Xylem Cell Development

Xylem Cell Development

Edited by
J. R. Barnett, PhD
Plant Science Laboratories, University of Reading

CHP CASTLE HOUSE PUBLICATIONS LTD

First published 1981 by
Castle House Publications Ltd.
Castle House, 27 London Road
Tunbridge Wells, Kent.

British Library Cataloguing in Publication Data

Xylem cell development.
1. Xylem
2. Botany—Ecology
I. Barnett, J. R.
581.4'1 QK646

ISBN 0 7194 0050 3

Production for the publisher by
Chambers Green Limited, Tunbridge Wells, Kent.

Printed in Great Britain by Henry Ling Ltd., at the Dorset Press, Dorchester, Dorset

Contents

Preface

Wood is a raw material of such importance that its physical properties and chemical composition have been investigated extensively. These properties are the manifestation of the combined properties of the various cell types comprising xylem and are a consequence of the changes which took place during the conversion of a meristematic derivative into a mature xylem element. Variation in wood properties within and between species is therefore a result of variability of the differentiation process, and if we hope to be able to control wood properties efficiently we should at least understand the process by which the xylem cells are formed. Unfortunately we are still far from this goal, and relatively few works have appeared which have been devoted exclusively to xylem differentiation.

The problem has been approached from a wide range of viewpoints, among which are studies of the nature of the stimulus causing a cell to commence the differentiation process, studies of the nature of the process itself, and of the effect of environmental and other factors in determining the form and size of the final product. The intention in preparing this book was to assemble as full a picture as possible of current thought on these aspects. It was felt that this might be achieved most effectively if a specialist in each field of interest were to prepare a chapter dealing with his own speciality. In this way, a depth and range of treatment beyond the scope of a single author ought to be obtained. The result of the decision to approach the subject in this way is presented here.

It is naturally not possible to ensure that in such a treatment every aspect of a subject will be covered, and an individual will always impose a personal slant on a topic. I feel however, that the authors have achieved a similar level of approach, each providing a useful review of the topic combined with an up-to-date summary of current thinking and recent developments in the field.

Emphasis is placed throughout at the cellular, rather than the whole tissue level, and the book concentrates on differentiation as far as xylem element completion. Secondary changes such as heartwood formation are not included.

Professor Preston's introductory chapter describes the wide range of problems to be solved if the process of xylem differentiation is to be understood. These problems are not, of course, unique to xylem, although useful pointers to the mechanism of differentiation in other tissues may come from a study of this material. Chapters in this book describe the actual changes taking place in cells during differentiation, the effect of the cell cycle and growth regulators in stimulating differentiation, the biochemical changes within differentiating cells, and the variability in size and form of the mature xylem cells induced by environmental and other factors.

I would like to thank the authors whose contributions have made this book possible. Their positive and enthusiastic response to my request for chapters has made my task a pleasant one. I must also, of course, acknowledge that I am responsible for any errors which may have crept in during the editing process.

J R Barnett
February 1981

Introduction

R D PRESTON
The University, Leeds, England

THE TOPICALITY OF XYLEM DIFFERENTIATION

It can be no accident that the first microscopic examination of an organised tissue, probably made by Henshaw in 1661, was of the xylem (in walnut trees), that the first illustration to be published was of the corky tissue of a woody plant (Hook's *Micrographia* of 1667) or that the first book dealing in a comprehensive way with the tissues of higher plants was devoted very largely to xylem (Grew's *Anatomy of plants* of 1682). Xylem constitutes the major supporting tissue of all spermatophytes and pteridophytes and is of equal significance to all plants in which it occurs, but its importance to man derives from its prodigious development in both dicotyledonous and coniferous trees. Man's origins, his past history and even his present existence centre around trees and it is not inconceivable that his survival, even or perhaps especially in this age of high technology, will depend upon what he is now doing, and will in the future do, to his trees. Throughout man's social and economic history there twists a thread in the progressively more intense use of trees. In prehistoric times, worship of trees as gods did not prevent the use of wood for fuel and for primitive tools. In historic times, clearance of forests for agriculture and the growing use of wood in abundance for uses such as fuel, buildings, farm implements and ships put greater and greater pressure on a commodity which must, however, at that time, have still seemed inexhaustible. With small populations, the practice of pollarding and coppicing (now too expensive!) meant that a comparatively small forest area could provide for all village needs.

Technological advances over the past 200 years have not diminished the call upon this resource; indeed, new uses found for wood have added to, not replaced, the older needs which themselves have burgeoned with the rapidly increasing population. The impact of the older use as fuel may be recognised by the circumstance that currently about 70% of wood available is consumed in this way and that, even in developed countries, the high cost of other, more recent fuels is now causing a minor upsurge in the use of wood-burning stoves. The impact of the newer technology itself may be exemplified by recalling that in 1961 the New York Times of Sunday, December 6th., ran to 678 pages weighing 6 lbs. in 1,458,558 copies. This used up 77,000 trees each of 70 years average age and growing on 360 acres[3]. Globally, 10^9 tons of wood are consumed annually and potential further uses for wood—as a source of raw material for the chemical industry for instance—can only add to this already huge demand. Although some countries have now taken steps to reverse the depletion of their forests there can be no doubt that, on a world-wide basis, areas devoted to forest are decreasing.

In this situation it is without question mandatory upon us to grow, use, and preserve wood much more efficiently in the future than in the past and this necessitates approaches from several disciplines. In the first place it is necessary to reach a better understanding of the relationship between the properties of wood and its anatomical and molecular architecture and this has already been recognised by the holding of a NATO Conference[7]. It follows that more needs to be known about the differentiation of wood, anatomically, physically and biochemically, from the apical meristem *via* the procambium and cambium of the plant producing it in order to lay out the mechanisms whereby desirable structures, and therefore desirable properties, may be reached; leading to the devising of breeding programmes and forest management processes whereby these features may be fostered. For this reason alone the process of xylogenesis needs urgent study.

Further than this, however, the differentiation of meristematic cells into xylem is only one example of a process ubiquitous in living things and presumably all differentiation mechanisms have much in common. Xylem differentiation has much to offer to the general problem not least because, in any one plant at any one time, the whole process is set out in space so that all stages are at one and the same time available. This is intellectually the more important aspect of the study if not the one of more practical immediacy. This book pays attention to both sides of the story.

PROBLEMS IN XYLEM DIFFERENTIATION

Xylem differentiation basically presents problems and asks questions at two levels. The first set of questions centres around the process whereby a cell which is isodiametric with a thin, viscoelastic wall becomes a large, anisodiametric cell with a thick, rigid wall, composed in part of the same molecular species—the polysaccharides—as the initial wall but containing also a new substance, lignin. As ancillary questions, one has to ask how it comes about that in gymnosperms the only cells produced, other than parenchyma, are tracheids whereas in angiosperms not only are fibres also produced, for these could be regarded as only a variation on the tracheid theme, but vessel elements in which the end walls, but not the side walls, become perforated. Beyond these lies a whole hierarchy of questions concerning the development, not of individual cells, but of wood with its complex organisation. How does it come about that the cellular organisation of the root apex is so different from that of the shoot apex while each is performing the same biochemical task? Why do only some cells in the apical meristems elongate into procambial cells and not others? By what control measures is it assured that the vascular bundles of the shoot apex are arranged in a strictly defined way? What factors lie behind the different positions of the primary xylem in shoots and roots? What is it that causes some cambial initials to swell into vessel elements and not their immediate neighbours, and why then is the cell immediately below induced to produce another vessel element? What is it that breaks the dormancy in the cambium during spring and what induces dormancy in autumn? What lies behind the development of annual rings in temperate species but not in most tropical species? The questions are endless. They were all asked during the last century, beginning with Unger, Hofmeister, Nägeli and Sanio[15], and repeatedly during this century by, among others, Haberlandt, J. H. Priestley, A. J. Eames and Kathleen Esau. Rational answers have been possible, however, only during the past few decades following the introduction of new techniques and new insights. By and large it seems that we are still in the main asking the same questions and only answering ourselves in a different way. We now say that cells are programmed, that they read instructions carried by the genome and direct biochemical processes to be switched on or off, so that finally they become committed to some path or other. This amounts to asking the same question in a different language and it remains that answers in terms of what actually happens in a cell, both physically and chemically at each stage of differentiation, are still to find. Much of this book is taken up by the search for answers at this level.

Although differentiation in xylogenesis, as with all cell differentiation, is basically a progressive modification of the biochemical processes in the cytoplasm and the effects of neighbouring (and distant) cells upon them by means of solute gradients, transmission of hormones and so forth, the end effect, the result of differentiation, is a change in the dimensions of the cell and in the wall surrounding it. On the face of it, there should be no difficulty in tracing the events occurring solely in the wall but in practice this turns out to be not quite straightforward.

DEVELOPMENTS IN THE MERISTEM

As a cell leaves the apical meristem to become a procambial cell it elongates considerably but does not develop the extensive vacuoles of the cells surrounding it which eventually form the cortex and the pith. Whether the position at which this occurs is determined by events at the apex (as, for instance, by the location of leaf primordia at the shoot apex) or whether it is determined by the pattern of the vascular tissue already present in the more mature system further from the apex, is not yet decided. It is not certain either whether the procambial cell extends by virtue of stresses in the wall induced by its own turgor or whether extension is induced passively by the elongation of the vacuolating cells around it. The wall is, however, extended by the same extent in either case and the question is how does the wall accommodate itself to this extension? The structure of the wall is not yet known and one can argue only by analogy partly with the walls of the fusiform initials of the cambium, partly with the walls of parenchyma cells in which the details are known, but mostly with single cells growing in isolation, such as the internodal cells of *Nitella*. In these, the net orientation of the microfibrils averaged through the wall, as determined under the polarising microscope, remains transverse, or slowly helical, however long the cell becomes. Since the primary wall retains about the same thickness throughout, this presented a major puzzle during the late 1930's and the 1940's since it led away from the accepted view that during growth the wall is stretched passively, and away from the idea that wall deposition is by apposition towards intussusception.

The puzzle was resolved to the satisfaction of most observers by Roelofsen's multinet-growth hypothesis[13]. This hypothesis is now too well known to need describing in detail and one may merely say that it has two elements. Firstly, the wall is continually extended during cell growth by the stresses induced upon it. Secondly, wall deposition is by apposition and every new lamella deposited has the same structure in which, with elongating cells, the microfibrils lie in preferred transverse or slowly helical orientation (Fig. 1). It follows that, with cells to which both these conditions apply, on passing through the wall from inside to outside, the strain continually increases and the microfibrils become continuously reoriented. The type of reorientation will depend upon the constitution of the newest lamella; if the microfibrils lie more or less parallel to each other

Figure 1: *Electron micrograph of the primary wall of a wheat coleoptile, metal shadowed, magnification 21 000×. Cell length approx. parallel to longer edge of page.*

in a helix round the cell the helix will steepen on passing through the wall; if, on the other hand, the microfibrils show an angular spread around a transverse orientation the resulting 'net' will be progressively pulled out until (if the lamella has not by that time broken) the net orientation on the outside of the wall becomes longitudinal. Both conditions have been reported[8] for parenchyma cells of higher plants and for the algae *Nitella* and *Chaetomorpha*. If, on the other hand, the architecture of the newly formed layer varies from time to time then the need for progressive reorientation is lost and in such cases lack of a progressive change through the wall is not evidence that the multinet growth hypothesis has failed. Such a case occurs with epidermal cells for which it has been shown that a modification, not a denial, of the hypothesis is needed[16]. Roland and his colleagues[14] have criticised the hypothesis, claiming that the walls of parenchyma cells show crossed microfibrillar orientation so that the structure upon which the hypothesis is based is therefore an artefact, and that, further, neither set of microfibrils becomes reoriented. This claim has, however, been questioned[9, 17]. Wardrop[17] has confirmed that the classic multinet growth hypothesis does apply to parenchyma cells of oat coleoptiles even when using the methods of Roland. This controversy forms the subject of an article which will appear elsewhere[10]. Both Roelofsen's hypothesis and the 'ordered fibril' hypothesis suggested by Roland demand that microfibrils should slip past each other and should become mutually readjusted as a lamella extends. This implies changes in the bonding between matrix polysaccharides and perhaps between these and the microfibrils. The metabolic machinery of the cell must as a consequence be involved, presumably through the agency of growth substances.

It is to be expected that these considerations should apply also to the procambium and the cambium though supporting evidence is not so far forthcoming. The wall structure of procambial cells is not known and, though with the fusiform initials of the cambium the microfibril orientation, as judged both by polarisation microscopy and X-ray diffraction analysis, is certainly more or less transverse[8] and remains so even though the initial has elongated to many times its original length, the circumstance that these cells show typical multinet growth changes is a paradox. This is because the tangential walls, new at each tangential longitudinal division, do not partake in the full extension and because the radial walls extend very considerably in the transverse direction after each such division during differentiation into a xylem cell. This situation needs further examination.

The steady elongation of the fusiform initials, extending in time over a number of years, raises questions not found with cells growing in isolation. These cells form a cylinder over a woody core which is continually expanding in girth due to the steady differentiation of new wood. The consequent increase in girth of the cylinder is not, however, taken up by radial longitudinal division but instead by cell extension. There does occur some pseudotransverse division, followed by a swinging round of the pseudotransverse wall toward the longitudinal but this does not add much to the adjustments necessary for circumference increase. The initials must therefore in effect interpenetrate in order to convert a length increase into a diameter increase and three proposals have been made as to how this might be achieved. The first, and obvious one, was that the initials are free to slide over each other so that the tip of any one initial could insert itself between two other initials—the so-called sliding growth. Because it was thought, erroneously, that this would break the protoplasmic connection thought to be a necessary prerequisite for the subsequent formation of bordered pits found on the radial walls of tracheids, and because no initial moves upwards or downwards on the woody cylinder, this was replaced by Priestley[11] by what he called 'symplastic' growth in which all the walls of the cambium are supposed to extend together with no slip but in which the growth rate must vary from point to point. As a third variant, Sinnott and Bloch proposed that each initial intrudes between others by tip growth—the so-called intrusive growth. On the whole, opinion seems to be settling on the last proposal (p. 63) though a final decision must await definitive observations on the way wall structure changes through the wall at any point along an initial.

It must not be forgotten either that this orderly behaviour is from time to time interrupted as a fusiform initial divides by repeated transverse division to form a ray initial. This seems to have the effect that no point in the xylem is allowed to be more than a certain (undefined) distance from a vascular ray; the manner in which this is achieved is unknown except in so far as that nutrient gradients appear to be involved.

THE FINISHED PRODUCT

a. *Molecular architecture*

The lateral extension of a daughter cell of the cambium raises a host of questions, many of which are discussed later in this book. Perhaps the most intriguing concerns the differentiation of vessel elements in angiosperm wood. The blowing up of a daughter cell into a relatively enormous derivative, when this does not happen to neighbouring cells except those below it in the formation of a vessel, seems to demand a control mechanism of high order. Even the disappearance of the end walls is by no means understood. With a vessel of a ring-porous dicotyledon the fusiform initial may have increased in diameter 20× and therefore the surface area of the end wall by 400×. For this reason Priestley[12] long ago contended that the end wall must be torn by an explosive extension. The alternative is that the end wall sustains this extension and is later digested, but this leaves open to question why this part of the cell envelope is removed while the rest of the wall, particularly the membranes of the pits, remains structurally intact. Such localised digestion of a wall is not, of course, unique to vessel elements; it occurs wherever a spore-producing cell, for example, drills a hole through a wall for the release of spores, at a location which may be specific. There are clearly mechanisms to be sought which protect parts of a wall while allowing other parts to be enzymatically removed and perhaps an eye should be kept by those involved with xylem differentiation upon trends in fungal wall studies where evidence is being found[1] for the presence in a cell of spatially controlled digestive enzymes and inhibitors. The contention that with these end walls, as with the pit membranes, only matrix polysaccharides are dissolved and that the cellulose is subsequently displaced by the transpiration stream seems difficult to support. The pressures likely to be induced on the tenuous filaments of cellulose by a stream travelling no faster than 4 m per hour would seem hardly high enough for this to occur.

The end product of xylogenesis with tracheids, fibres and vessel elements, but not necessarily with parenchyma cells either of the vertical series or of the vascular rays is the thick secondary wall surrounding a lumen empty of cytoplasm. The thickening of this wall proceeds in a cell which has ceased, or almost ceased, to extend and presumably involves a change from high rates of protein synthesis to high rates of polysaccharide synthesis coupled with a high rate of intake of carbohydrates. Whether cessation of growth in size comes about because the wall becomes too rigid to allow the creep under turgor forces which the viscoelastic primary wall had hitherto permitted, or whether some other signal is obeyed, is not yet certain. Wall deposition is unquestionably by apposition, but this is probably no more than a continuation of the habit of the primary wall. In spite of what is commonly believed, the thickness which the secondary wall eventually reaches does not vary much over a growing season. The visible distinction between early wood and late wood in conifers is due mainly to the much larger lumen size in the former. The factors underlying this distinction are very complex, as discussed later (p. 236).

The onset of the secondary wall must, however, be taken to signal a pronounced change in the physical condition of the biochemical pathways leading to synthesis of the wall. Up to this point, the cellulose microfibrils have not been well oriented, they have been widely separated on an average, mostly by imbibed water, and the preferred direction has been approximately transverse. The average spacing between microfibrils may be calculated to be of the order of several microfibril widths. In the secondary walls (Fig. 2) the microfibrils are comparatively well ordered in rather precise parallel arrays, they lie in relatively steep helices round the cell (except with vessel elements) and are separated by only one or two

Figure 2: *Electron micrograph of the wall of a tracheid in* Pinus radiata *xylem, metal shadowed, magnification 14 000×. Tracheid length approx. parallel to longer edge of page. S1 to left; S2 to right.*

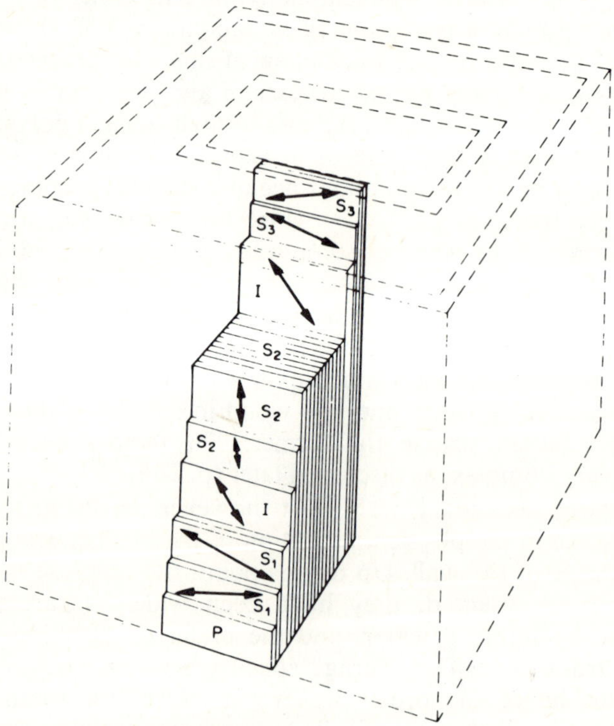

Figure 3: *Diagrammatic representation of the structure of a tracheid wall. Double arrows show the run of the microfibrils in the various layers. The intermediate layers I are not always present, and the layer S3 is absent in some species.*[8]

microfibril widths. There occurs, moreover, in all species a switch from a relatively slow helix in the first-formed S1 layer to a steep helix in S2, followed in most species by a reversion to a slow helix in S3 (always missing in tension wood, a peculiarity which has been speculated upon without any firm conclusions) (Fig. 3). The nature of this sophisticated end product of xylogenesis must be very fundamental to the whole process examined in this book. Parallel with this development, though somewhat displaced in its appearance with time, is the equally significant impregnation with lignin, also discussed in detail later (p. 115). Whether the phenylpropane derivatives forming the substrate for lignin synthesis are moved into the differentiating cell after secondary thickening has ceased, or whether they are already present in the cell seems to be moot, with the concensus of opinion somewhat in favour of the latter. It is certain that, as with the primary wall, the wall even at this late stage still contains synthesising enzymes. Lignification induces another difference between the layers of the secondary wall in that the lignin content per unit wall volume (but not the total lignin in any layer) decreases from the outer parts of the wall, where lignification begins, towards the inner side. This has important consequences for timber decay and its prevention which lie, however, beyond the bounds of this book.

There are still other differences between the wall layers. The evidence is that there occurs a gradation in the kind and quantity of the matrix polysaccarides from within outwards. It seems that at this period of xylogenesis there are profound biochemical adjustments in process or, in modern terms, that the cell is receiving a controlled series of detailed messages modifying the original message setting off the whole process. These deal with the solid framework of the wall. Whether they also involve the porosity of the wall, which it is vital to maintain, is not certain. Pore diameter in the secondary wall seems to be about 4 nm on an average and few pores exist which are larger than this except in the pit membrane[8]. Pore diameters in the walls of fusiform initials are not known but an estimate for other primary walls has been given[2] also at about 4 nm. This latter figure has been reached by a solute exclusion technique and may be somewhat in error for a variety of reasons. If it is correct, then the same pore size is maintained over the whole wall of the xylem elements in spite of a pronounced gradation in chemical composition and structure.

The features of the wall in the secondary xylem continue in a modified way a relationship with cell length already found in the primary xylem. In the primary situation, this takes the form that the secondary wall remains discontinuous so long as it develops in an internode which is elongating and in which therefore the developing xylem is being extended, presumably passively. This is evidently an adaptation to the need to preserve open water channels during elongation but the genesis of this condition is not clear. The form of the secondary thickening whether annular or helical, is sometimes preceded by a similar orientation of bands of vesicles, or microtubules, in the cytoplasm, but the precise significance of this remains to be worked out. It could be that an understanding of cellulose biosynthesis in xylogenesis needs to await clarification of the similar situation found in cell types in which it can be more conveniently studied.

With the secondary xylem the relation between cell length and structure for tracheids and fibres is more specific but perhaps continues the same trend though with a different expression. For any cell of length L the angle between the helically wound microfibrils and cell length, θ, is given by the relation $L = a + b.cot\theta$[8] though perhaps in some species the connection may be between L and θ instead of $cot\theta$. There is a similar relation between θ and cell breadth so that, just as the cytoplasm in a primary vessel element 'knows' that the cell is elongating, so the cytoplasm at any point even in a very long tracheid 'knows' how long and wide the cell is. This is a peculiarity which cannot be explained by invoking a need for the number of turns of the helix to be kept constant or by any other simple consideration of helix geometry. There seems to be some subtle influence here of cell form on the biosynthesis of microfibrils, and it seems probable that this is exercised through the plasmalemma. The relationship is found in phloem fibres as well, and in monocotyledons, and is indicated even with some algae. It harmonises with the observation that, in general, the longer and thinner a cell of any kind is the steeper the microfibrillar helix, and vessel elements are no exception. Whether the relation holds within vessel elements is not known since small changes in flat helices are difficult to observe when the run of the microfibrils is so disturbed by the abundant pitting.

The consequences of this relation are clear for the secondary xylem just as they are for the primary xylem. Since at any point of an inner annual ring the tracheids are shorter on average than are those of an outer ring, over about 60 years for some conifers though for a shorter period with hardwoods, then the wall helices are flatter in the inner ring than they are in the outer ring. It may be the effects of this which have led to the description of the innermost rings as juvenile wood. Since cells with steep helices have a greater Young's modulus, greater breaking strength and lower extensibility than do shorter cells[8], the outer wood is stronger in tension parallel to the grain than is inner wood. This has clear relevance in timber utilisation and for tree breeding but the implications for the tree itself are not straightforward.

The increasing average length of tracheids across the annual rings is due, of course, to the increasing length of the appropriate fusiform initials, as first established by Sanio and confirmed by I. W. Bailey, as a consequence of the continual cutting off from the initial of replicas of itself. Nevertheless the final tracheid so produced is often if not always somewhat longer than the initial that produced it. This post-cambial increase in length may well be an accommodation to the increased circumference of the woody cylinder caused by the radial expansion of new tracheids whose differentiation is just beginning. For, since differentiating tracheids do not increase in tangential width, the circumference increase must be taken care of by the pushing apart of cells through the insertion between them of cell apices. Otherwise the increase in length of the differentiating tracheids would induce a compression in the very outermost parts of the xylem and a tension within the bulk of the wood; whereas, for reasons discussed in this book (p. 275) the outer wood ends up in tension and the inner wood in compression.

It seems reasonable to assume, therefore, that some consequences of xylem differentiation arise through the simple geometry of the tree, others perhaps through the direction of flow of nutrients and other necessary ingredients but that the whole is sensitively controlled by a series of biochemical events influenced by messages released from the information store in the genetic complex.

b. *Cell wall models*

Whenever the development of a structure is being looked for it is clearly an advantage if the structure to be produced is known in all its details. Unfortunately this advantage cannot be claimed for xylem tissues. As indicated above, the architecture of xylem cells is well recognised as far as the cellulose component is concerned but very little is known of the way in which the microfibrils are held together by the matrix substances or how these themselves are bonded. We need, both for the purposes of developmental investigations and for an understanding of timber properties, a coherent model of the wall, but this is so far lacking. It is known that some molecular chains of xylan and mannan are bound to the microfibrils and therefore lie parallel to them. Beyond this, it has been claimed that the molecular chains of all hemicelluloses lie parallel to the general run of microfibrils. To this extent, a model of the wall could consist merely of a set of parallel lines, the lines within microfibrils being regularly spaced but not those outside them (since in the native wall the hemicelluloses are crystalline in the optical but not the X-ray diffraction sense). There is a problem here with branched chains such as arabinogalactans; but since it is now known that in branched polysaccharides from bacteria the branches do not stand out at right angles from the main chain but instead are rotated parallel to the main chain and hydrogen bonded to it[6] there is conceptually no difficulty in arranging the consequent rod-shaped molecule parallel to the nearest microfibril. What remains is to determine the distribution of the various hemicelluloses and the bonding between them, and there can at present be no certainty about this. Albersheim and his colleagues[4] some time ago presented a detailed model of the primary wall upon which a secondary wall model might have been based were it not for doubts that the model is firmly based or acceptable in terms of known wall properties[9]. A much more acceptable partial model has been presented by Monro, Penny and Raymond[5] and a somewhat similar model, based both on the known conformation of polysaccharides and on known wall properties, is also available[9]. This model—based on the primary wall—is presented in figure 4. Apart perhaps for the links to a protein, and with the *caveat* that bonding to lignin is lacking, this might serve as a potential though only partial model for the secondary wall.

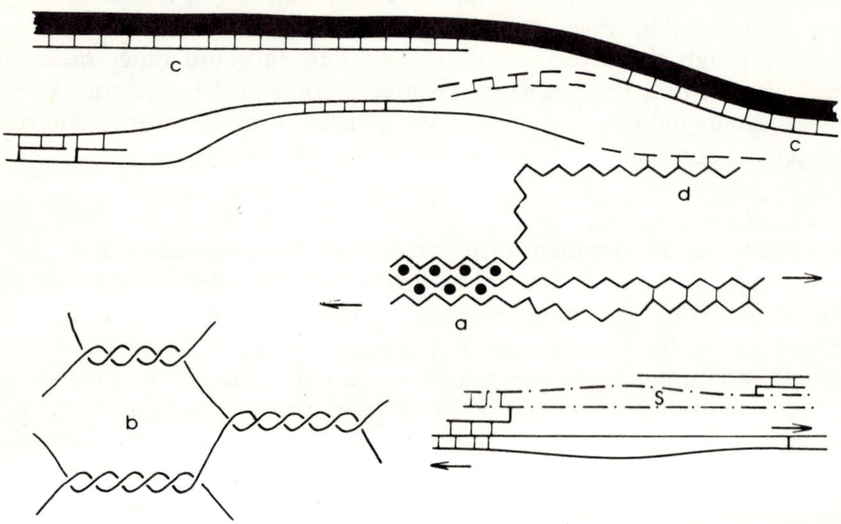

Figure 4: *Diagrammatic representation of some possible bonding schemes in primary walls. Thick rod—microfibril. Solid straight, or smoothly curved, lines—ribbon-like polysaccharides. Broken lines—twisted ribbon-like polysaccharides e.g. xylan with a three-fold axis. Zigzag lines—pectic acid. Broken and dotted lines—protein. Helices—helical chain molecules such as 1-3 linked xylans. Full circles—calcium ions. Shorter thinner lines—hydrogen bonds. (Not to scale).*[9]

REFERENCES

1 Burnett, J. H., Trinci, A. P. J. (Eds) *Fungal walls and hyphal growth*. Symp. British Mycol. Soc. C.U.P. Cambridge, London, New York, Melbourne (1979)

2 Carpita, N., Sabularse, D., Montezinos, D., Delmer, D. P. Determination of pore size of living plant cells. *Science* **205,** 1144-1147 (1979)

3 Feininger, A. *Trees*, Viking Press, New York (1978)

4 Keegstra, K., Talmadge, K. W., Bauer, W. D., Albersheim, P. The structure of plant cell walls III. A model of the walls of suspension-cultured sycamore cells based on

the interconnections of the macromolecular components. *Plant Physiol.* **51,** 188-197 (1973)

5 Monro, J. A., Penny, D., Raymond, W. B. The organisation and growth of primary cell walls of lupin hypocotyls. *Phytochem.* **15,** 1193-1198 (1976)

6 Moorhouse, R., Winter, W. T., Arnott, S. Conformation and molecular organisation of fibres of the capsular polysaccharide from *Escherichia coli* M41 mutant. *J. Mol. Biol.* **109,** 373-391 (1977)

7 N.A.T.O. *Properties of wood in relation to its structure* (E. G. Kovach, Ed.) Scientific Committee Conference, Scientific Affairs Division, N.A.T.O. Brussels, Belgium.

8 Preston, R. D. *The physical biology of plant cell walls.* Chapman and Hall, London (1974); and references therein.

9 Preston, R. D. Polysaccharide conformation and cell wall function. *Ann. Rev. Plant Physiol.* **30,** 55-78 (1979)

10 Preston, R. D. To be published.

11 Priestley, J. H. The meristematic tissue of the plant. *Biol. Revs.* **3,** 1-20 (1928)

12 Priestley, J. H., Scott, L. I., Malins, M. E. Vessel development in the Angiosperm. *Proc. Leeds Phil. Soc.* **3,** 42-58 (1935)

13 Roelofsen, P. A. Orientation of fibrils in the cell wall of growing cotton hairs and its bearing on the physiology of cell wall growth. *Biochem. Biophys. Acta.* **7,** 43-53 (1951); Roelofsen, P. A., Houwink, A. L. Architecture and growth of the primary wall in some plant hairs and in the *Phycomyces* sporangiophores. *Acta Bot. Neerl.* **2,** 218-225 (1953)

14 Roland, J. C. The wall of the growing plant cell; its three dimensional organisation. *Int. Rev. Cytol.* **61,** 129-166 (1979); and references therein.

15 Sachs, J. von. *History of Botany* (1530-1860). Clarendon Press, Oxford (1890)

16 Sargent, C. Differentiation of the crossed fibrillar outer epidermal wall during extension growth in *Heracleum vulgaris* L. *Protoplasma* **95,** 309-320 (1978)

17 Wardrop, A. B., Wolters-Arts, M., Sassen, M. A. A. Changes in microfibril orientation in the walls of elongating plant cells. *Acta Bot. Neerl.* **28,** 313-333 (1979)

1. The Primary Xylem

T P O'BRIEN
Department of Botany, Monash University, Victoria, Australia

INTRODUCTION

The bodies of most vascular plants are derived ultimately from embryos (which may be produced sexually or apomictically) or from some form of vegetative propagule. In all cases, and in the case of apogamously produced sporophytes, all major organs become vascularized during their development by primary vascular tissues. Although the pattern in which the two tissue systems—primary phloem and primary xylem—are arranged varies in fascinating ways in different organs in the same plant and among similar organs in different species, in the great majority of species that have been studied to date, the vascular system forms a completely interconnected network that links all major organs of the body to one another. Remembering that the primary bodies of plants display indeterminate growth (new buds, leaves, internodes, nodes and roots are produced more or less continuously throughout the life of the plant) there must be mechanisms that allow the vascular system produced within the newly produced organs to connect smoothly with that already differentiated within the mature tissues of the same body. A point which is less often stressed in this connection is that organ loss (abscission of buds, leaves, roots, flowers and fruits, whole branch systems) normally accompanies plant development; partial die-back of leaves is also commonplace. Clearly there must be mechanisms which seal off the severed vascular connections, minimising loss of sap, and rendering the surface resistant to invasion by micro-organisms which could otherwise enter the ruptured vascular tissue.

Vascular tissues have fascinated botanists since the early descriptions by Grew (1682)[37] and as the techniques of microscopy and histology advanced to a high level of sophistication by the end of the 19th Century (see Bracegirdle[7] for an excellent history of micro-technique), a voluminous literature accumulated. Among a number of more minor themes, it is possible to detect at least three major themes in this literature. First, comparative vascular anatomy was used as a major tool, supporting and extending morphological observations, in attempts to unravel taxonomic affinities and phylogeny in extant and fossil vascular plants. Many modern students, raised on a diet rich in biochemistry, physiology and molecular biology, will not be aware of the intensity with which such matters were studied in the past 150 years, but the books by Sporne[91], Foster and Gifford[32], and Bierhorst[5] and the modern research literature all testify to the interest that this subject once held and still holds in modern times. The second theme is a functional theme, in search of answers to such questions as: how does vascular tissue form? What does it do? How do its cells act? In his book Physiological Plant Anatomy, Haberlandt[44,45] epitomises this approach in a way that reflects to some extent the approach of his predecessor, Schwendener, to similar questions. Thirdly, there is a developmental theme, characterised by the work of Professor Esau and her students over the last 45 years and so ably enshrined in her books.[22,23,24] Here, the concern is not only with comparative anatomy and function but with the more difficult question: how does the tissue come into being?

The rise in knowledge in biochemistry and in cell and molecular biology has begun to revivify a fourth theme. The detailed structural knowledge now available and the wealth of tissue-culture facilities in various laboratories around the world have lead to a resurgence of interest in the question: how is the tissue and cellular pattern controlled? It is interesting to note what has sparked off this resurgence of interest in this ancient question. The introduction of electron microscopy as a serious tool in the study of plant bodies started about 1950, gained momentum in 1963 with the introduction of glutaraldehyde as a primary fixative[83] and, pushed along by a catena of technical advances (by-pass ultramicrotomes, diamond knives, epoxy resins and other plastics as embedding media: see Feder and O'Brien[27], Bracegirdle[7]), ushered in the New Anatomy. Resolved images at 5–10 nm resolution of the majority of plant cells and tissues, backed up with high-quality light microscopy at 0.25 μm resolution, are now widely available.[43,57,68] Vascular tissues, partly due to their historical importance, partly due to the characteristics of their cells and partly because of spectacular advances made in regulation of vascular-tissue differentiation in culture, have received an enormous amount of very detailed attention in the last 15 years. The factors regulating differentiation of tracheary elements and the development of their beautifully sculptured walls have been intensively investigated and several major reviews have appeared (see O'Brien[67], Roberts[81], Esau and Charvat[25], Shininger[89], Berlyn[4], Fukuda and Komamine[33]).

This chapter briefly reviews some basic anatomical facts about the primary xylem, including those details of structure that have been established by the New Anatomy. There follows a discussion into what is known about the development of tracheary elements, the control of the pattern in their extracellular skeletons, and the regulation of the pattern in which they are formed in tissues. Finally, a discussion of xylem parenchyma cells, until recently, a rather neglected category of cell type.

CELL TYPES OF THE PRIMARY XYLEM

Two types of cells are present universally in the primary xylem: tracheary elements and xylem parenchyma cells. Supportive tissue elements (fibres and variously lignified parenchyma cells) are also common but not always present. Tracheary elements constitute the final products of one of the most remarkable sequences of cellular differentiation known. At maturity they consist solely of pipes of extracellular material; the wall of the pipe is fabricated chiefly by the cytoplasm of the tracheary elements but part of the wall is also contributed by adjacent parenchyma cells. Adjacent mature tracheary elements may be interconnected both laterally and/or vertically and the nature of these interconnections is used to distinguish two broad categories of tracheary element. During the construction of tracheary elements, each cell is separated initially by a layer of primary wall and eventually by secondary lignified walls whose pattern of deposition where the cells lie adjacent may be quite complex. When the cytoplasm dies at maturity, the lignified secondary walls remain, but the primary wall is partially degraded. In tracheids, a residue of this partially degraded wall persists as a so-called 'pit membrane' (the term is unfortunate for there is nothing in common between pit membranes and any of the lipo-protein membranes of the cytoplasm) which persists across the gaps in the lignified walls where adjacent tracheids are in contact. While the pit membrane probably has minimal effect upon the passage of solutes and water between adjacent tracheids, gas bubbles do not pass across it. In contrast, where vessel members are in contact at their ends, a perforation plate is created out of the adjacent walls of the vessel members. The structure of perforation plates varies enormously in the organs of different species and can vary widely in the same plant but all have one feature in common: somewhere, the perforation plate is truly perforated by a pore across which no pit membrane extends. Sometimes the pores may be just a small percentage of the plate area but more commonly most of the plate is involved and the area of the pore may be almost equal to the total cross-section of the adjacent vessel members. Series of interconnected vessel members are called vessels and may extend many centimetres through primary xylem. Bubbles of gas will therefore pass unhindered through adjacent vessel members in a vessel and may break the transpiration stream.

In both types of tracheary element, tracheids and vessel members, the wall of the pipe consists chiefly of polysaccharides, rich in cellulose microfibrils and with lignin deposited between the adjacent chains of polysaccharide. The pattern in which the wall is deposited varies in remarkable ways with the growth rate of the tissue at the time the cells are being formed (see p. 19) but in addition to the elaborate patterns of secondary wall deposition that characterise these cells, the pipe wall may be made up in part by segments of the walls of adjacent xylem parenchyma cells. Here the walls are always unlignified, low in cellulose, rich in non-cellulosic polysaccharides that form open, porous gels of coarse texture. These gaps in the lignified pipe-surface are almost certainly the preferred pathways across which nutrient exchange takes place between the moving transpiration stream and the living cells of the xylem and eventually, more distant cells of cortex, mesophyll and epidermis. Tracheary elements must be viewed then as a most remarkable feat of internal engineering, from which the plant has fabricated an elaborate internal plumbing system that interconnects all major organs in the primary plant body. The technique used is to construct extracellular skeletons, then to partially digest the skeletons while removing all of the cytoplasm of the tracheary elements. Contact with living cell surfaces is maintained at the pits where the pipe wall is composed at least in part of the walls of living cells.

The parenchyma cells of the primary xylem had been studied very much less than the tracheary elements until the last 20 years (see O'Brien[67]). However, following recognition that such cells in coleoptile xylem were rich in mitochondria and rough endoplasmic reticulum (ER)[69], it was established that such cells were very active in amino acid accumulation in the internodes of mature pea plants[73]. This led indirectly to the discovery of what are now called 'transfer cells', a specialised modification of a wide variety of cell types in which the cell surface area is enhanced by as much as 20-fold with the development of a wall/membrane apparatus (see p. 39). This apparatus appears to be induced in any site in which unfavourable surface area restricts short-distance, high-intensity solute fluxes. Modifications of this kind are widespread in the xylem parenchyma in the nodes of monocots and dicots[41] and are seen especially clearly in all nodes of all grasses that have been studied (see references in Busby and O'Brien[10]).

While it is true that the discovery of vascular transfer cells shone a spotlight of world interest briefly on the xylem parenchyma cell, it is still true that this cell type is largely unexplored. Studies on xylem parallel to those carried out by Kuo *et al.*[54] on phloem are urgently needed to gain insight into the symplastic pathways that connect the transpiration stream metabolites to the rest of the tissues. Detailed stereological analyses similar to those recently published by Briarty *et al.*[8] on wheat endosperm would be invaluable for assessing the significance of the exchange surface available at the surface of tracheary elements, how this varies in different tissues of the same plant and in the same tissues of different plants. In particular there is urgent need of studies that analyse the surface and volume relationships between parenchyma cells, their neighbouring tracheary elements, and the extent of pit contact with tracheary elements.

The Procambium

In normal organs, primary xylem differentiates in a very predictable sequence from a precursor tissue, the procambium or provascular tissue.

Since vascular differentiation is restricted in normal organs (as distinct from tissue cultures and wounded systems) to the cells of the procambium, it follows that the pattern in which this tissue differentiates and the links forged between adjacent strands of procambium determine the broad outline of the vascular system. Vascular cells do not normally form anywhere else. One might have hoped that the structure of procambial cells would contain some detectable clues to their future behaviour. Unfortunately, it is not possible to cite any single cytological feature that will distinguish procambial cells reliably from their neighbours. In general, if these cells are viewed in cross-section, they appear to be somewhat less-vacuolated, to stain more intensely with cytoplasmic stains and to consist of cells of smaller diameter than those adjacent cells which will produce non-vascular tissues (Fig. 1.2). This last difference, the smaller diameter in cross-section, is by no means strict since the procambial cells which will form the largest vessels of the primary xylem are often the largest diameter cells in the procambium.

However, the plain fact of the matter is that, while this tissue was studied extensively prior to 1950, in the last 30 years it has not received the attention it deserves. Since it is the natural precursor of vascular tissues, and since mature vascular tissues are vital to long-distance metabolite transfer in the primary body, one might have hoped that procambial cells would have been the objects of a great deal of developmental study. D'Amato[20] and Lai and Srivastava[55] have summarised the evidence that shows that, at least for some species, and especially in roots, partial or full genome amplification by endopolyploidy occurs in the procambium that gives rise to tracheary elements. Very recently, Gunning *et al.*[39,40] have studied the structure of procambial cells in the roots of the aquatic fern, *Azolla*. These roots display determinate growth and elaborate their structure by a very precisely regulated sequence of divisions and growth from their apical cell, generating a series of units (merophytes) in and across which it is possible to follow cytological features with great precision. It appears from this study that procambial cells undergo distinctive modifications to their populations of cortical microtubules during the early stages of their formation. What is needed are comparable studies on shoot and leaf procambium and in procambium of roots which display indeterminate growth.

It is disappointing, and a little frustrating, to be going into the 1980's with so little that is new to say about the cells of this vitally important meristem. A good deal of recent evidence implicates this tissue in hormone transport that in turn may regulate the development of the vascular network (see Sachs[85,86] and references cited therein). These exciting ideas cannot be grounded upon a solid, structural foundation because there is so little known about the structure of procambial cells. It is time that the degree of symplastic connection between procambial cells, and between them and their neighbours, was measured accurately. A detailed comparative study of the fine structure of procambial tissues from different organs of the one species, and between comparable organs of different species, is long overdue. As Shininger[89] states: "We do not have a meaningful concept of procambium at a biochemical or ultrastructural level, and the development of concepts at these levels is important."

Protoxylem and Metaxylem

In all organs that part of the procambium that is destined to produce the xylem does not do so all at once: some cells complete their differentiation ahead of others. It is customary to distinguish the tissues which complete their differentiation first—the protoxylem—from those which form later—the metaxylem. These are terms of convenience, for no sharp boundary delimits the two, either in time or in space. However, in all parts of the plant body where vigorous extension accompanies xylem differentiation, the structure of the wall deposited by the developing tracheary elements reflects the rate of extension taking place in a most remarkable way. In internodes, petioles, leaf blades, and flower stalks, protoxylem tracheary elements commonly deposit a lignified secondary wall that consists solely

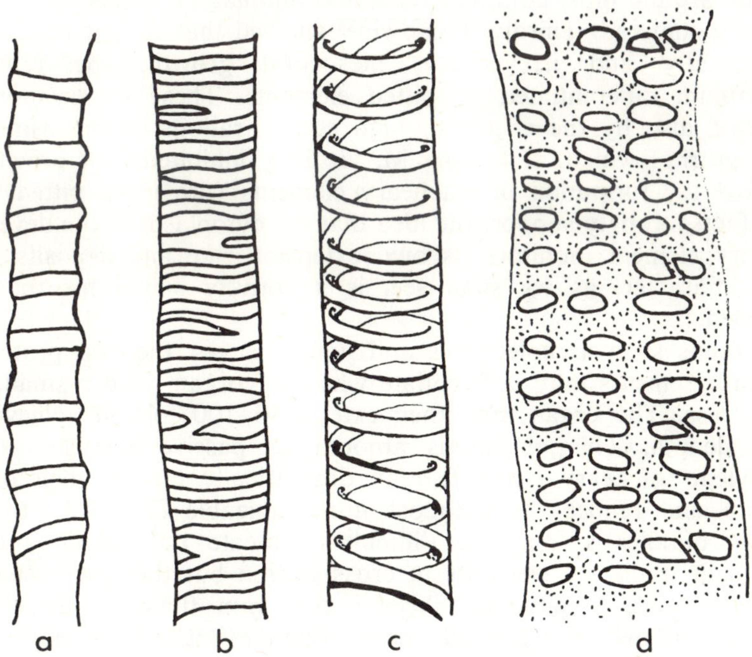

Figure 1.1: *Selected examples of primary xylem tracheary elements. (a) stretched annular; (b) reticulate; (c) helically thickened; (d) pitted element.*

of annular bands, separated from one another by regions that consist solely of primary wall (see Figs. 1.1, 1.3 and 1.9). Alternatively, such tracheary elements may show gyres of lignified wall thickening, deposited as a helix (Fig. 1.1). As the growth rate begins to decline, the later formed tracheary elements deposit walls in which the lignified layers are reticulate, and finally, when extension stops, only pitted elements are produced (Fig 1.1). Here, the primary wall is covered by a continuous matrix of lignified secondary wall, perforated only at pits which may correspond to regions where adjacent tracheary elements are in contact (Figs 1.3, 1.4, 1.6), or where xylem parenchyma cells or xylem fibres contact the lumen of the tracheary elements (Figs 1.3, 1.12-1.14). The adaptiveness of this behaviour will become clearer when the detailed development of the tracheary element has been considered.

Three facts strongly support the argument that it is the growth rate of the tissue, in the region and at the time at which tracheary element differentiation is occurring, which determines the pattern of wall deposited. First, in most roots the bulk of extension is completed before the protoxylem tracheary elements complete their differentiation, and root protoxylem usually contains reticulate, not annular or helically thickened tracheary elements. Secondly, in nodes, where extension is minimal or absent, pitted elements are the most common type, and annular or helical elements are rarely encountered. Thirdly, Goodwin[34] showed that if light was used to stop rapidly and irreversibly, mesocotyl growth, the remaining procambium differentiated as pitted elements. These observations are supported also by the effects of high concentrations of the osmoticum pentaerythritol which, at 0.4-0.5 M, virtually inhibited root growth and provoked the formation of tracheary elements with more pitted walls[9]. These facts strongly support the idea that the cytoplasm of the developing tracheary element monitors its own extension rate and deposits a wall whose character, as we shall see, is adaptively suited for the tissue concerned.

In organs such as coleoptiles and young leaves, the first protoxylem tracheary elements may differentiate when the organ is very small, i.e. 1 mm or less. The organ may then extend 50–100 fold in achieving its mature length. This enormous amount of passive extension of the protoxylem tracheary elements may dramatically modify their original structure and render them almost unrecognizable (Fig 1.3). In some vascular bundles it may be difficult to locate the positions of the protoxylem tracheary elements in cross-section but they can usually be seen in cleared, longitudinal slices (0.5–1 mm thick), taken through herbaceous stems, or in cleared whole mounts of mid-ribs of leaves. Such preparations offer an ideal, convenient and very simple way of studying the progressive alteration in structure that adjacent files of tracheary elements

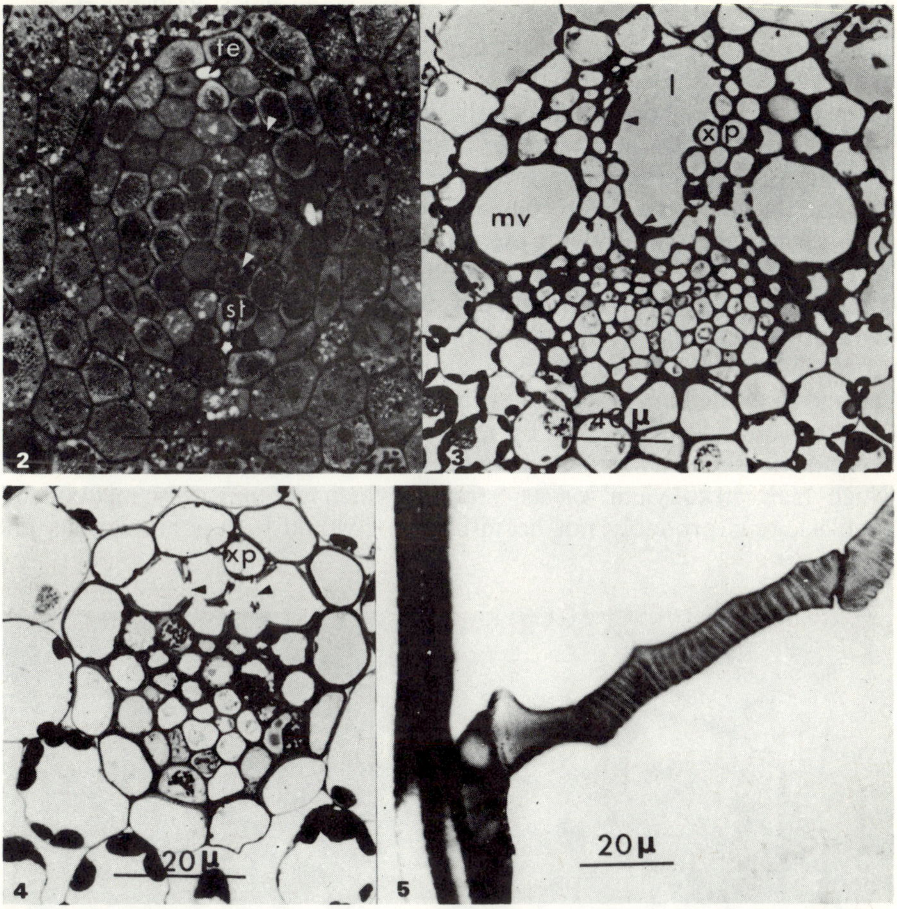

Figure 1.2: *Procambial strand cut in cross-section in an early stage of vascular differentiation from the mid-rib bundle of the primary leaf of a wheat seedling. Note numerous cells in division (arrows); st, sieve tube; te, tracheary element. Reproduced with permission from J. G. Swift and T. P. O'Brien.* Aust. J. Bot. **19,** 63-71, 1971.

Figure 1.3: *T.S. of midrib bundle from a wheat leaf. Two large metaxylem vessels (mv) flank a lacuna (l), created by the disruption of protoxylem tracheary elements, portions of whose secondary bands of thickening are visible at arrows. Note bilateral symmetry of the bundle about its midline and the xylem parenchyma cells (xp).*

Figure 1.4: *T.S. of a large intermediate bundle from a wheat leaf. All tracheary elements tend to be approximately the same diameter and very much smaller than the large vessels of Fig. 1.3. Xylem parenchyma cells (xp) are few. Note numerous intervessel pits (arrows). Figs. 1.3 and 1.4 stained with toluidine blue 0 at pH 9, and reproduced with permission from J. Kuo, T. P. O'Brien and M. J. Canny,* Planta **121,** 97-118 (1974).

Figure 1.5: *Part of a vessel in a transverse vein that interconnects the longitudinal veins of the leaf. The leaf was fed 1% ferric chloride and cleared in sodium sulphide. Note the curious patterns in which wall thickenings are deposited, with pits confined to selected areas of the wall where these cells abut xylem parenchyma cells (see Fig. 1.14). The remainder of the wall is thickened uniformly throughout, and lacks apertures. Reproduced with permission from J. Kuo, T. P. O'Brien and S. Y. Zee.* Aust. J. biol. Sci. **25,** 721-737, 1972.

undergo as organ growth-rate declines, and such preparations can be made quite readily by students in a class.

As Esau[22] has pointed out, the use of the terms protoxylem and metaxylem as used here, i.e. as terms to describe the early- and late-forming xylem, is the earliest established usage of these terms. Perhaps because the structure of the tracheary elements in internodes tends to vary in a predictable manner with the time of their differentiation, these terms have acquired other shades of meaning and other implications. Thus it is common to find the term 'metaxylem tracheary element' used in the literature as a synonym for 'pitted element'. Again, if a particular strand of procambium differentiates almost entirely after growth of an organ has finished (e.g. the smallest bundles in grass stems and leaves: Kuo *et al.*[53]), the tracheary elements will be only reticulate or pitted (Fig 1.4). It is common to see such strands referred to in their mature state as 'strands which lack protoxylem' or as 'strands which contain only metaxylem'. Such usage is probably not harmful but it would be less ambiguous and

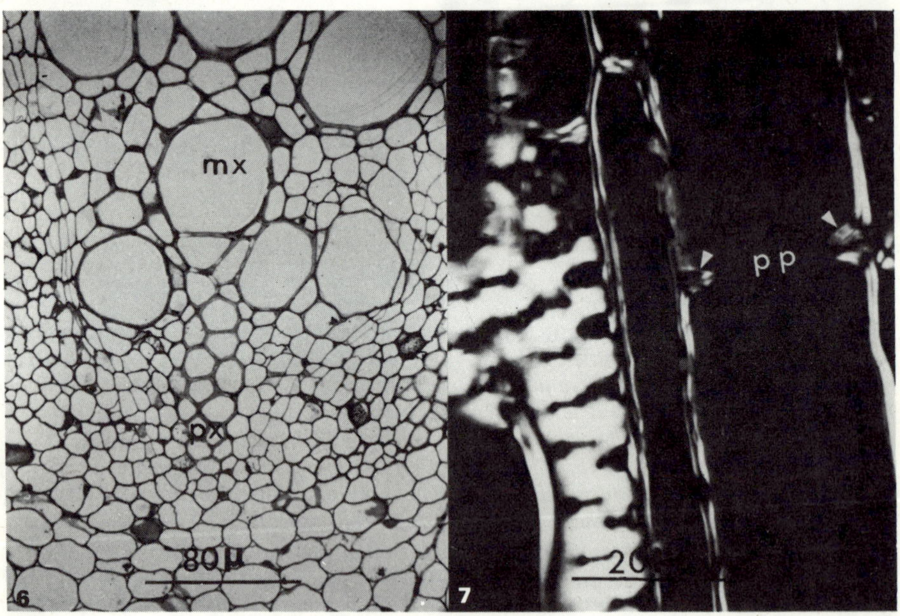

Figure 1.6: *Protoxylem (px) and metaxylem (mx) tracheary elements and associated xylem parenchyma cells in the cross-section of a pea root. Reproduced with permission from T. P. O'Brien and M. E. McCully.* Plant Structure and Development (London: Collier-Macmillan 1969).

Figure 1.7: *Longitudinal section of part of vessel perforation plate, pp, viewed between crossed polarizer. The rim of the plate consists of lignified secondary wall (arrows).*

therefore preferable to stick to the original use of the terms and to state that 'the later formed strands contained only pitted elements'.

The Differentiation of Tracheary Elements within the Procambium

The production of a mature primary-xylem tracheary element involves three major steps: the deposition of a characteristic pattern of wall thickening, the removal of the cytoplasm, and modification of the cell walls between adjacent elements to form interconnected units. Many of the basic facts about these processes were established long ago and before the advent of electron microscopy (see Esau[22], Roelofsen[82]), but some aspects have been further clarified in the last 20 years and many additional questions raised.

At the time it begins to deposit secondary wall material, the procambial cell is encased in an unlignified primary wall. It seems to be a general feature of primary-xylem tracheary elements that this original primary wall remains unlignified, in sharp contrast to the situation in fibres or in tracheary elements of the secondary xylem where all or most of this wall becomes lignified as differentiation continues. At this stage in their development, the fine structure of the cytoplasm is not particularly noteworthy, the cells containing the usual complement of cell organelles including leucoplasts, mitochondria, ER, dictyosomes and polyribosomes. Nuclei are large and contain prominent chromatin masses and one or more nucleoli. If, as is commonly the case, the cell is long and narrow, the nuclei are also elongated and may extend 30–50% of the cell length. When present, cortical microtubules are disposed around the cell periphery, the long axis of the tubules oriented circumferentially around the long axis of the cell (Fig 1.8, inset). Studies on *Azolla* indicate that the tubules are probably disposed as arrays of overlapping arcs in which the average tubule-length is probably a few microns[46].

Pickett-Heaps[76] was the first to show that one of the earliest signs that a procambial cell was about to embark upon differentiation as a tracheary element was a change in the distribution of cortical microtubules. The tubules grouped into bands (36 consecutive, regularly-spaced groups of tubules were found in one cell). He felt that initially, the bands of secondary wall-thickening formed *between* the groups of tubules which later came to overlie the bands, the position they continue to occupy during later stages of band development. This initial location of band development to the position between the tubule groups is yet to be confirmed by other investigators. However, other studies have confirmed that the tubules do group into bands. Hardham and Gunning[46], in a study that includes an analysis of tracheary element differentiation in the root of the water fern *Azolla*, show that, before the wall thickenings develop, the microtubules become grouped into bands whose number depends on the length of the

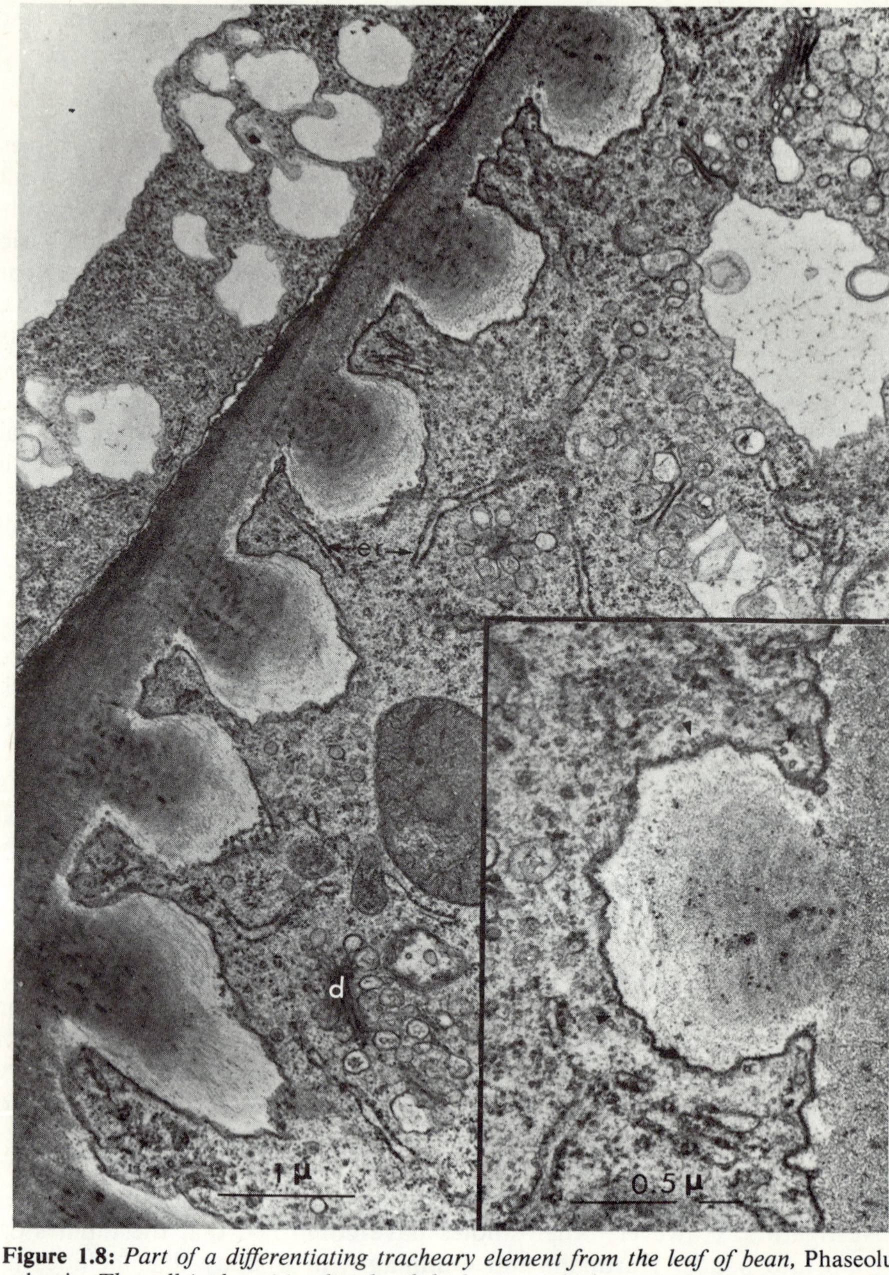

Figure 1.8: *Part of a differentiating tracheary element from the leaf of bean,* Phaseolus vulgaris. *The cell is depositing bands of thickening, which are probably partly lignified. The cytoplasm is rich in dictyosomes (d), and profiles of ER (er), some elements of which appear to penetrate between the bands of thickening. The inset shows microtubules (arrows) grouped above a band of thickening and absent from the interband regions.*

cell. During this period, there is an increase in the number of microtubules per μm of wall length, an increase which precedes the initiation of the wall bands. The maximum 'density' of microtubules, about 7-8 μm^{-1}, occurs when the deposition of the wall thickenings is half to two-thirds complete, the number reducing as maturation of the element nears completion. In this *Azolla* system, xylem differentiation advances acropetally towards the root apex in such a way that, at different stages of root growth, formation of wall thickenings is initiated in cells of widely different length. In all cases, the spacing between bands is approximately 2 μm and the maximum 'density' of microtubules found in the cell is the same.

The association of microtubules with bands of wall thickening, first reported by Hepler and Newcomb[48] in wounded *Coleus* internodes undergoing xylem regeneration, has been confirmed in all cases for which it has been looked with appropriate methods and can be safely accepted as a generalisation that probably holds true for all primary xylem tracheary elements that develop banded wall-thickenings (e.g. see Cronshaw and Bouck[15], Esau *et al.*[26], Hepler and Fosket[47], Esau and Charvat[25], among others). As pointed out earlier[67] Bierhorst and Zamora[6] suggested that all primary xylem tracheary elements are constructed initially as annular or reticulate elements and that pitted elements are derived ontogenetically from such elements (before cytoplasmic destruction of course!) by the addition of more wall material between the gyres of thickening. This question does not seem yet to have been studied, the only report of additional wall material being added to an annular element being contained in the work of Hardham and Gunning[46]. In their Fig. 11, they illustrate the addition of irregular wall ingrowths, reminiscent of the wall ingrowths of transfer cells, both to the original gyres of thickening and to the wall surface between the gyres. Unlike the wall ingrowths of transfer cells, these ingrowths become lignified. The remarkable tracheary elements found in minor veins of wheat leaves[53] would be particularly interesting to study in this connection since here, one face of the cell develops gyres of thickening, while the other face of the same cell has a completely solid wall (see Figs. 1.5 and 1.14).

One of the greatest difficulties that confronts electron microscopists is establishing cause and effect relationships between cellular structures and the functions they carry out. Since the discovery of the cortical microtubule by Ledbetter and Porter[56] this organelle has been widely implicated in regulating the pattern in which cell-wall microfibrils are deposited (see Newcomb[62], O'Brien[66], Hepler and Palevitz[49], Gunning and Steer[43]). The strong correlation between the orientation of wall microfibrils and cortical microtubules in differentiating primary xylem tracheary elements, is among the strongest circumstantial evidence that supports this hypothesis. And yet, despite many observations and a number of interesting experiments, the functions of the microtubules in these cells are tantalisingly obscure. Key observations were made by Pickett-Heaps[77] who showed that if colchicine was administered to procambial cells before the microtubules grouped into bands, no microtubules were detectable and no bands of wall thickening were formed. If the colchicine was applied to cells after the thickenings had begun to form, most of the microtubules disappeared and the regular pattern of wall deposition was destroyed. Wall deposition continued (eliminating the possibility that the microtubules were essential for wall synthesis *per se*), but the thickenings that were deposited were irregular in size and sometimes were combined into giant bands. Interestingly, there was no evidence that the orientation of the cellulose microfibrils within the irregular bands became disorganised. However, Hepler and Fosket[47] in a study of wound-vessel formation in colchicine-treated *Coleus* internodes, showed that colchicine treated cells lacked microtubules and also led to disoriented cellulose deposition within the irregular bands. They also made the remarkable observation that the presence of a neighbouring tracheary element (even a mature, dead element) was able to 'stabilise' the formation of normal thickenings, directly across the primary wall, in a colchicine-treated differentiating element that lacked microtubules. This 'influence' of an adjacent, mature tracheary element did *not* spread to sites on the opposite side of the differentiating element.

Some of the problems in interpreting studies that use inhibitors are raised by the work of Brower and Hepler[9] and others have been discussed before[66]. Brower and Hepler[9] applied colchicine or the herbicide isopropyl N-phenylcarbamate (IPC) to wheat and onion seedlings and then sampled the roots. 4×10^{-4}M colchicine for 24 hours destroyed the normal pattern of secondary wall deposition (evaluated in squash preparations by Nomarski or polarizing optics), while at the end of 5 days exposure, no normal tracheary elements were present. IPC, even at 10^{-3}M, had little effect at the end of 24 hours, but at 5 days, the elements produced were more like pitted elements. However, such roots experienced a profound growth inhibition and growth inhibition with the osmoticum pentaerythritol, produced similar, though not identical results. In onion, when IPC treated cells were examined after conventional fixation in the EM, the effects of the drug on microtubules were very variable. Some cells appeared to lack microtubules completely, others showed no reduction while in others the number of tubules associated with adjacent bands of thickening within the one cell was highly variable. However, in wheat, IPC at 10^{-3}M behaved just like colchicine, lower concentrations yielding variable results just like those seen in onion.

It may turn out to be impossible to distil useful information about the role of microtubules from inhibitor studies of this kind. The recent work of

Gunning and his colleagues upon the *Azolla* root show that cells entering upon differentiation arrive with a population of microtubules that consists of 'an initial quota' and 'subsequent interpolations'. In the case of the endodermal cell file in the *Azolla* root, they have shown that these two components of the cell's microtubule population are differentially sensitive to treatments that affect microtubules (high-pressure, D_20, colchicine, low temperature). In the case of tracheary element differentiation in *Azolla*[46] it is known that the onset of wall deposition is accompanied by a marked rise in the number of interpolated tubules. Studies at this level of detail have yet to be carried out for other types of meristem, but it is obvious that, if different components of the microtubule population are differentially sensitive to drugs, it will be impossible to evaluate drug action by examining random cells in tissues treated for 1–5 days. The stage of development reached by the cell at the time the drug arrived cannot be determined.

The role of adjacent wall thickenings in causing wall deposition is equally difficult to evaluate. If there was a general influence of wall thickenings in older cells upon the induction of pattern in their procambial neighbours, one might expect to find it expressed all the time. However, as Gunning and Hardham[38] point out for *Azolla*, and as Esau and Charvat[25] do likewise for bean, this association, though not uncommon, is by no means obligatory.

Since the early work of Cruger[16] and Sinnott and Bloch[90] there have been suggestions that the cytoplasm is 'pre-patterned' in some way that reflects the future pattern in which the wall material will be deposited. It is not entirely plain just what features of the cytoplasm may have generated these suggestions (see Goosen de Roo[36]). Electron microscopy of a great many cells of higher plants[43] shows that, in addition to cortical microtubules, the cell cortex is usually populated by a very complex network of ER. The disposition of this network, whose presence was first emphasised by Porter and Machado[78] in onion root-tip cells, has yet to be analysed from serial sections in any developing plant cell. Nonetheless, it is plain from the many published images of this system that it consists of an array of cisternal rough ER attached to more tubular and/or sac-like elements of semi-smooth and smooth ER, which penetrate on occasions very close to the cell membrane. The temporal stability of this network is not known; in particular it is not known if it is exposed to gross displacement during cytoplasmic streaming, or whether it remains anchored in a more gelled region of the cell cortex, free of major disturbance from streaming. In tracheary elements that have begun to deposit bands of thickening, parts of this ER network are often visualised penetrating between the bands (Fig. 1.8 and inset). One of the most dramatic illustrations of this phenomenon is contained in an early paper by Pickett-Heaps[76]. Although some authors claim not to be able to detect this arrangement of the ER, it is my view that this network of ER is probably always present between bands of thickening, and sometimes overlying them, as long as one allows for the difficulties that arise in interpreting different planes of section (see discussion in O'Brien[67]). More recently, Esau and Charvat[25] illustrate in beans a particularly clear example of this association, which is also detectable in *Azolla* roots (Fig. 10, Hardham and Gunning[46]).

In looking through the literature on this topic, it is also true that most authors express the view that the amount of ER and dictyosomes seen in the cortical cytoplasm of differentiating tracheary elements increases during the onset of development of the bands. There have not yet been thorough stereological analyses of this question of the type carried out by Briarty *et al.*[8] on wheat endosperm. Nonetheless, this general feeling is supported by the data obtained by Goosen de Roo[35] who measured ER and dictyosome abundance in a 'standard sampling region' of the cell cortex in cells at different stages of development. A full stereological analysis will remove the possibility that the cell redistributes its population of ER and dictyosomes, rather than actually increasing it.

The composition of the bands of thickening is not known from direct biochemical analysis of isolated band material, but there is no doubt that they are rich in polysaccharides, including cellulose microfibrils whose average orientation (determined by polarised light), is parallel to the band axis (Roelofsen[82], p. 118). Since the late 1950's, stimulated initially by observations on root-cap and root-epidermal cells, the notion that the ER, dictyosomes, and their vesicles constitute the sub-cellular sites of synthesis and transport of cell-wall polysaccharides has been a central tenet of most who have published in this field (see reviews by Northcote[63, 64], O'Brien[66], Chrispeels[13], and references therein). Recent reviews of the biochemistry of polysaccharide synthesis (Elbein[21], Clarke *et al.*[14]) and of the difficulties of interpretation that beset the results obtained with cell-free systems[79] continue to suggest the need for caution in accepting this widely-held idea. It is an idea of attractive simplicity, but hard-core evidence to support it, especially for the case of the wall-thickenings of tracheary elements, is sparse indeed, and fiendishly difficult to obtain. There has always been a strong penchant to assign wall-forming role to any evident differentiation of the cytoplasm of cells known to be engaged in wall formation, ignoring other possible roles for the cytoplasmic state observed[66]. The ER-dictyosome activity seen in cells engaged in wall formation (Fig. 1.8) might well be engaged in the events associated with ion and nutrient exchange between the free space and the cytoplasm, in comparable events between cytoplasm and vacuole, in the formation of vacuoles, in the production and secretion of cell-wall enzymes or in the elaboration and transport of phenolic and other glycosides, to name but a selection. Radio-autographic and EM histochemical attempts to implicate the ER and dictyosomes in tracheary element-wall formation were not entirely conclusive; autoradiography of tissue treated with glucose-6-^{3}H showed only weak incorporation into the wall up until the time the bands began to appear, and marked incorporation afterwards[76]. In the cytoplasm, a small percentage of the dictyosomes was labelled, and a significant amount of label appeared in the ER, though the resolution of the method does not permit one to be sure that label is really in the ER cisternae between the bands. In any event, attempts to demonstrate that the ER label was in polysaccharide by the use of the silver hexamine reaction were negative, leaving the significance of the observations somewhat in doubt (see discussion in O'Brien[66]).

Dramatic alterations in nuclear volume and structure often accompany the early stages of differentiation of tracheary elements in both the primary[59] and secondary xylem[2], and these events have recently been investigated in corn leaves by Lai and Srivastava[55]. It is clear from List's study[59] that varying degrees of nuclear enlargement accompany cell growth during differentiation of metaxylem elements in the roots of different species. In one case *(Marsilea quadrifolia)* the differentiating tracheary elements became multinucleate, displaying synchronous mitosis; cells were detected with eight metaphases present. In others (e.g. corn) nuclear volume increases dramatically and appears to be accompanied by endopolyploidy, generating nuclear DNA contents up to the 64C level. In others *(Dennstaedtia cicutaria)*, nuclei did not enlarge beyond the volume they occupied in the apical cell. Endoduplication, leading to 4, 8 and occasionally 16C levels, has been reported during tracheary element differentiation in cultured pea-root segments by Torrey and his colleagues[58,74,92] but Phillips and Torrey[75] did not find a preferred level of DNA content/nucleus that was associated specifically with tracheary element differentiation.

In a light and EM study of vessel element differentiation in corn leaves, Lai and Srivastava[55] found a distinct increase in the volume of the nucleus and nucleolus. DNA and histone levels doubled, and the nucleolus differentiated to show fibrillar, granular and vacuolar regions. These changes preceded deposition of the secondary wall. Histone levels doubled again during wall deposition, to be followed by the development of lobed nuclei, clumping of chromatin, shrinkage of nucleolar volume, and loss of nucleolar zonation. The authors interpret these observations to indicate that early in vessel element differentiation, there is a high template activity of DNA both in the nucleoplasm and in the nucleolus. This work supports the conclusions of Avanzi *et al.*[1] who obtained evidence for amplification of ribosomal cistrons during maturation of metaxylem in the root of onion, *Allium cepa*.

Endopolyploidy and bi- and multinuclear states are common in the somatic tissues of plants (see D'Amato[20], Oksala and Therman[70], and references therein) and appear to be the outcome of uncoupling DNA synthesis from karyokinesis and/or cytokinesis, generating varying degrees of genome amplification within the cell. These phenomena are probably closely associated with the problems of administering a large cytoplasmic volume; it is unlikely that they are associated in any very specific manner with the events of tracheary element differentiation.

Destruction of the Cytoplasm and Modification of the Cell Wall

According to Lai and Srivastava[55] degeneration of the nucleus precedes that of the cytoplasm and occurs before cell-wall development has been completed. Whether this is generally true is not known since it is rare to encounter cells with partially degenerate protoplasts. One imagines that the reason this is so is that autolysis of the cytoplasm is rapid once the enzymes that bring it about are released, and that they rapidly degrade the macromolecules of the cell to units which are not stabilized by fixation. It is therefore difficult, and probably unwise, to place too much reliance on the appearance of degraded cytoplasm seen in late developmental stages. Nonetheless, such evidence as existed at the time led O'Brien[65] to propose that the cell membrane was among the last to be destroyed and that it was not until it was destroyed that the unprotected parts of the primary wall were exposed to the hydrolases that degrade them. Such a view assigns a very passive role to the cytoplasm of the cell in the matter of wall digestion: released enzymes simply degrade all parts not protected from attack. Against this view must now be set the observations of Esau and Charvat[25], who show very clearly that the primary walls between gyres of lignified thickening, as well as that part of the cross wall that will form the pore by its removal, swell noticeably in maturing tracheary elements that have an intact protoplast. One might dismiss this swelling of the primary wall as due to deposition of more wall material were it not for the results of the labelling experiments performed by Pickett-Heaps[76]. These results show that the labelling of the primary wall in the interband region is minimal, as is labelling of the wall in the future pore region (his Fig. 5). While not conclusive, these results suggest that the swelling of the primary wall regions which will be digested marks an early action of wall degrading hydrolases which have been added to the wall prior to protoplast collapse, an action which is completed as the cytoplasm is destroyed. Such a view would assign a more active role in the release of the wall-degrading hydrolases by the cytoplasm than I envisaged earlier.

The pattern in which the wall material is removed is most interesting. All lignified walls appear to be totally resistant to digestion[65]. At lateral walls, the result depends on the nature of the associated element. Where the

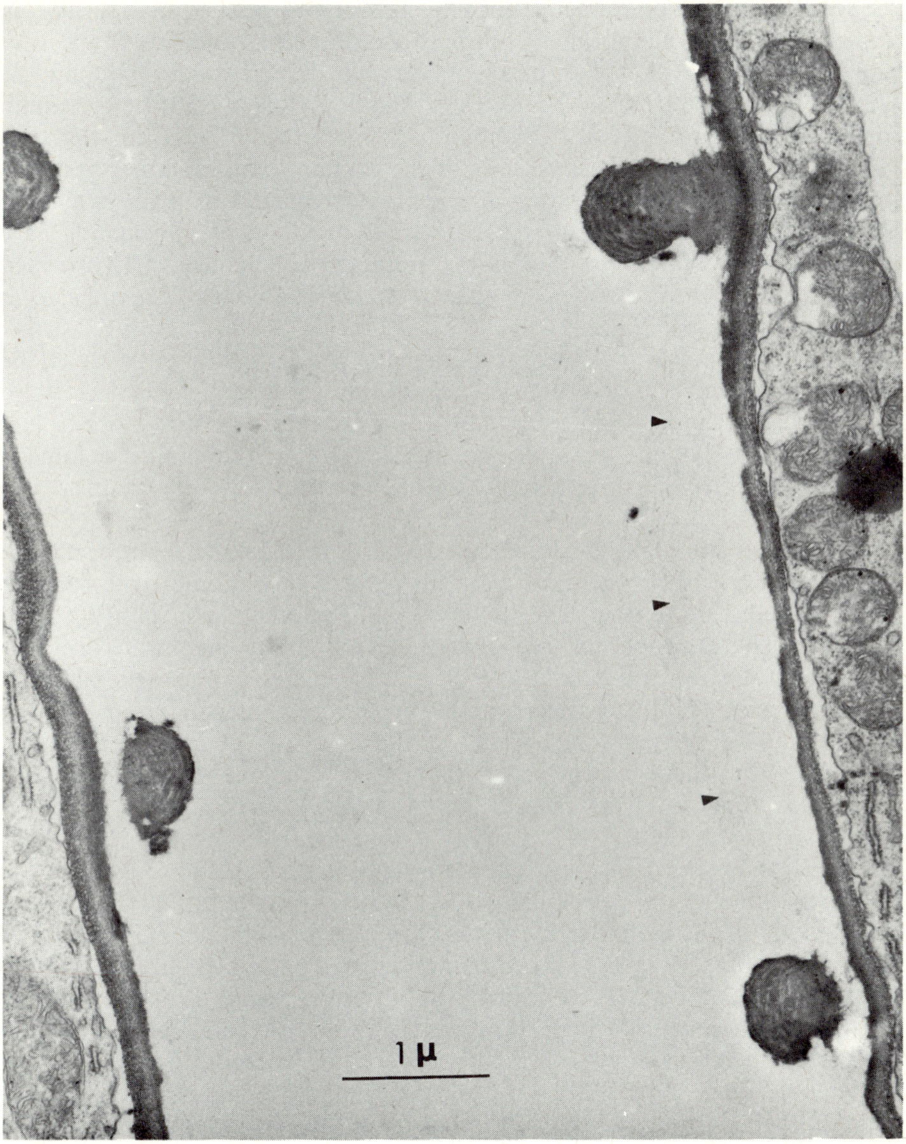

Figure 1.9: *Longitudinal section of part of a stretched annular tracheary element and adjacent parenchyma cells from the vascular bundle of an oat coleoptile. Remnants of a stretched, hydrolysed wall can barely be detected in the interband region (arrows). Note that in the same element, primary wall beneath the bands of thickening may be present or absent, and that where it is absent, the ring is tilted.*

adjacent element is another tracheary element, the two primary walls are reduced to a mat of fibrils, termed a 'hydrolysed wall'[69], comparable in appearance to pit membranes seen between adjacent tracheids (Figs. 1.10 and 1.11). However, those parts of the two primary walls and middle lamella which are overlain by gyres of lignified thickening usually remain intact, anchoring bands of thickening to one another in adjacent elements (Figs. 1.9-1.11). Why these regions of the primary wall are not digested is not known but it is clear that if they are protected in some way, the protection is not absolute. Occasionally, rings of thickening are encountered in which the underlying primary wall has been hydrolysed (Fig. 1.15); when this happens in an element with annular thickenings, such a ring often tilts (see Fig. 1.9).

Where the tracheary element is adjacent to a xylem parenchyma cell, the interband region of the primary wall of the tracheary element is again converted to a hydrolysed wall and digestion appears to stop more or less at the middle lamella region. The extent to which the middle lamella material is degraded is variable within one cell and between adjacent cells (of Figs. 1.12 and 1.13).

The composition of the hydrolysed wall is not known with certainty. Such walls are birefringent, and fail to stain with cationic dyes such as toluidine blue 0, and are negative in the Periodic acid/Schiff's (PAS) reaction. These properties are shared with the pit membranes of the intervessel pits in the secondary xylem of willow[65] and several other woody species[19] and, most interestingly from a phylogenetic point of view, by the 'oblique primary end-walls' of the hydroids of certain mosses[87]. It is likely that these hydrolysed walls contain cellulose microfibrils which give them their birefringence. It might at first be thought that this is inconsistent with their PAS negativity since cellulose is usually believed to be PAS positive[52]. However, under the standard conditions used in the PAS reaction (15 min periodic acid, 15–30 min in Schiff's reagent) cell walls with high cellulose contents are PAS negative (O'Brien and Thimann[69], Figs. 5-7). Prolonged oxidation with periodic acid (45–60 min) does render such walls PAS positive (unpublished results), so that cellulose in walls is not inherently PAS negative, but rather it is relatively unreactive. The electron density of the fibrillar mat seen in hydrolysed walls is quite variable (of Figs. 1.10, 1.11, 1.12 and 1.13), and it is not safe to conclude that it is the cellulose which is stained. As Heyn[51] points out, uranyl acetate does not stain crystalline cellulose and Ray[80] showed that the α-cellulose residue of chemically extracted outer epidermal walls had almost no affinity for the uranyl acetate and lead salts that give most of the contrast in stained EM sections. These facts suggest that other constituents are present in hydrolysed walls, constituents that may be coating the cellulose microfibrils in the wall remnant, and thereby negatively staining

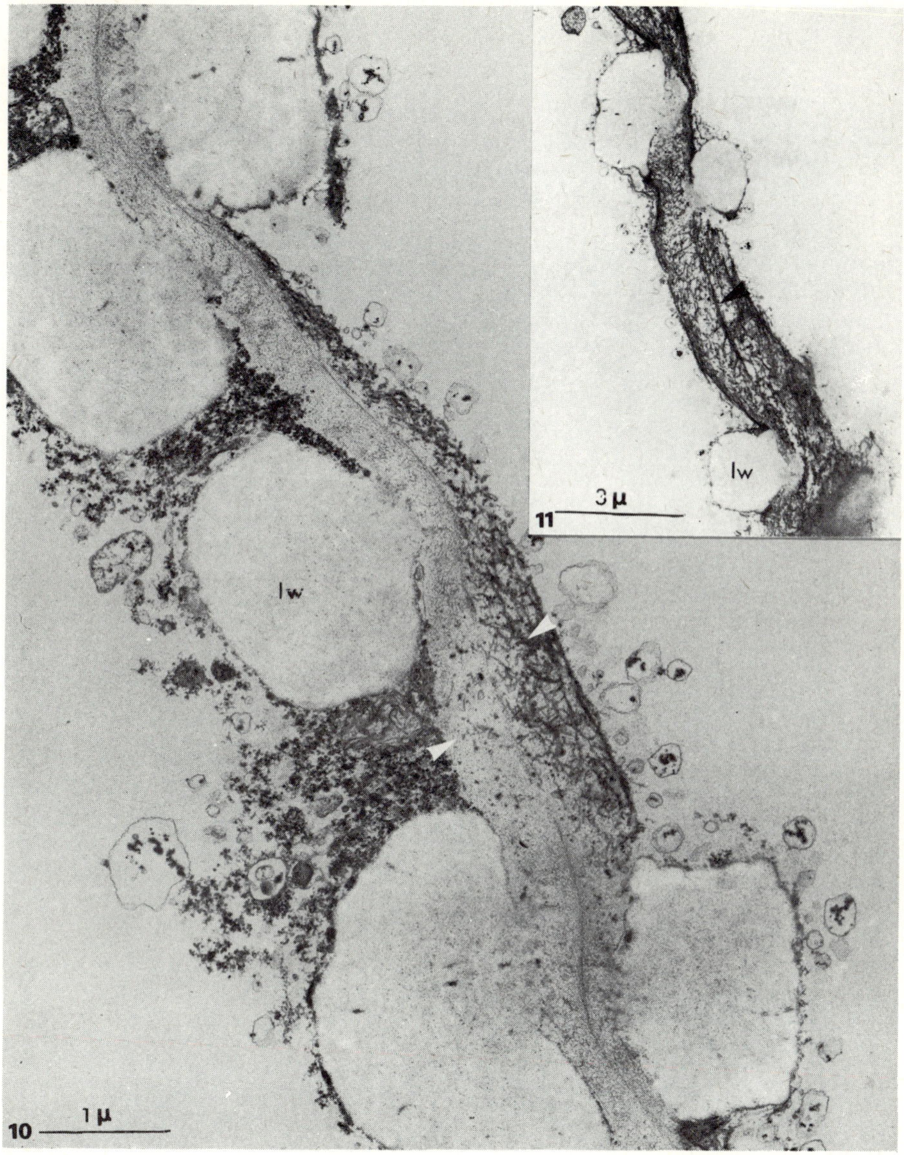

Figures 1.10 and 1.11: *Side walls between vessel members with completely developed secondary walls in xylem from petiole of* Phaseolus vulgaris, *seen in L.S.*

Figure 1.10: *In the vessel member to the right, degradation of the primary wall is almost complete (arrow) and little cytoplasmic residue is left in the cell. To the left, the primary wall is partly hydrolysed between the coils of the helix (arrow), and a considerable amount of cytoplasmic debris is present in the cell. lw, lignified wall.*

Figure 1.11: *Wall between two vessel members that were much stretched and probably no longer functional. The secondary wall is attached to the coarse set of fibrils that remained after the primary wall was degraded. A dense strand of fibrils indicates the region of the middle lamella (arrow). lw, lignified wall. Figures 10 and 11 reproduced with permission from K. Esau and I. Charvat.* Ann. Bot. **42,** 665-667, 1978.

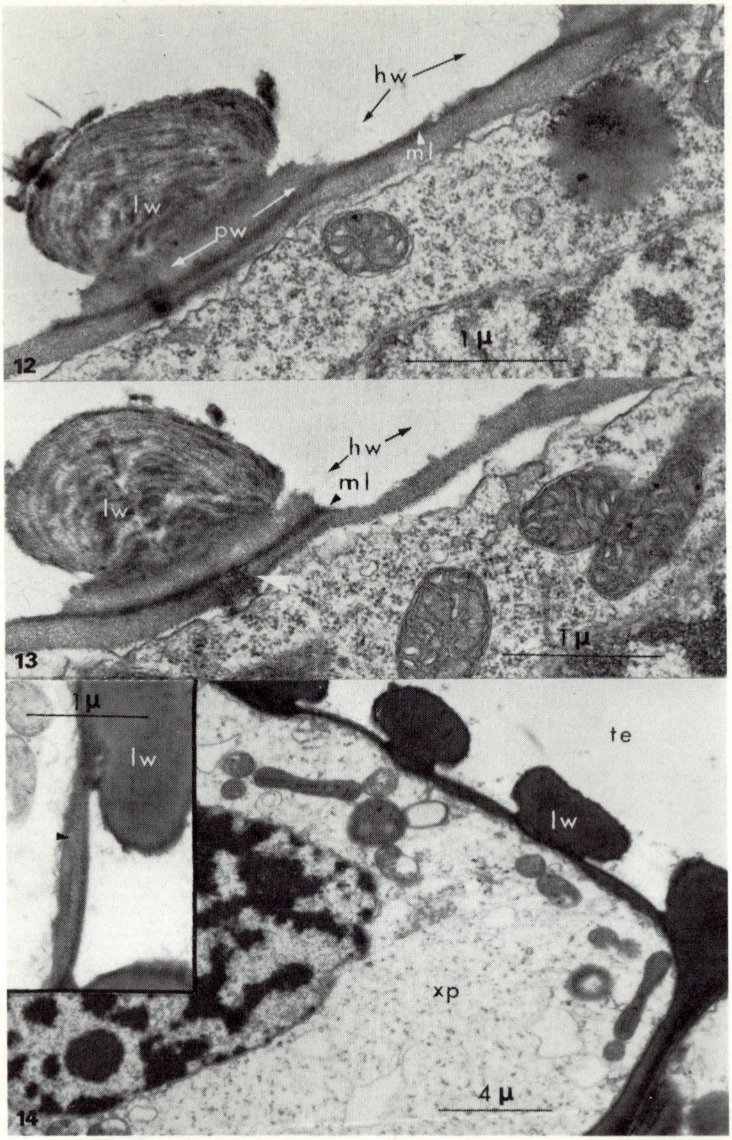

Figures 1.12 and 1.13: *Two views of the hydrolysed region (hw) in the interband region, showing the varying degree to which middle lamella material (ml) is removed. pw, primary wall of tracheary element; lw, lignified band of secondary wall. Note occluded plasmodesmata in Figure 13 (arrow). Adjacent cell is xylem parenchyma cell (oat coleoptile xylem).*

Figure 1.14: *Part of tracheary element (te)/xylem parenchyma (xp) cell boundary in transverse vein of a wheat leaf. Such tracheary elements have pits only where they abut the xylem parenchyma cells (see Figure 1.5). Note the swollen wall of porous texture (inset, arrow) in the pit. lw, lignified wall of tracheary element.*

the cellulose in the traditional EM sections. The wall residues in bean[25] are particularly strongly stained (see Figs. 1.10 and 1.11).

The presence of a cellulose residue in the hydrolysed walls makes it unlikely that cellulase is responsible for the complete removal of the two primary-wall segments in the pore region of the perforation plate, unless the secretion of the enzyme is strongly restricted to the wall region(s) that are to be removed. It seems more likely that the unlignified parts of future perforation plates are converted first to hydrolysed walls and that the pore(s) of vessels (Fig. 1.7) is created by the collapse of the relatively unsupported finely fibrous walls. This collapse may be initiated by the transpiration stream. However, images of the steps in pore formation in primary xylem tracheary elements are elusive, though Esau and Charvat[25] claim to have encountered an image of a finely fibrous mat of wall material at the site of a future pore, but the results are not yet published. Published work on wall hydrolysis in primary xylem is still based on very few species, and many of the observations are casual and have been made incidental to some other study. Perhaps, this situation may soon be rectified and the list of material examined increased, to include, if possible, those species in which Bierhorst and Zamora (1965) report that even lignified wall can be removed during the formation of some perforation plates. Digestion of wall material to create the pores of vessels and the pit membranes of inter-tracheid and intervessel pits has obvious adaptive advantages, but it is interesting to note the added adaptiveness that lateral-wall hydrolysis has in the case of the annularly-thickened tracheary element. Such cells form characteristically in tissues undergoing rapid elongation. To form a pipe, these cells must lose their contents, but at the same time they lose their capacity to plasticise the extension of the unlignified primary walls in the regions between the bands of thickening. The view I had of this problem in 1970 remains the same:

> What would happen in a rapidly elongating organ to the primary wall of a protoxylem tracheary element once its protoplast was destroyed? It is unlikely that the adjacent living cells could plasticise the wall of the dead tracheary element. At best, the dead protoxylem would act to check the growth rate of the extending organ, and at worst the wall might rupture, perhaps tilting the rings of thickening and collapsing the all-important water conduit. But this problem is solved because the non-cellulosic polysaccharides are removed from the primary wall *between* the bands of thickening, releasing the cellulose microfibrils in those regions from interaction with the non-cellulosic substances. Thus, precisely at the time when the vital activity of the tracheary element is unable to plasticise the wall, the 'concrete is removed from the reinforcing steel' and the resistance of the wall to passive

stretching must largely disappear. The lignified bands of thickening remain, anchored to similar bands in neighbouring cells or to walls of living vascular cells. These bands prevent collapse of the passively stretching element, giving time for the development of a new file of tracheary elements (see Figs. 1.4, 1.9).

In summary, it appears that some form of genome amplification is commonly associated with the early stages of tracheary element differentiation, but it is unlikely that it is a *sine qua non* for such differentiation. The development of a population of grouped microtubules seems to be a key morphogenetic event in the transition from procambial cell to differentiating tracheary element, but the interpretation of experiments in which grouped microtubules are removed from cells by colchicine is not free of difficulties, and the ontogeny of pitted tracheary elements is still not sufficiently well-known to assess whether grouping of microtubules is always associated with tracheary element induction. In cells which deposit a banded wall, the initial grouping of tubules into bands and their subsequent association with the bands of thickening during wall development seems to be a major feature that determines the controlled pattern of wall deposition, but there is evidence that the wall pattern can also be determined in part by the pattern of pre-existing wall thickenings, though this is not always so. The ER shows a distinct spatial distribution in the cell cortex, being more abundant between the bands of thickening, and the amount of ER and Golgi vesicles in the cell cortex increases significantly during the deposition of the wall. It is widely accepted that the ER and Golgi vesicles contribute polysaccharides to wall deposits in this and other cell types but hard-core evidence to support this view is still slim. Towards the closing stages of maturation, those parts of the non-lignified primary wall that are not overlain by lignified thickenings are attacked and converted to a hydrolysed wall which consists probably of cellulose microfibrils coated by varying amounts of other materials. Whether the pattern in which this wall digestion takes place is regulated in an active way by the controlled release or activation of hydrolases, or whether it originates in a more passive manner by a generalised action of the autolysing protoplast on unprotected parts of the wall, cannot yet be decided unequivocally. The total loss of part of the wall from the pore region of vessels probably involves an initial production of a hydrolysed wall, with passive removal of the pore segment; published micrographs to support this view are not yet to hand, and a localized action of cellulases cannot be excluded.

Control of the Distribution and Nature of the Cell Types Differentiating as Xylem from the Procambium

One cannot long study the anatomy of organs of vascular plants without realising that the pattern in which the xylem differentiates is subject to a number of controls which must act in a precise manner in both space and time. Thus, in roots, the protoxylem tracheary elements differentiate on the outside of the procambial cylinder (adjacent to the pericycle; exarch differentiation, Fig. 1.6), while in dicot stems the protoxylem commonly forms towards the inner surface either of the procambial annulus (in stems with a continuous ring of procambium) or of the isolated procambial strands (in stems with distinct primary bundles). Other patterns are known. A moment's reflection on the variety of stelar types found in lower vascular plants also shows that the patterns can be specific for particular groups. However, the control processes at work here may have subtler aspects for it seems that the positions occupied by the tracheary elements within the xylem, as well as their diameter, may also be precisely regulated. Precision control of cell position is shown at its best in the work of Gunning *et al.*[39, 40] on the *Azolla* root, and the unexpected effects of precise control of vessel diameter in the work of Kuo *et al.*[53, 54], studying the vascular tissues in wheat leaves (Figs. 1.3 and 1.4). The venation of such leaves is striate, with the veins mostly longitudinal but connected one to the other 3–4 times per cm by tiny cross-bridges (Fig. 1.5), 0.02 cm long, containing a single narrow vessel (3–5 μm in radius). The longitudinal veins are of two main kinds: about one-third of them contain vessels of large diameter, while the other two-thirds contain vessels of much smaller diameter (cf. Figs. 1.3 and 1.4). Calculations showed that the 20 large-diameter vessels present in the cross section of the leaf accounted for 96% of the water flow through the leaf, the remaining 126 tracheary elements accounting for the balance. Experiments with iron salts[53] and with dyes (unpublished) confirm that the large vessels in the few veins that have them do indeed do what the calculations suggest, for the vessels in the smaller longitudinal veins are filled laterally through the short cross-bridges by tracers moving rapidly up the large vessels, not by flow from below within the smaller veins. This system was seen by Kuo *et al.*[54] as a device to help separate the outwardly directed water-flux from the inwardly directed sugar-flux, nearly 100 times smaller than the water-flux. One cannot help but wonder if the small diameter of protoxylem tracheary elements in roots, and the fact that they lie on opposite arcs to the primary phloem (Fig. 1.6) may be associated with a comparable need to separate the inwardly directed water-flux from the outwardly directed sugar-flux in the root tip. Since flow in a tracheary element varies with the fourth power of the radius, it provides a fine control on the flow rate, and whether tracheary elements are acting as high-flow conduits or as low-flow backwaters with perhaps greater opportunity for diffusive exchange with their neighbouring cells, will depend sensitively upon their radius. The physiological significance of the precise patterns in which tracheary elements are produced is therefore probably much greater than we currently realise and provides additional motivation, if any were needed, for studying the control mechanisms. Patterns of vascular development in roots have been most intensively investigated in recent years[89]. It is plain that pattern control resides in the tissues of the quiescent centre and in those immediately adjacent to it in what Feldman and Torrey[29] call the 'proximal meristem' (see Feldman and Torrey[30], Feldman[28], Charlton[12] for references). Barlow[3] has offered an interesting model for the behaviour of the root apex but the detailed mechanisms which produce vascular pattern in the root apex still remain obscure.

A successful manipulation of the kinds of cells differentiating in the secondary xylem from cambial tissues has been reported by Hess and Sachs[50], working with *Phaseolus vulgaris*. It was possible to control the type of xylem formed in the stem by the selective removal of leaves of different ages. Many vessels formed below young leaves, while a single mature leaf caused the formation of a large volume of xylem consisting almost exclusively of fibres (see also Sachs[84]). It appears that the stimuli emanating from young and mature leaves differed by at least one component, the activity of a mature leaf being imitated by a mixture of auxin and gibberellin, while that of a young leaf could be imitated by auxin alone.

Xylem Parenchyma Cells

Situated in close proximity to the contents of the transpiration stream, xylem parenchyma cells are strategically located for metabolite exchange between the apoplast and the symplast and, by the development of outgrowths, to modify the available pathway for the transpiration stream. The role of these cells in the storage of reserves has always been evident in ray parenchyma cells of the secondary xylem of sapwood where the cells are often filled with starch. What, then, is the nature of the apoplast/symplast junction?

In the secondary xylem, vessel-associated parenchyma cells often develop, in addition to their normal primary and secondary walls, an extra layer, originally called the 'protective layer' by Schmid[88]. In oak[61] and willow[65] this layer of wall material is deposited asymmetrically in the cell, being thicker on the side adjacent to the vessel lumen. In willow, this wall material had an open porous texture and its staining reactions were consistent with those of primary wall material, even though parts of the layer were clearly deposited internal to a lignified secondary wall. At the vessel/parenchyma cell pits, it appeared that no sharp boundary existed between the original primary walls of the vessel and parenchyma cell and the 'protective layer'. However, as Schmid[88] and Czaninski[19] show plainly, this is not always so, the 'protective layer' in some species being sharply differentiated from the rest of the wall at the pit, while in other species, it can be absent[17]. Czaninski[19], using EM histochemistry, has confirmed within the limits of those methods, that when the layer is present at the pit, it is unlignified, rich in acidic polysaccharides, and in hemicelluloses.

Whether such wall material in the parenchyma cell/vessel pit plays any role in protecting the vessel-associated xylem parenchyma cell from attack during the closing stages of tracheary element differentiation is a moot point. In the primary xylem, there is no doubt that comparable regions of the walls of xylem parenchyma cells sometimes show evidence of partial degradation (Fig. 1.13). On occasions, one encounters images that can be interpreted as the addition of extra wall material by the cytoplasm to just that area of the xylem parenchyma cell-wall which is adjacent to the tracheary element, the whole being strongly reminiscent of the secondary xylem image (Fig. 1.15). Unfortunately, such images are not encountered all of the time in one species, nor in all species and their interpretation has to remain open. Perhaps such reaction walls have more to do with the induction of tyloses (Figs. 1.16 and 1.17), wall outgrowths from xylem parenchyma cells long known to be involved in the partial or total plugging of tracheary elements in both primary and secondary xylem (see Foster[31], Meyer[60], Meyer and Côté[61], Czaninski[17], O'Brien[67] for references). The ability of xylem parenchyma cells to secrete copious amounts of gums and resins, especially during the formation of heartwood[23], or when challenged by infections[11] should also be borne in mind when interpreting such images.

Experiments with ferric chloride showed that the non-lignified wall of the xylem parenchyma cells in the transverse veins of wheat leaves was apparently more porous to this tracer than adjacent regions of the lignified wall[53]. The cytochemical properties of this type of wall in the veins are very similar to those found in xylem parenchyma cells when they differentiate the wall-membrane apparatus characteristic of wheat transfer cells[18]. The taxonomic distribution of transfer cells in leaves and nodes, and the many sites at which these cells are probably involved in solute accumulation and secretion, have been reviewed in detail[41, 42, 71], and only a few points need to be noted here. Pate *et al.*[72] showed that vascular transfer cells were particularly well-developed where differentiating vascular tissue of the plumule was in close proximity to the cotyledonary traces in a dicot seedling. Autoradiographic evidence was obtained for a pathway from the xylem of the cotyledonary traces to the tissues of the plumule via the xylem transfer cells. Xylem transfer cells are strongly developed in all nodes of wheat and probably in all other grasses[93, 94]. In the vegetative node, it has been shown[10] that the differentiation of these transfer cells is initiated soon after the appearance of the first tracheary elements in the departing leaf-trace bundles and reaches its peak just as the leaf to which the bundle

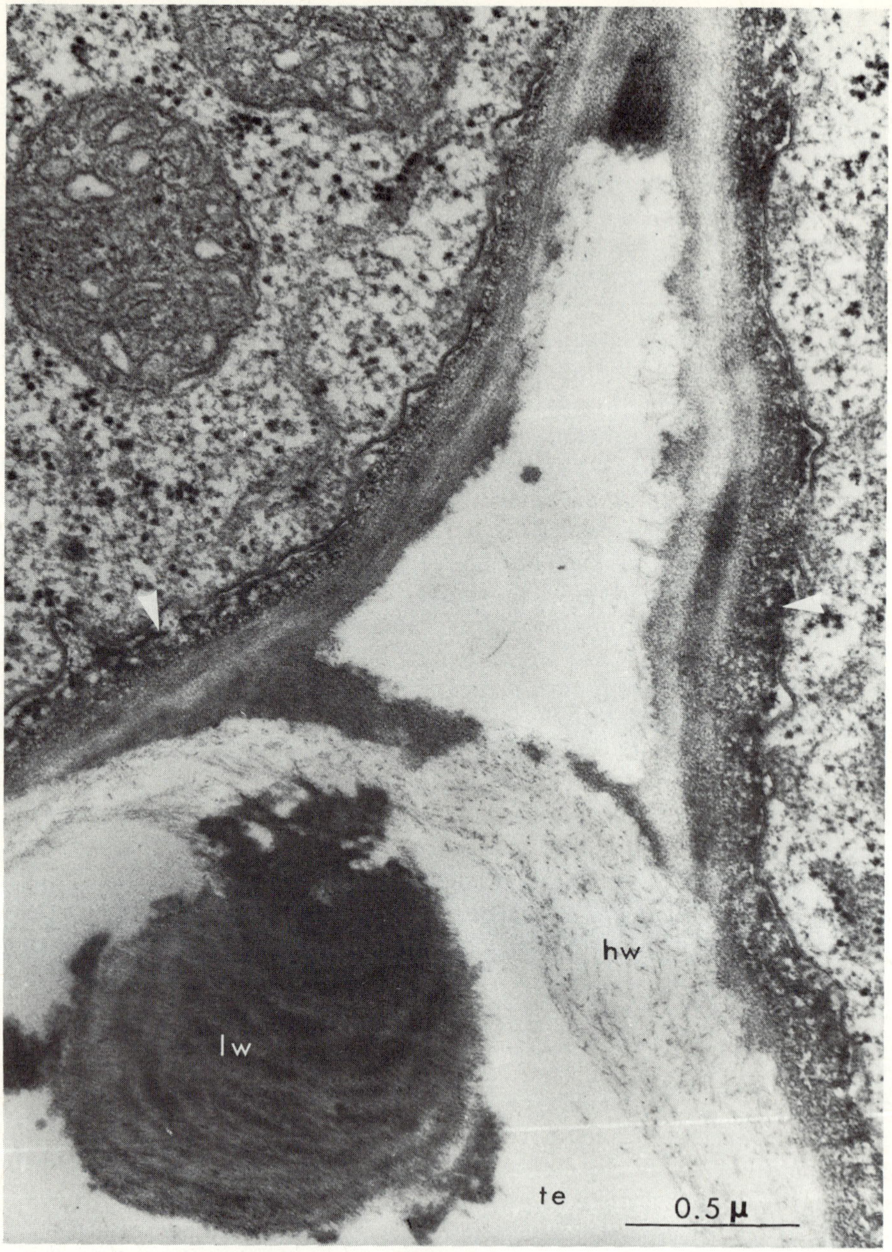

Figure 1.15: *The junction between a tracheary element (te), and two xylem parenchyma cells in rye coleoptile xylem. Note the hydrolysed wall (hw) of the tracheary element, which in this case extends beneath the band of lignified thickening (lw). The intercellular material at the cell junction is partly digested and the wall of the adjacent parenchyma cells has a curious appearance (arrows) consistent with the view that additional wall is being added to the cell surface adjacent to the tracheary element.*

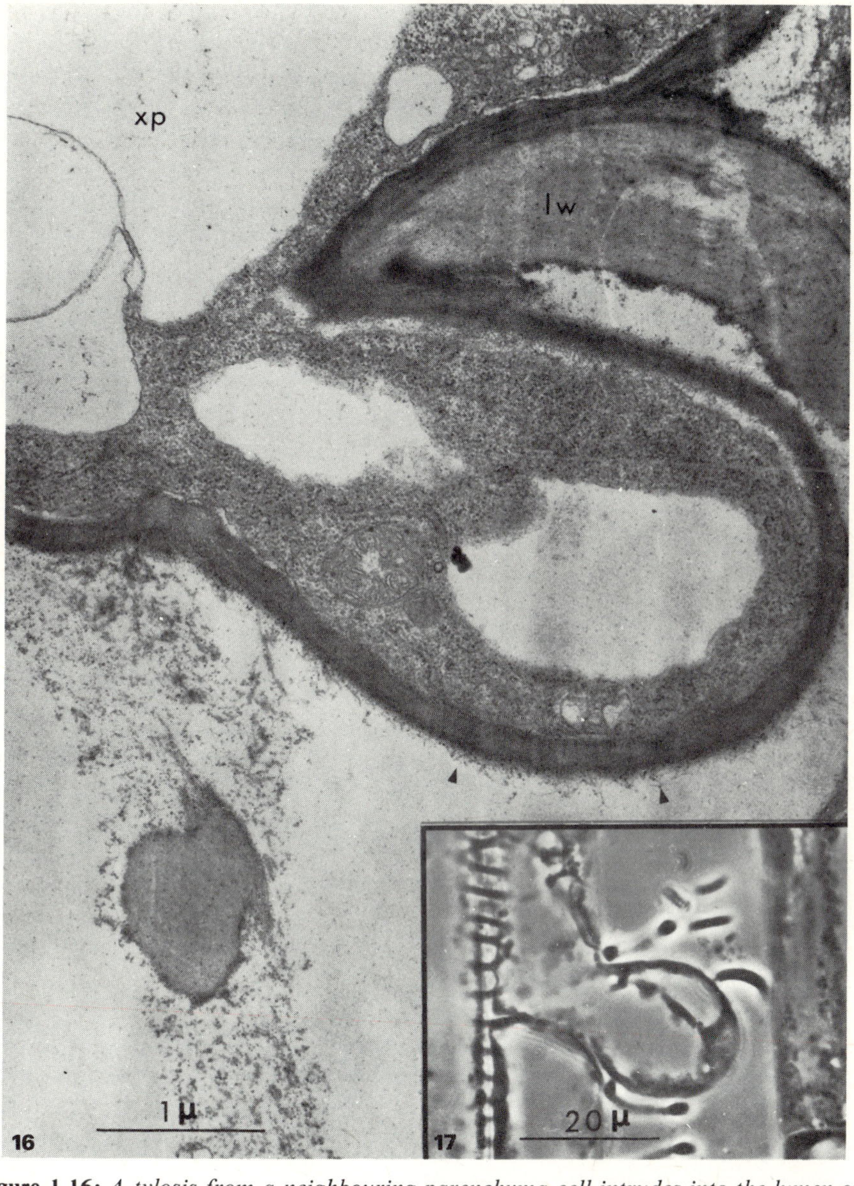

Figure 1.16: *A tylosis from a neighbouring parenchyma cell intrudes into the lumen of a stretched annular tracheary element in oat coleoptile xylem. Phase contrast. Reproduced with permission from T. P. O'Brien and K. V. Thimann,* Protoplasma **63,** 443-478, 1967.

Figure 1.17: *Tylosis from a xylem parenchyma cell (xp) invades the lumen of a tracheary element in xylem from the leaf of bean,* Phaseolus vulgaris. *Note the band of helical thickening (lw), and the finely fibrillar residue (arrows) of the hydrolysed lateral wall of the tracheary element.*

belongs becomes fully expanded. In this system also, the authors feel that xylem transfer cells play an important role in redirecting solutes travelling in the xylem of the mature leaf to the developing leaves at the shoot apex.

The distribution of xylem parenchyma cells, modified to the transfer cell state, with wall ingrowths whose cytochemistry mimics that of non-lignified walls of comparable cells in other sites, provides strong circumstantial evidence for a major role of these cells in xylem physiology. It is to be hoped that the next decade will see further experimentation that will give greater insights into the activities of these cells.

ACKNOWLEDGEMENTS

Thanks are due to Mrs F. M. Cole for assistance with figures, to Mr B. Fuhrer for photography, and Ms A. Lucarelli for typing. The work has been supported by facilities provided by the Australian Research Grants Committee and the Rural Credits Development Fund of the Reserve Bank.

REFERENCES

[1] Avanzi, S., Maggini, F., Innocenti, A. M. Amplification of ribosomal cistrons during maturation of metaxylem in the root of *Allium cepa*. *Protoplasma* **76,** 197-210 (1973)

[2] Bailey, I. W. The cambium and its derivative tissues. III. A reconnaissance of cytological phenomena in the cambium. *Amer. J. Bot.* **7,** 417-434 (1920)

[3] Barlow, P. W. Towards an understanding of the behaviour of root meristems. *J. Theor. Biol.* **57,** 433-451 (1976)

[4] Berlyn, G. P. Physiological control of differentiation of xylem elements. *Wood and Fiber* **11,** 107-126 (1979)

[5] Bierhorst, D. W. *Morphology of Vascular Plants.* (New York: The Macmillan Co. 1971)

[6] Bierhorst, D. W., Zamora, P. M. Primary xylem elements and element associations of angiosperms. *Amer. J. Bot.* **52,** 657-710 (1965)

[7] Bracegirdle, B. *A History of Microtechnique.* (London: Heinemann Educational Books 1978)

[8] Briarty, L. G., Hughes, C. E., Evers, A. D. The developing endosperm of wheat—a stereological analysis. *Ann. Bot.* **44,** 641-658 (1979)

[9] Brower, D. L., Hepler, P. K. Microtubules and secondary wall deposition in xylem: the effects of isopropyl-N-phenylcarbamate. *Protoplasma* **87,** 91-111 (1976)

[10] Busby, C. H., O'Brien, T. P. Aspects of vascular anatomy and differentiation of vascular tissues and transfer cells in vegetative nodes of wheat. *Aust. J. Bot.* **27,** 703-711 (1979)

[11] Catesson, A. M., Czaninski, Y., Péresse, M., Moreau, M. Sécrétions intravasculaires de substances 'gommeuses' par les cellules associées aux vaisseaux en réaction à une attaque parasitaire. *Bull. Soc. bot. Fr.* **123,** 93-107 (1976)

[12] Charlton, W. A. Primary vascular patterns in root meristems of *Pontederia cordata* and their relevance to studies of root development. *Canad. J. Bot.* **58,** 1351-1369 (1980)

[13] Chrispeels, M. J. Biosynthesis, intracellular transport and secretion of extracellular macromolecules. *Ann. Rev. Plant Physiol.* **27,** 19-38 (1976)

[14] Clarke, A. E., Anderson, R. L., Stone, B. A. Form and function of arabinogalactans and arabinogalactan-proteins. *Phytochemistry* **18,** 521-540 (1979)

[15] Cronshaw, J., Bouck, G. B. The fine structure of differentiating xylem elements. *J. Cell Biol.* **24,** 415-431 (1965)

[16] Crüger, H. Zur Entwicklungsgeschicte der Zellenwand. *Bot. Zeit.* **13,** 601-613; 617-629 (1855)

[17] Czaninski, Y. Observations sur une nouvelle couche pariétale dans les cellules associées aux vaisseaux du Robinier et du Sycamore. *Protoplasma* **77,** 211-219 (1973)

[18] Czaninski, Y. Etude cytochimique ultrastructurale de la paroi des cellules de transpert dans la tige du Blé. *Biologie Cellulaire* **29,** 221-224 (1977)

[19] Czaninski, Y. Cytochimie ultrastructurale des parois du xylème secondaire. *Biologie Cellulaire* **35,** 97-102 (1979)

[20] D'Amato, F. Nuclear Cytology in Relation to Development. (Cambridge: Cambridge University Press 1977)

[21] Elbein, A. D. The role of lipid-linked saccharides in the biosynthesis of complex carbohydrates. *Ann. Rev. Plant Physiol.* **30,** 239-272 (1979)

[22] Esau, K. *Plant Anatomy* (New York: John Wiley and Sons 1953)

[23] Esau, K. *Plant Anatomy*, 2nd Ed. (New York: John Wiley and Sons 1965)

[24] Esau, K. *The Anatomy of Seed Plants*, 2nd Ed. (New York: John Wiley and Sons (1977)

[25] Esau, K., Charvat, I. On vessel member differentiation in the bean (*Phaseolus vulgaris* L.) *Ann. Bot.* **42,** 665-677 (1978)

[26] Esau, K., Cheadle, V. I., Gill, R. H. Cytology of differentiating tracheary elements. II. Structures associated with cell surfaces. *Amer. J. Bot.* **53,** 765-771 (1966)

[27] Feder, N., O'Brien, T. P. Plant microtechnique: some principles and new methods. *Amer. J. Bot.* **55,** 123-142 (1968)

[28] Feldmann, L. J. The generation and elaboration of primary vascular tissue patterns in roots of *Zea. Bot. Gaz.* **138,** 393-401 (1977)

[29] Feldman, L. J., Torrey, J. G. The quiescent centre and primary vascular tissue pattern formation in cultured roots of *Zea. Canad. J. Bot.* **53,** 2796-2803 (1975)

[30] Feldman, L. J., Torrey, J. G. The isolation and culture *in vitro* of the quiescent centre of *Zea mays. Amer. J. Bot.* **63,** 345-355 (1976)

[31] Foster, R. C. Fine structure of tyloses in three species of *Myrtaceae. Aust. J. Bot.* **15,** 25-34 (1967)

[32] Foster, A. S., Gifford, E. M. Jr. *Comparative Morphology of Vascular Plants*, 2nd Ed. (San Francisco: W. H. Freeman and Co. 1974)

[33] Fukuda, H., Komamine, A. Direct evidence for cytodifferentiation to tracheary elements without intervening mitosis in a culture of single cells isolated from the mesophyll of *Zinnia elegans. Plant Physiol.* **65,** 61-64 (1980)

[34] Goodwin, R. H. On the development of xylary elements in the first internode of *Avena* in light and dark. *Amer. J. Bot.* **29,** 818-828 (1942)

[35] Goosen-de-Roo, L. The relationship between cell organelles and cell wall thickenings in primary tracheary elements of the cucumber. II. Quantitative aspects. *Acta bot. neerl.* **22,** 301-320 (1973)

[36] Goosen-de-Roo, L. The structure of protoplast in primary tracheary elements of the cucumber after plasmolysis. *Acta bot. neerl.* **22,** 467-485 (1973)

[37] Grew, N. *The Anatomy of Plants* (New York: Johnson Reprint Corporation 1965)

[38] Gunning, B. E. S., Hardham, A. R. Microtubules and morphogenesis in plants. *Endeavour* **3,** 112-117 (1979)

[39] Gunning, B. E. S., Hughes, J. E., Hardham, A. R. Formative and proliferative cell divisions, cell differentiation and developmental changes in the meristem of *Azolla* roots. *Planta* **143,** 121-144 (1978)

[40] Gunning, B. E. S., Hardham, A. R., Hughes, J. E. Evidence for initiation of microtubules in discrete regions of the cell cortex in *Azolla* root-tip cells and an hypothesis on the development of cortical arrays of microtubules. *Planta* **143,** 161-179 (1978)

[41] Gunning, B. E. S., Pate, J. S. Transfer cells. In *Dynamic Aspects of Plant Ultrastructure* (A. W. Robards, ed.) (London: McGraw-Hill 1974)

[42] Gunning, B. E. S., Pate, J. S., Green, L. W. Transfer cells in the vascular system of stems: taxonomy, associations with nodes and structure. *Protoplasma* **71,** 147-171 (1970)

[43] Gunning, B. E. S., Steer, M. W. *Ultrastructure and the Biology of Plant Cells.* (London: Edward Arnold 1975)

[44] Haberlandt, G. *Physiologische Pflanzenanatomie.* (Leipsig: W. Engelmann 1909)

[45] Haberlandt, G. *Physiological Plant Anatomy.* Reprint Edition. (New Delhi: Today and Tomorrow Book Agency 1965)

[46] Hardham, A. R., Gunning, B. E. S. Interpolation of microtubules into cortical arrays during cell elongation and differentiation in roots of *Azolla pinnata. J. Cell Sci.* **37,** 411-442 (1979)

[47] Hepler, P. K., Fosket, D. E. The role of microtubules in vessel member differentiation in *Coleus. Protoplasma* **72,** 213-236 (1971)

[48] Hepler, P. K., Newcomb, E. H. Microtubules and fibrils in the cytoplasm of *Coleus* cells undergoing wall deposition. *J. Cell Biol.* **20,** 529-533 (1964)

[49] Hepler, P. K., Palevitz, B. A. Microtubules and microfilaments. *Ann. Rev. Plant Physiol.* **25,** 309-362 (1974)

[50] Hess, T., Sachs, T. The influence of a mature leaf on xylem differentiation. *New Phytol.* **71,** 903-914 (1972)

[51] Heyn, A. N. J. The microcrystalline structure of cellulose in cell walls of cotton, ramie and jute fibres as revealed by negative staining of sections. *J. Cell Biol.* **29,** 181-197 (1966)

[52] Jensen, W. A. *Botanical Histochemistry.* (San Francisco: Freeman Co. 1962)

[53] Kuo, J., O'Brien, T. P., Zee, S-Y. The transverse veins of the wheat leaf. *Aust. J. biol. Sci.* **25,** 721-737 (1972)

[54] Kuo, J., O'Brien, T. P., Canny, M. J. Pit-field distribution, plasmodesmatal frequency, and assimilate flux in the mestome sheath cells of wheat leaves. *Planta* **121,** 97-118 (1974)

[55] Lai, V., Srivastava, L. M. Nuclear changes during differentiation of xylem vessel elements. *Cytobiologie* **12,** 220-243 (1976)

[56] Ledbetter, M. C., Porter, K. R. A 'microtubule' in plant cell structure. *J. Cell Biol.* **19,** 239-250 (1963)

[57] Ledbetter, M. C., Porter, K. R. *Introduction to the Fine Structure of Plant Cells.* (Berlin: Springer-Verlag 1970)

[58] Libbenga, K. R., Torrey, J. G. Hormone-induced endoreduplication prior to mitosis in cultured pea root cortex cells. *Amer. J. Bot.* **60,** 293-299 (1973)

[59] List, A. Some observations on DNA content and cell and nuclear volume growth in the developing xylem of certain higher plants. *Amer. J. Bot.* **50,** 320-329 (1963)

[60] Meyer, R. W. Tyloses: development in white oak. *Forest Products Journal* **17,** 50-56 (1967)

[61] Meyer, R. W., Côté, W. A. Formation of the protective layer and its role in tylosis development. *Wood Science and Technology* **2,** 84-94 (1968)

[62] Newcomb, E. H. Plant microtubules. *Ann. Rev. Plant Physiol.* **20,** 253-288 (1969)

[63] Northcote, D. H. The chemistry of the plant cell wall. *Ann. Rev. Pl. Physiol.* **23,** 113-132 (1972)

[64] Northcote, D. H. The involvement of the Golgi apparatus in the biosynthesis and

secretion of glycoproteins and polysaccharides. *Biomembranes* **10,** 51-76 (1979)

[65] O'Brien, T. P. Further observations on hydrolysis of the cell wall in the xylem. *Protoplasma* **69,** 1-14 (1970)

[66] O'Brien, T. P. The cytology of cell-wall formation in some eukaryotic cells. *Bot. Rev.* **38,** 87-118 (1972)

[67] O'Brien, T. P. Primary vascular tissues. In *Dynamic Aspects of Plant Ultrastructure* (A. W. Robards, ed.) (London: McGraw-Hill 1974)

[68] O'Brien, T. P., McCully, M. E. *Plant Structure and Development.* (London: Collier-Macmillan 1969)

[69] O'Brien, T. P., Thimann, K. V. Observations on the fine structure of the oat coleoptile. III. Correlated light and electron microscopy of the vascular tissues. *Protoplasma* **63,** 443-478 (1967)

[70] Oksala, T., Therman, E. Endomitosis in tapetal cells of *Eremurus* (Liliaceae). *Amer. J. Bot.* **64,** 866-872 (1977)

[71] Pate, J. S., Gunning, B. E. S. Transfer cells. *Ann. Rev. Plant Physiol.* **23,** 173-196 (1972)

[72] Pate, J. S., Gunning, B. E. S., Milliken, F. F. Function of transfer cells in the nodal regions of stems, particularly in relation to the nutrition of young seedlings. *Protoplasma* **71,** 313-334 (1970)

[73] Pate, J. S., O'Brien, T. P. Microautoradiographic study of the incorporation of labelled amino acids into insoluble compounds of the shoot of a higher plant. *Planta* **78,** 60-71 (1968)

[74] Phillips, R., Torrey, J. G. DNA synthesis, cell division and specific cytodifferentiation in cultured pea root cortical explants. *Develop. Biol.* **31,** 336-347 (1973)

[75] Phillips, R., Torrey, J. G. DNA levels in differentiating tracheary elements. *Develop. Biol.* **39,** 322-325 (1974)

[76] Pickett-Heaps, J. D. Incorporation of radioactivity into wheat xylem walls. *Planta* **71,** 1-14 (1966)

[77] Pickett-Heaps, J. D. The effects of colchicine on the ultrastructure of dividing plant cells, xylem wall deposition and distribution of cytoplasmic microtubules. *Dev. Biol.* **15,** 206-236 (1967)

[78] Porter, K. R., Machado, R. D. Studies on the endoplasmic reticulum. IV. Its form and distribution during mitosis in cells of onion root tip. *J. Biophys. Biochem. Cytol.* **7,** 167-180 (1960)

[79] Quail, P. H. Plant cell fractionation. *Ann. Rev. Plant Physiol.* **30,** 425-484 (1979)

[80] Ray, P. M. Radioautographic study of cell wall deposition in growing plant cells. *J. Cell Biol.* **35,** 659-674 (1967)

[81] Roberts, L. W. *Cytodifferentiation in Plants.* (London: Cambridge University Press 1976)

[82] Roelofsen, P. A. The plant cell-wall. In *Encyclopaedia of Plant Anatomy.* III. Pt. 4. (Berlin-Nikolassee: Gebruder Borntraeger 1959)

[83] Sabatini, D. D., Bensch, K., Barrnett, R. J. Cytochemistry and Electron Microscopy. *J. Cell Biol.* **17,** 19-58 (1963)

[84] Sachs, T. The induction of fibre differentiation in peas. *Ann. Bot.* **36,** 189-197 (1972)

[85] Sachs, T. The control of the differentiation of vascular networks. *Ann. Bot.* **39,** 197-204 (1975)

[86] Sachs, T. Patterned differentiation in plants. *Differentiation* **11,** 65-73 (1978)

[87] Scheirer, D. C. Hydrolysed walls in the water-conducting cells of *Dendroligotrichum* (Bryophyta): histochemistry and ultrastructure. *Planta* **115,** 37-46 (1973)

[88] Schmid, R. The fine structure of pits in hardwoods. In *'Cellular Ultrastructure of Woody Plants'* (W. A. Côté, ed.) 291-304. (Syracuse, New York: Syracuse University Press 1965)

89 Shininger, T. L. The control of vascular development. *Ann. Rev. Plant Physiol.* **30,** 313-337 (1979)
90 Sinnott, E. W., Bloch, R. The cytoplasmic basis of intercellular patterns in vascular differentiation. *Amer. J. Bot.* **32,** 151-156 (1945)
91 Sporne, K. R. *The Morphology of Pteridophytes.* (London: Hutchinson and Co. 1962)
92 Torrey, J. G., Fosket, D. E. Cell division in relation to cytodifferentiation in cultured pea root segments. *Amer. J. Bot.* **57,** 1072-1080 (1970)
93 Zee, S-Y. Distribution of vascular transfer cells in the culm nodes of bamboo. *Canad. J. Bot.* **52,** 345-347 (1974)
94 Zee, S-Y., O'Brien, T. P. Vascular transfer cells in the wheat spikelet. *Aust. J. biol. Sci.* **24,** 35-49 (1971)

2. Secondary Xylem Cell Development

J R BARNETT
Plant Science Laboratories, University of Reading, Whiteknights, Reading, England

INTRODUCTION

The secondary xylem of a woody stem is a heterogeneous mixture of cell types whose presence or absence, relative abundance, and arrangement with respect to one another is typical and diagnostic for any species. The cells may be considered as making up two distinct groups. The first comprises elements such as fibres, tracheids, vessels, parenchyma and the cells of resin canals. These are distinguished by having their longest axis oriented parallel to the main axis of the stem. The second group consists of those cells which make up the rays and usually, but not always, have their longest axis oriented radially with respect to the axis of the stem. They are most often parenchymatous, although tracheary elements occur in the rays of some conifers.

The axially oriented elements arise from superficially identical fusiform initials in the vascular cambium, while ray cells arise from ray initials which are in turn formed by differentiation of fusiform initials. The development from apparently similar fusiform cambial cells of such diverse elements as long, narrow, thick walled, lightly pitted fibres; and short, large diameter, heavily pitted and perforated vessel elements, is achieved through variation in the extent and nature of wall growth and elaboration during differentiation. The control mechanism for this is still far from being understood, as indeed are the processes themselves. In this chapter it is intended to describe and discuss what is known of the physical changes taking place in a xylem cell during its formation and to consider in particular the contribution that has been made by electron microscopy.

Cambium and developing secondary xylem are enclosed by phloem and bark, making it effectively impossible to study the process of differentiation *in situ*. Thus the only way to observe the *in vivo* process is to remove tissue from the stem and examine it microscopically. This unfortunately involves fixation and sectioning and provides a static image of what is a dynamic process. The observer is thus obliged to extrapolate in an attempt to relate this image to the changes which occur during differentiation. Some attempt has been made to observe cambial cells alive, however. Bailey[4] developed techniques for removal and maintenance of living cambial cells from woody stems. By careful manipulation of the support medium he was able to study these cells by light microscopy for up to 500 hours. The environment in which they were kept was very different from that inside the stem, however, and he does not record having observed a tendency for these cells to differentiate.

THE ULTRASTRUCTURE OF FUSIFORM CAMBIAL CELLS

The most notable contributions to our knowledge of the cytoplasmic structure in cambial cells, prior to the development of electron microscopy, were made by Bailey[1,4]. He pointed out how little was known of these cells at that time (1919) owing to the difficulty of preparing the tissue for examination. Among the fundamental observations that he made was that each cambial cell possessed one nucleus. Up until that time it was thought that the sphere of influence of a nucleus was limited, so that large cells, like those of the cambium, would require several nuclei. He described the process of mitosis in cambium, and the subsequent formation of the cell plate by movement of so-called kinoplasmosomes towards the cell tips. In living cells he described seasonal changes in vacuolation[4]. The vacuome of dormant cambial cells was found to consist of many small vacuoles, while in active cambium, it comprised a single large one. This observation was later confirmed by Catesson[19] in an extensive light microscope study of cambium in *Acer pseudoplatanus*.

Hohl[52] was the first to apply electron microscopy to cambial cytoplasm, in his study of *Datura stramonium*. In the twenty years since that study, descriptions of the cambia of a small number of other species have been published. This work has been reviewed recently by Catesson[21]. Apart from size, there appears to be little to distinguish cambial initials of one species from those of another. In addition, nothing has yet been found to distinguish structurally a cambial initial about to give rise to one type of axial element from one about to give rise to another type. Similarly, in studies of cambium from which compression wood tracheids were developing, Timell[121,122] demonstrated that the cambial cells resembled those from which normal wood was developing in the same species. It appears, therefore, that the factors responsible for directing the fate of a cambial derivative will need to be sought at the molecular level, beyond the present resolution limits of electron microscopy (chapters 5, 6, 7).

The ultrastructure of cambial cells, and in some cases, its seasonal variation, has been examined in both gymnosperms[7,8,9,58,76,115,121,122,123] and angiosperms[20,21,36,65,72,100,114]. In those cases in which the ultrastructure of dormant and active cambia have been compared, the most consistently noted feature has been the seasonal change in vacuolation first described by Bailey[4] in living cells. It is reassuring to find that electron microscope preparation techniques preserve this fundamental characteristic. The only other universally reported seasonal change is in the form of the endoplasmic reticulum (ER) which is smooth in dormant cells and rough in active ones. Other seasonal changes described by some authors, but not by others, relate to the presence or absence of starch in plastids, activity or otherwise of dictyosomes, the condition of the plasmalemma and the presence or absence of protein bodies. Catesson[21] has recently shown that the nucleolus apparently changes in appearance from a more open structure in active cells to a more compact structure during dormancy.

The cambium of *Aesculus hippocastanum* L. as seen with the electron microscope is illustrated in Figs. 2.1-2.6. A typical transverse section through a group of fusiform cells is shown in Fig. 2.1. Four radial files of cells may be seen, with the cambium cells (*sensu* I.A.W.A.[54]) in the central region of each file. Partially differentiated phloem cells are present at one end of each row, with xylem cell precursors commencing enlargement at the opposite end. Those cells with larger tangential diameters have been sectioned nearer their centres than those with smaller tangential diameters. The actual initial cell in any file cannot be identified with certainty, although Murmanis[75] demonstrated that careful examination of the tangential cell walls could be used to reveal its approximate location. This technique is based on the fact that a new layer of primary wall is deposited around the entire cell at each division, so that walls which have been in existence the longest will also be the thickest. Consequently, the thinnest tangential wall will belong to an initial or its immediate derivative.

Fusiform cells in active cambium are highly vacuolate with cytoplasm confined to a thin parietal layer. This is often less than 0.1 μm thick except where the presence of organelles causes an inward deformation of the tonoplast. There is a tendency for organelles to be congregated in the cytoplasm alongside the radial walls and around the nuclei (Fig. 2.2). The organelles are similar to those described in other cambia and meristematic tissues. Nuclear chromatin is in a dispersed form; ER, mitochondria and proplastids are abundant. In some cases, proplastids are in the process of accumulating starch and differentiating into amyloplasts, while dictyosomes are actively producing vesicles (Fig. 2.4). Many individual ribosomes are present in the cytoplasm (Fig. 2.3) and there is, at this stage no tendency for these to form polysomes. Plasmodesmata are occasionally observed in radial walls between fusiform cells (Fig. 2.5) and between fusiform cells and ray cells (Fig. 2.6). Despite extensive searching, these structures could not be found in radial walls of cambial cells in *Pinus radiata*, however[11]. The significance of these observations for pit development will be discussed later. Recent work has suggested that the composition of the radial walls of cambial cells is very different from that of tangential walls[110], and it is possible that this is related to subsequent differential growth (see later). Catesson[21] has also suggested that this difference in composition may reflect the presence of an apoplastic pathway for transport of nutrients from the phloem along the radial walls.

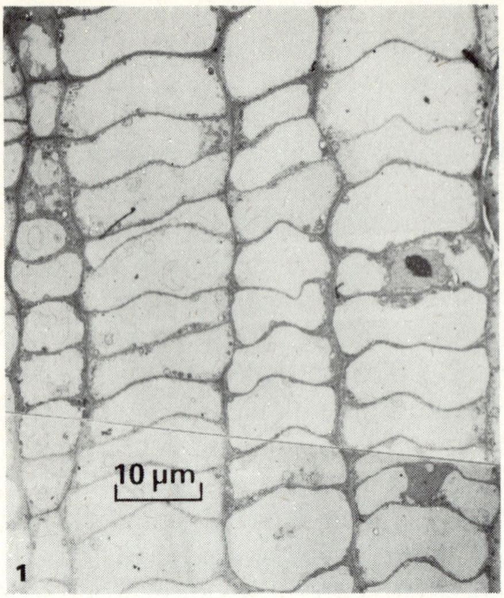

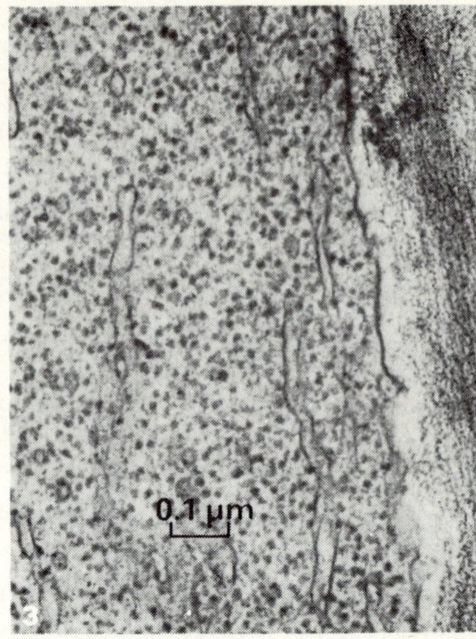

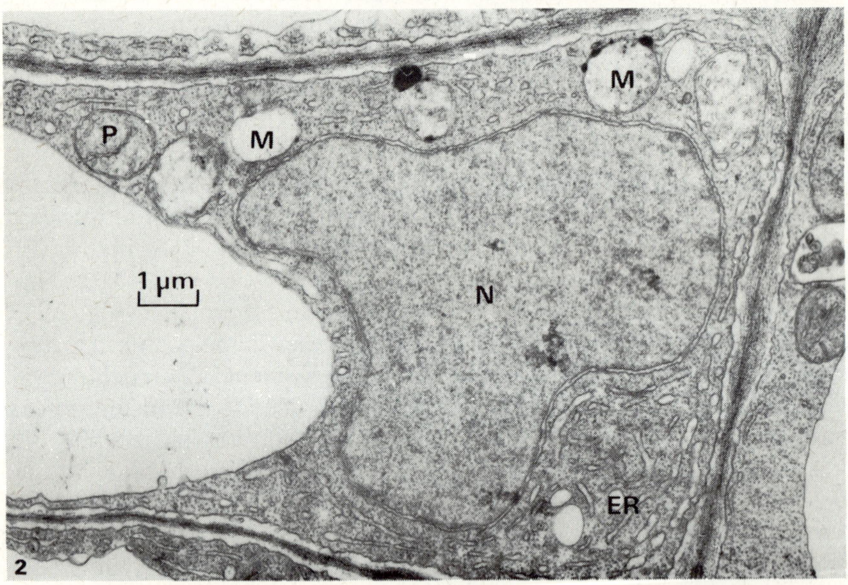

Figure 2.1: *Transverse section (TS) through active cambium of* Aesculus hippocastanum. *Four files of fusiform cells are visible. The initial cells occupy the central region of the micrograph, while the cells at the top are differentiating into phloem, and those at the bottom are in the enlargement stage of differentiation into xylem. The cells are highly vacuolate with cytoplasm confined to a narrow peripheral layer.*

Figure 2.2: *The nucleus (N) of a fusiform cambial cell seen in TS. Note that the chromatin is present in a dispersed form. Mitochondria (M), proplastids (P) and endoplasmic reticulum (ER) are abundant in the cytoplasm enclosing the nucleus.*

Figure 2.3: *Greater magnification of the cytoplasm of a cambial cell showing the presence of numerous free ribosomes.*

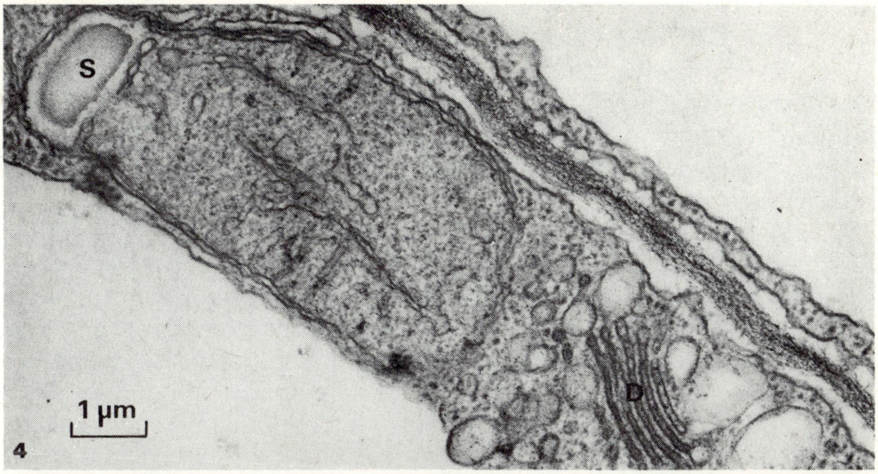

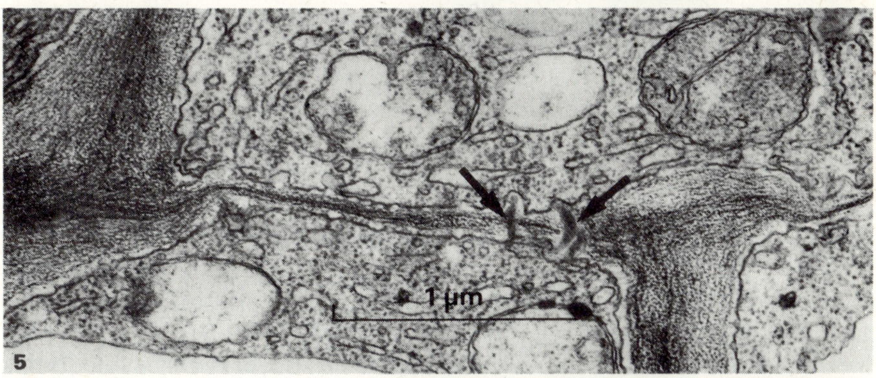

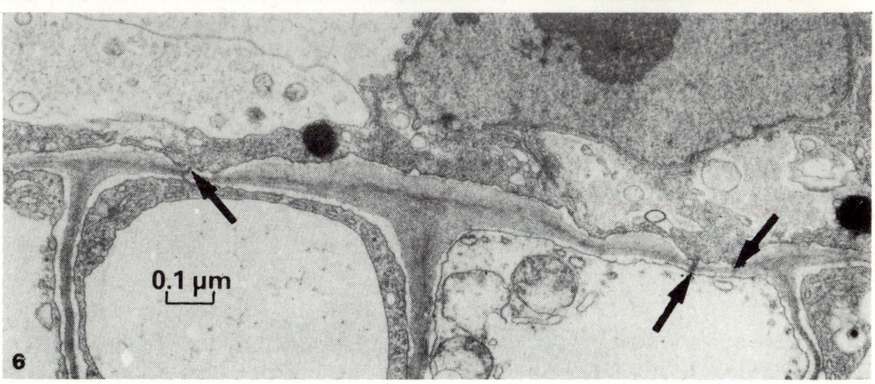

Figure 2.4: *Starch (S) accumulating in a developing amyloplast in a cambial cell. An active dictyosome (D) may also be seen.*

Figure 2.5: *Plasmodesmata (arrows) in a radial wall between two fusiform cambial cells.*

Figure 2.6: *Plasmodesmata (arrows) in a radial wall between a ray cell and fusiform cambial derivatives.*

RAY CELL ULTRASTRUCTURE

Ray cells have received less attention than fusiform cells in descriptions of cambial ultrastructure. This is unfortunate since the rays form a major pathway for flow of nutrients to the zone of xylem differentiation. Robards[98] has described ray parenchyma cells of *Salix fragilis*, while Timell[121] has done so for *Picea abies*.

A ray passing through the cambium of *Aesculus hippocastanum* is illustrated in Fig. 2.7. The ray cells are typical of those described for other species in that they contain a higher proportion of cytoplasm to vacuole than do the adjacent fusiform cells. The nucleus occupies a large part of the total cell volume and is, like the nuclei of fusiform cells, enclosed by cytoplasm containing an aggregation of organelles (Fig. 2.8). As in fusiform cells there is an abundance of mitochondria, ER, active dictyosomes and free ribosomes. Plastids frequently contain starch and appear to be at a more advanced stage of differentiation into amyloplasts than are the plastids in tangentially adjacent fusiform cells (Fig. 2.9). An additional feature not observed in fusiform cells is the presence of large osmiophilic bodies of unknown composition, lying free in the cytoplasm (Figs. 2.7 and 2.9).

Tangential walls between ray cells are perforated by numerous plasmodesmata, usually aggregated into pit fields (Fig. 2.9). The wall in the pit fields is thinner than in those regions which are unperforated by plasmodesmata. Microtubules have been observed grouped close to the thicker regions of the wall (Fig. 2.10). The presence of plasmodesmata through radial walls between ray and fusiform cells in this species has already been mentioned, although as indicated above, such plasmodesmata are absent from the radial walls of ray cells in *Pinus radiata*. These cells are similar in most other respects to those in *Aesculus* as described above.

In the case of *Pinus radiata*, as with some other conifers, ray initials differentiate into ray parenchyma and ray tracheids. In ray tracheids the walls develop irregular thickenings and bordered pits form between one ray tracheid and another, and between ray tracheids and axial tracheids (Figs. 2.11 and 2.12). It is difficult to predict the fate of a ray initial or its derivative from electron microscope sections, but a clue as to whether such a cell will form a parenchyma cell or a tracheid may lie in the presence of bands of microtubules in some cells (Fig. 2.13). Such microtubules are known to be associated with developing cell wall thickenings in many plant cells (Chapter 1, p. 25), and it is possible that their presence in the cell illustrated in Fig. 2.13 is an early indication that it will become a ray tracheid. The high level of activity in this cell is indicated by the presence of many polysomes. The wall thickenings which develop must leave areas of primary wall free to form pit membranes. An early stage is illustrated in Fig. 14 in which thickenings may be seen overlying and obscuring

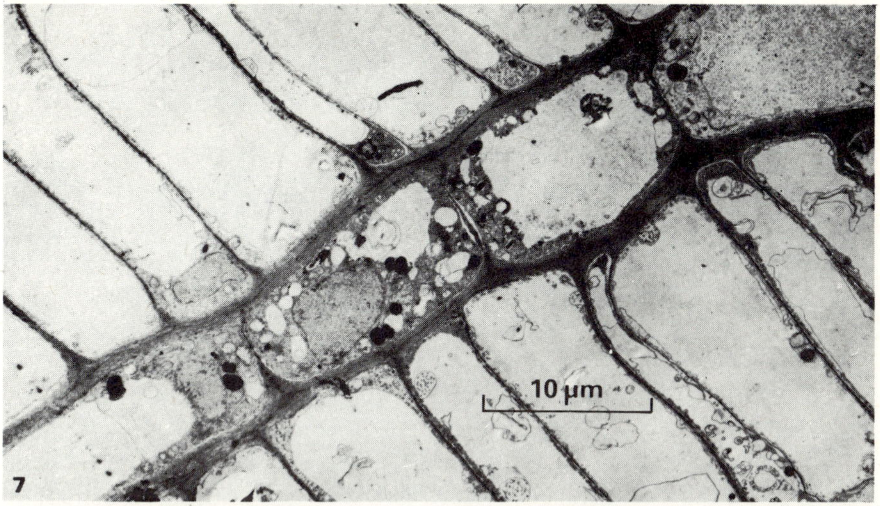

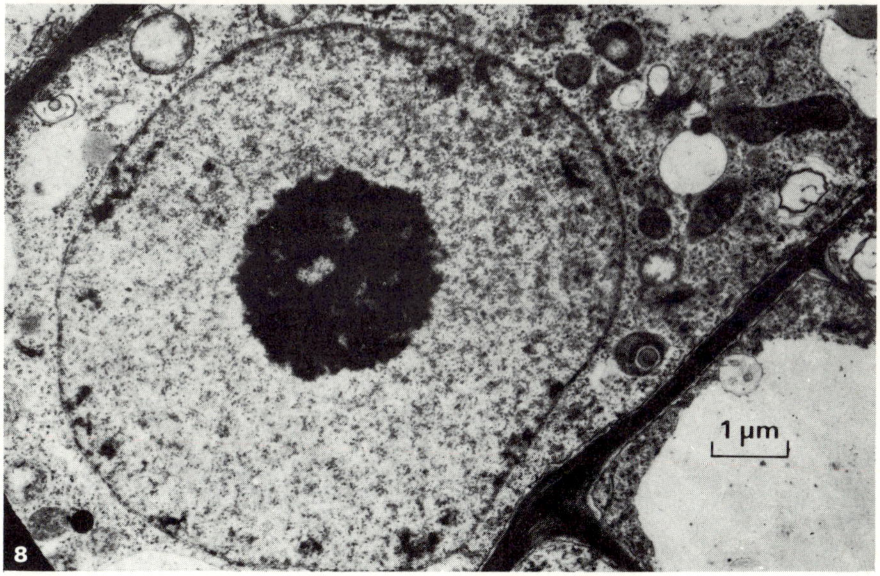

Figure 2.7: *A ray in the cambial zone of* Aesculus hippocastanum. *The cells of the ray contain a higher proportion of cytoplasm to vacuole than do the adjacent fusiform cells.*

Figure 2.8: *Nucleus of a ray cell in* Aesculus hippocastanum. *The cytoplasm enclosing the nucleus is rich in organelles and ribosomes.*

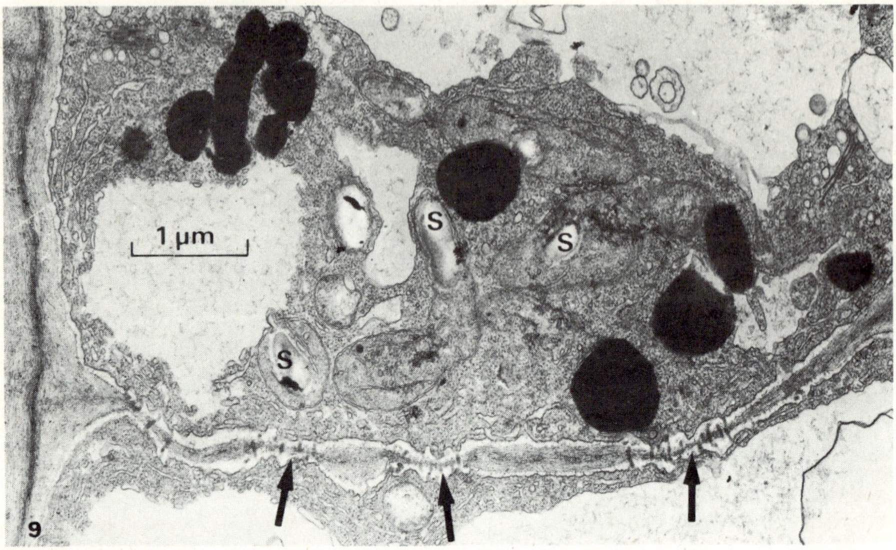

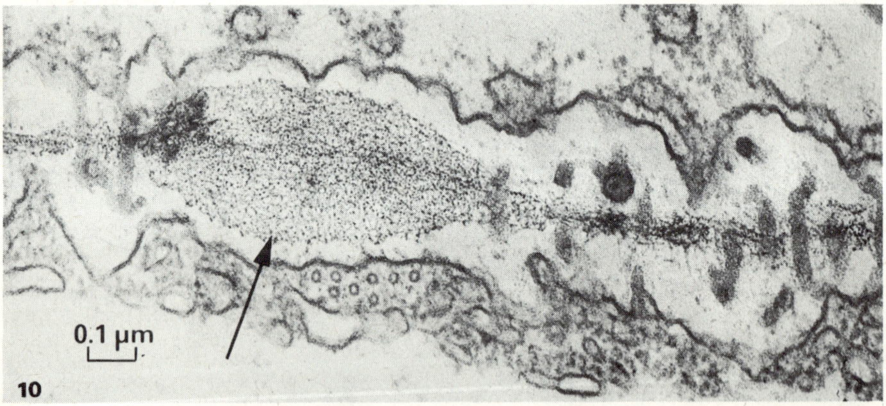

Figure 2.9: *Part of a ray cell in* Aesculus hippocastanum. *Note the presence of starch in the plastids (S), and the organisation of plasmodesmata into pit fields (arrows) in the tangential wall between this cell and the next. Large, heavily stained osmiophilic bodies are free in the cytoplasm of ray cells (see also Figure 2.7). These are not found in fusiform cells.*

Figure 2.10: *Two pit fields in a tangential ray cell wall in* A. hippocastanum. *The groups of plasmodesmata are separated by a thicker zone of primary wall (arrow). Note the group of microtubules adjacent the thickened primary wall.*

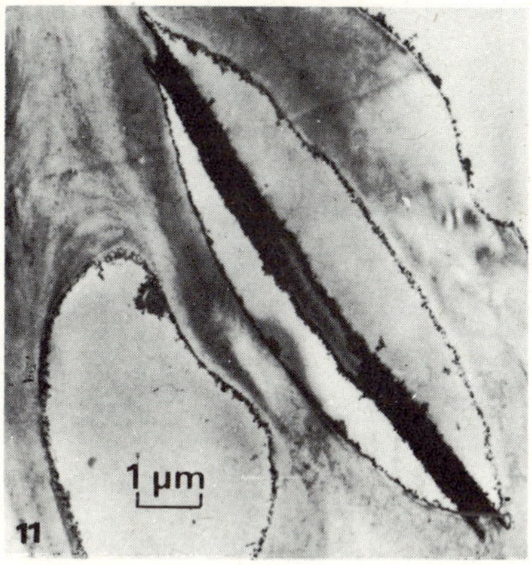

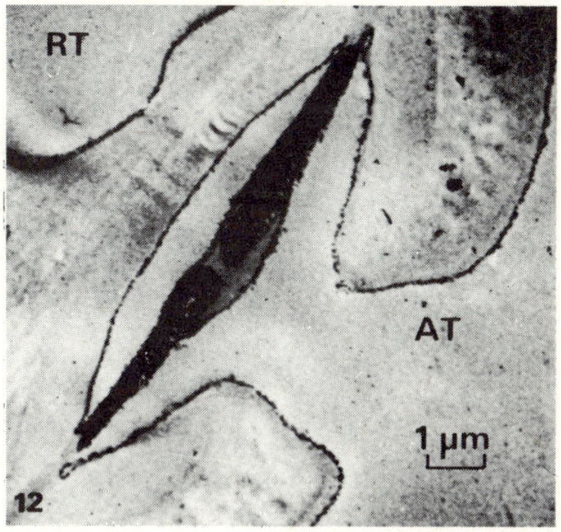

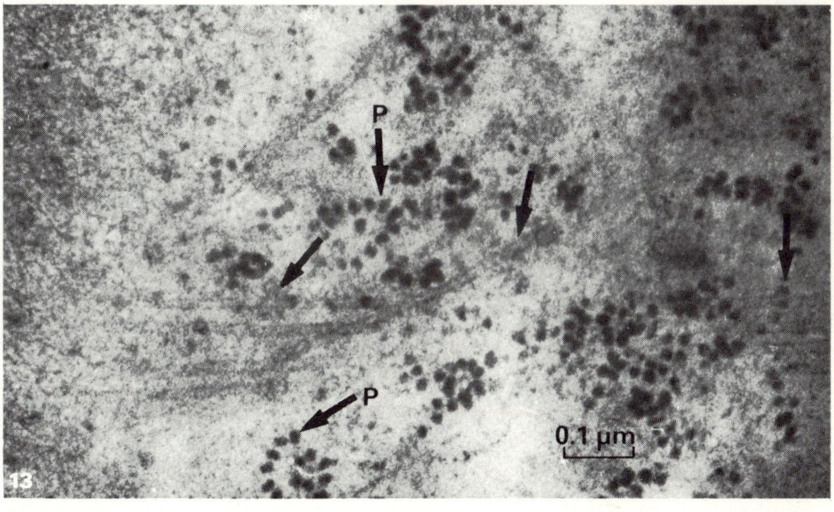

Figure 2.11: *Bordered pit between ray tracheids in* Pinus radiata.

Figure 2.12: *Bordered pit between a ray tracheid (RT) and an axial tracheid (AT) in* P. radiata.

Figure 2.13: *Microtubules (arrows) in a differentiating ray cell in* P. radiata. *These possibly indicate that wall thickenings of the type found in a ray tracheid are about to develop in this cell. Ribosomes are aggregated into polysomes (arrows marked P).*

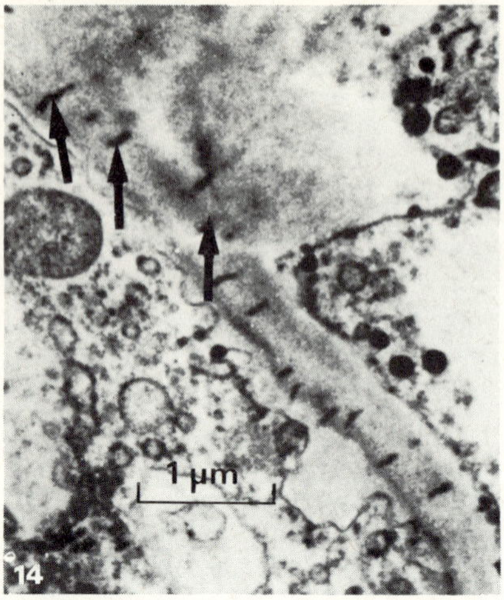

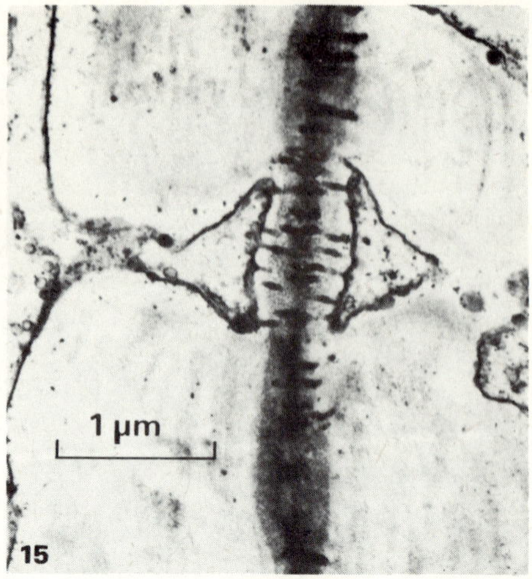

Figure 2.14: *Tangential wall between differentiating ray cells in* P. radiata. *The cell on the right already has much wall thickening, indicating that it is developing into a ray tracheid. Note that the presence of plasmodesmata through the primary wall has not prevented deposition of secondary wall, so that several now end blindly in the wall thickening (arrows).*

Figure 2.15: *Tangential wall between developing ray parenchyma cells in* A. hippocastanum. *As in* P. radiata *(Figure 2.14), wall thickenings have developed without regard to the location of plasmodesmata. As many have been obscured by wall deposition as have persisted through the pit membrane.*

plasmodesmata. A similar phenomenon may be observed in the parenchyma cells of *Aesculus*, which develop a more regular secondary wall and leave simple pits for maintenance of the symplast (Fig. 2.15).

As ray parenchyma cells develop, they elongate radially, and some changes take place in their cytoplasmic organisation. In *Aesculus*, plastids accumulate a paracrystalline assembly of phytoferritin-like material (Fig. 2.16), while single-membrane bound crystal-containing bodies also appear (Fig. 2.17). The latter were found in *Acer rubrum* by Cronshaw[31] and are of unknown function and composition. Phytoferritin in cambium was described in *Acer pseudoplatanus* by Catesson[20].

It is in and along the rays that heartwood substances or their precursors accumulate and pass inwards. Parameswaran and Bauch[84] observed what appeared to be the synthesis of phenolics in sapwood rays of *Abies alba*. This process was occurring in ER derived vesicles, where particles, four times larger than ribosomes were assembled and merged to form osmiophilic globules. It was shown that these globules contained a different material from the phenolics of heartwood[84]. Similar structures are present in ray parenchyma of *Pinus radiata* (Fig. 2.18).

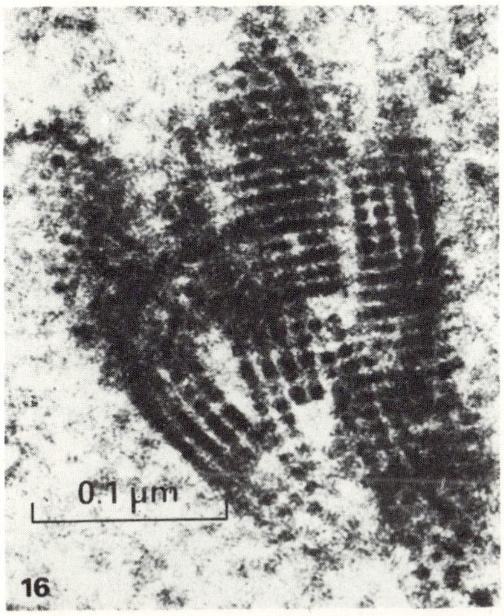

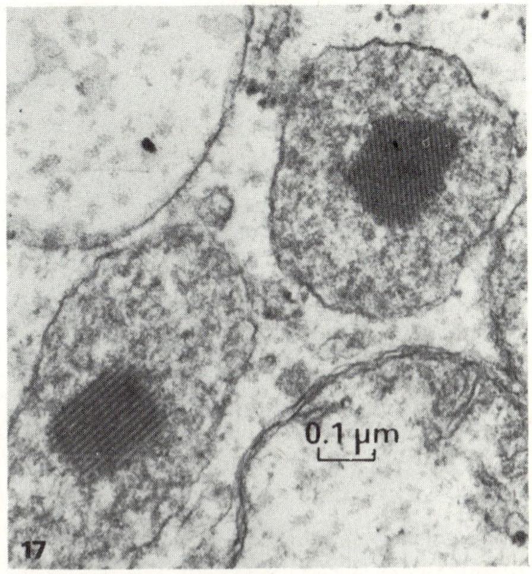

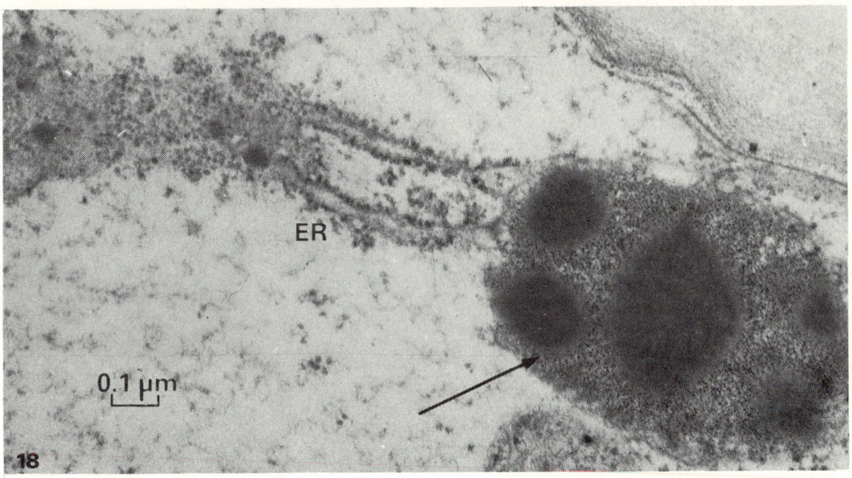

Figure 2.16: *Paracrystalline material in a maturing ray cell of* A. hippocastanum. *This resembles phytoferritin as described for these cells in* Acer pseudoplatanus[20].

Figure 2.17: *Paracrystalline material of unknown composition contained by single-membrane bound organelles in ray cells of* A. hippocastanum.

Figure 2.18: *Osmiophilic material (arrow) associated with endoplasmic reticulum (ER) in ray parenchyma of* P. radiata. *It has been suggested that this represents the synthesis of phenolics by the reticulum*[84].

FUSIFORM CELL DIVISION

This process is the first step in differentiation. The role of division and the cell cycle in differentiation is discussed elsewhere in this volume (chapter 5). Division *in vivo* follows a regular pattern which may vary from one species to another. Mahmood[70] found that in *Pinus radiata* the xylary derivative of division by an initial divided periclinally twice more to produce four daughter cells. These cells then underwent enlargement and differentiation to form tracheids. A similar pattern was observed in *Pinus strobus*[75]. The pattern of division has not been determined unequivocally for any other tree species.

The elongated form of fusiform cambial cells means that true periclinal and anticlinal division requires accurate alignment of the cell plate over long distances, particularly in conifers where cells may be several millimetres long. Bailey[1] first described pseudotransverse division in conifers, showing that the cell plate is inclined with respect to the vertical, reflecting the tilted division plane. His observation that the cell plate is formed by kinoplasmosomes which travel towards the cell tips from the centre of the cell has been mentioned above.

Electron microscope studies of cambial division have been carried out by Evert and Deshpande[36] and Goosen de Roo and van Spronsen[43]. The mitotic spindle was shown to consist of microtubules[36] between which cytoplasmic organelles could be observed. The new cell plate develops from dictyosome vesicles having either electron lucent or electron opaque contents after osmium fixation. Coated vesicles, also derived from dictyosomes were found associated with the developing plate at later stages. As the cell plate forms in one area, the microtubules associated with it disperse and are reassembled to reform the phragmoplast (Bailey's kinoplasmosome) which thus advances towards the cell tip. Plasmodesmata arise at this time by the persistence of strands of ER through the developing plate.

Some stages of cambial division are illustrated in Figs. 2.19-2.24. A section through a dividing fusiform cell of *Pinus radiata* shows the condensed form of the chromosomes (Fig. 2.19). Mitochondria and plastids appear to be excluded from the spindle region in this case. Fig. 2.20 shows a section through the phragmoplast in which the developing cell plate and associated microtubules are visible. The edge of the phragmoplast consists of osmiophilic vesicles in an apparently random arrangement (Fig. 2.21). Osmiophilic material is also abundant along the developing cell plate of the *Aesculus* cell in Fig. 2.22. Large numbers of ribosomes may also be seen in association with this material (Fig. 2.23) while there is ample evidence of dictyosome activity (Fig. 2.24).

The way in which the developing cell plate is guided through the length of cambial cells is not understood. It is possible that the microtubules have a role in this.

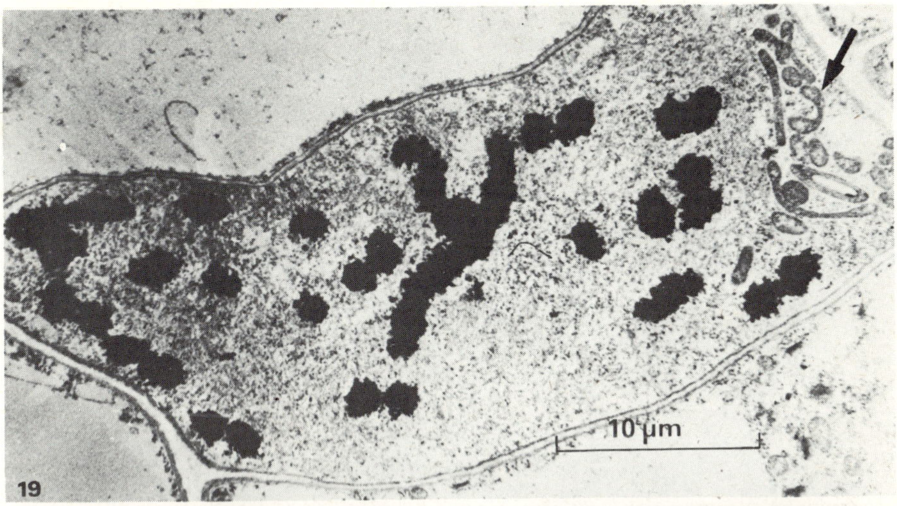

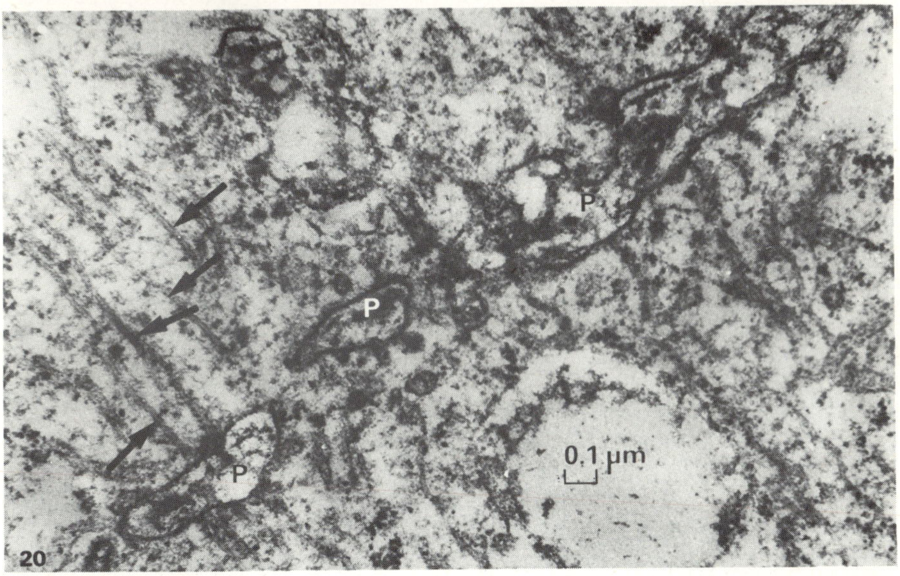

Figure 2.19: *Mitotic nucleus in the cambium of* P. radiata. *Note that the larger organelles (mitochondria and plastids) appear to be excluded from the spindle region, being aggregated at the periphery of the cell (arrow). This is contrary to the observations of Evert and Deshpande*[36].

Figure 2.20: *Part of a developing cell plate in* P. radiata. *The microtubules of the phragmoplast may be seen (arrows). The cell plate still appears to be in a fragmentary state with sections of the future plate enclosed by membranes (P).*

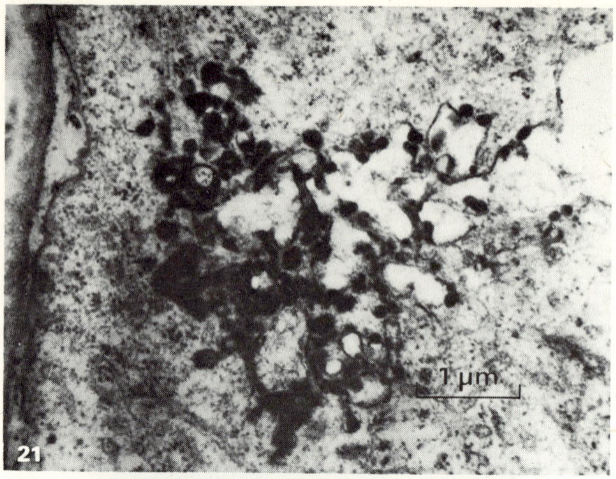

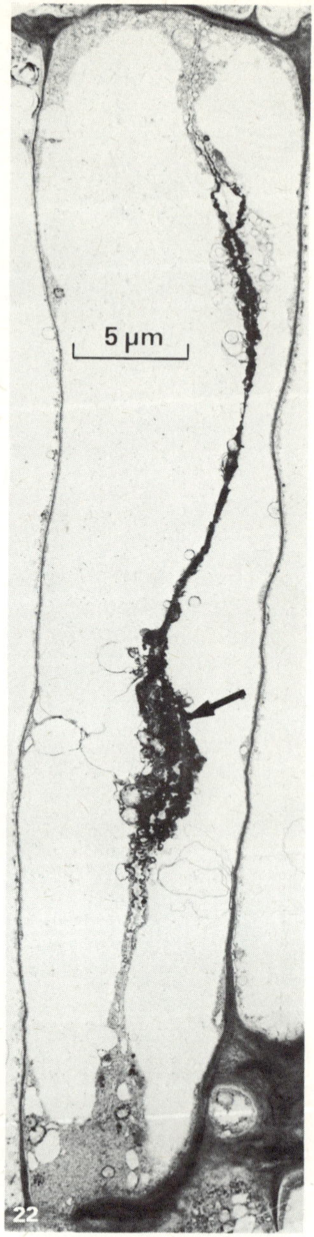

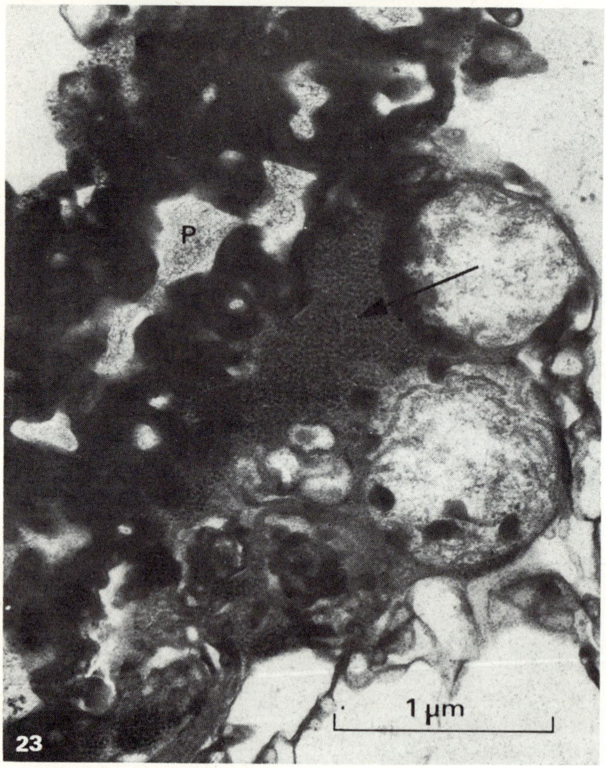

Figure 2.21: *The edge of a phragmoplast in* P. radiata.

Figure 2.22: *Transverse section of a dividing fusiform cell in* A. hippocastanum. *The developing cell plate is associated with highly osmiophilic material (arrow).*

Figure 2.23: *Greater magnification of part of the cell plate illustrated in Figure 2.22. An accumulation of ribosomes (arrow) is present among the osmiophilic material associated with the developing cell plate (P).*

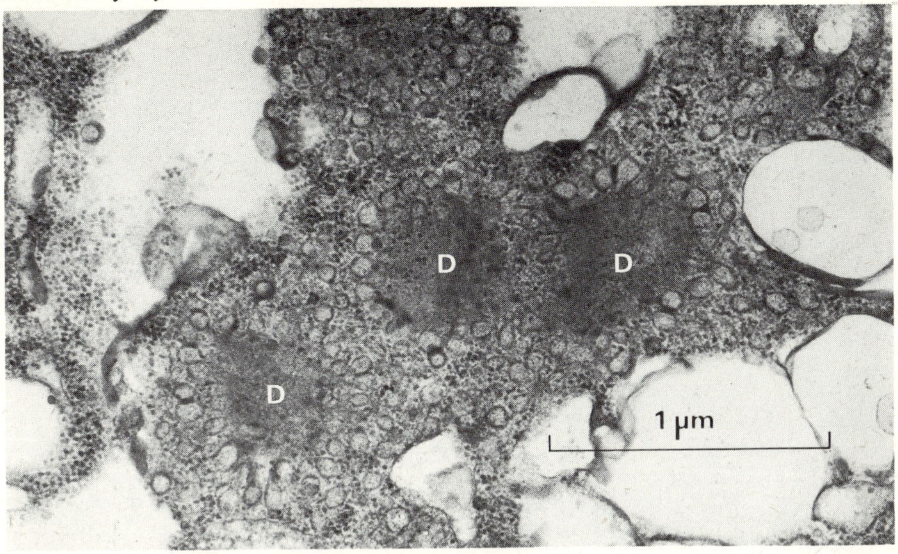

Figure 2.24: *Active dictyosomes in the cytoplasm near the phragmoplast illustrated in Figures 2.22 and 2.23 (D).*

CELL ENLARGEMENT

The form, and therefore to some extent, the function of a xylem element is determined by the direction and amount of growth which the cambial derivative undergoes prior to secondary wall deposition. Bailey[2] attempted to determine the extent to which fusiform cells grow during differentiation. He found that gymnosperm tracheids are 5–15% longer than their parent cambial cells; angiosperm fibres and fibre tracheids several hundred per cent longer; while vessel elements remain about the same length. Radial expansion is a major factor in volume increase in all types of cell, while vessel elements also expand tangentially. In some cases, notably those cells adjacent vessels, there may be tangential contraction to accommodate vessel growth (Fig. 2.25).

The zone of enlarging cells is a cylinder of tissue in which cells are growing in various directions and at the same time accommodating the growth of their neighbours. They do this under considerable physical constraint, enclosed as they are between xylem, which is virtually incompressible, and a rigid bark. Cells are also subject to downward pressure from the mass of tissue above them. The walls of these growing cells are very fragile as witnessed by the ease with which the tissue may be damaged during attempts to remove it from the tree. It must, therefore be turgor pressure within the cells which withstands the pressures to which the cells are subjected, and which provides the driving force for growth. In fact,

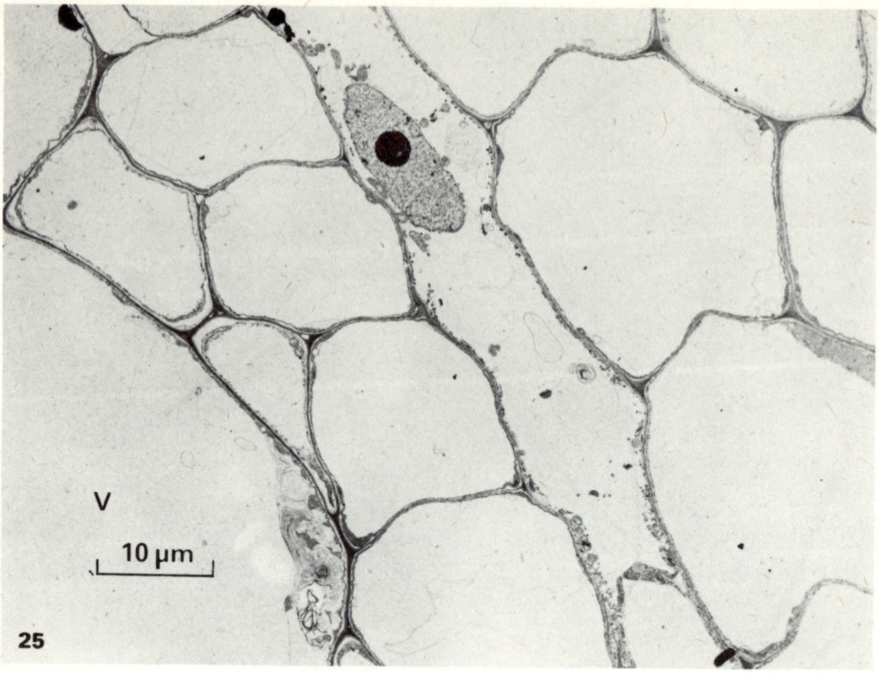

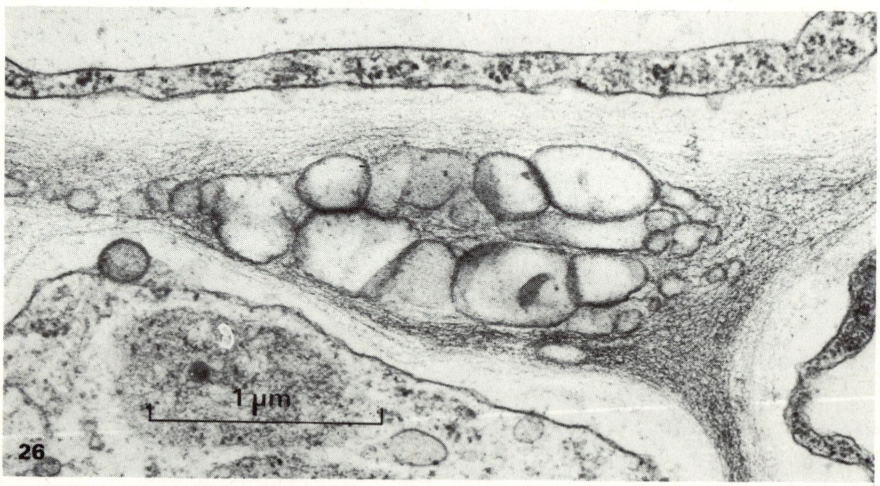

Figure 2.25: *Zone of cell enlargement in developing xylem of* A. hippocastanum. *The enlarging vessel (V), part of which is visible in the bottom left hand corner of the micrograph, has caused distortion of those cells which abutt onto it. Note that the cytoplasm is still confined to a narrow parietal layer in all the cell types.*

Figure 2.26: *Vesicular material contained within the middle lamella between enlarging cells. This material may be involved in loosening the wall to permit growth of a cell tip between the presently adjacent cells.*

this turgor pressure has long been considered the prime driving force in cell growth. Krabbe[67] suggested that an increase in turgor associated with vacuolation in particular cells would allow them to expand and form large diameter vessel elements. This enlargement, he supposed, would occur at the expense of adjacent cells which would elongate as a result of the lateral compressive force to which they were subjected. To permit this, it would be necessary for cells so squeezed to slide past one another, and for each cell to possess its own wall. This type of growth was called sliding growth, and was generally accepted as the most plausible mechanism until it was reviewed by Priestley[96,97]. He found the hypothesis unsatisfactory and suggested instead that cambial derivatives grow by a mutual adjustment of common walls. This 'symplastic-growth' hypothesis could be applied to expansion of the new wall between a pair of cambial cells formed by pseudotransverse division of an initial, but could not explain satisfactorily the type of growth in which an enlarging cell, such as a fibre, increases the number of cells with which it is in contact as it grows. On the basis of *in vivo* studies of growing root tips, Sinnott and Bloch[113] concluded that sliding growth was not occurring, and that a symplastic growth mechanism could not account for the extensive adjustments which they observed between cells. They proposed instead that existing wall contacts were maintained, and that during growth, new wall contacts were made as wall material was deposited by a cell tip extending between cells which had been separated along their middle lamella. By this mechanism, growth would always be an active process originating within the cell, rather than a passive one caused by external pressures as proposed in the sliding growth hypothesis. Bannan[5,6] in describing the results of his painstaking analysis of the fate of cambial cells and their derivatives supported the intrusive growth hypothesis.

Intrusive growth by cell tips means that cells ahead of the tip must be separated in some way to make room for the expanding cell. Neeff[81] proposed that a growing tip secretes an enzyme ahead of itself to loosen the middle lamella. This idea was restated recently by Wenham and Cusick[131]. They noted a swelling of the middle lamella ahead of growing fibre tips in *Salix viminalis* and other species. This was cited as evidence for a change in the structure of the lamella preceding its separation. Vesicle-containing swollen regions of middle lamella have been recorded for enlarging tracheid precursors in *Pinus radiata* by Barnett and Harris[11] and are present in enlarging cells in *Aesculus hippocastanum* (Fig. 2.26). The loose texture of the middle lamella, and the swelling apparently induced by the presence of these vesicles, strongly suggests that they are involved in wall loosening ahead of an expanding cell tip. Whether they are formed as a consequence of separation, or whether they are the causative agent is not certain. Hejnowicz recently suggested that tangential tension in the radial walls of enlarging cells would assist enzyme attack in the separation of walls along the middle lamella[49]. The probable source of this tension is to be found in radial expansion of cambial derivatives. A ring of cambium cells with, typically, an initial radial diameter of 6 μm, expanding to form tracheids with a radial diameter of 30 μm, will effectively increase stem circumference by approximately 157 μm. Thus the cells of the next outermost ring are subjected to a tangential tension which may be relieved by tangential expansion of these cells, anticlinal division and tip growth, or by tip growth alone. Several layers of cells undergo radial enlargement simultaneously and cells in this phase do not divide further or, in the case of gymnosperms, expand tangentially, so tip growth must be the main mechanism for accommodating the tension in the zone of cell enlargement.

Cells of the enlargement zone in *Aesculus hippocastanum* are illustrated in Fig. 2.25. While the radial walls of cambial zone cells are much thicker than the tangential walls, in the enlargement zone the radial walls become, as it were, stretched, until their thickness is reduced to that of the tangential walls. Where they are some distance from vessels, fibres enlarge until their radial diameter is approximately the same as their tangential diameter. In contact with a vessel they exhibit considerable lateral distortion, however (Fig. 2.25). As in fusiform cambium cells, cytoplasm in enlarging cells is confined to a thin parietal zone. Most of the increasing cell volume is taken up by expansion of the vacuole. Localised concentrations of organelles along radial walls such as are common in the cambium, are no longer found.

CELLULOSE MICROFIBRILS AND WALL GROWTH

Using polarisation microscopy, van Iterson demonstrated that the orientation of cellulose micelles in cambial walls is predominantly at more than 45° to the long axis of the cell[57]. This was supported by Preston and Wardrop[93] using X-ray diffraction, when they also showed that the micellar orientation followed a shallow spiral around the cell.

With the advent of electron microscopy and its application to the study of plant cell walls came the first observations of the microfibrillar structure of cellulose[41,91]. Cambium of *Pseudotsuga taxifolia* was one of the first tissues in which microfibrils were subsequently examined[50]. Frey-Wyssling and Stecher[42] examined the walls of enlarging parenchyma cells and found areas which were thinner, and contained fewer microfibrils than normal. These were thought to arise as a consequence of intrusion by cytoplasm into the wall, causing a localised loosening of its texture and permitting expansion. Deposition of fresh wall material in these extended areas would then restore their original thickness. This process was known as mosaic growth. Frey-Wyssling[39] later pointed out that dried primary walls prepared for electron microscopy present a much more compact appearance than is the case in natural hydrated walls, and that there was therefore plenty of space for addition of new wall material by intussusception. The small number of microfibrils to be found near the tips of growing cells also led him to suggest that the advancing tip was naked, and that in this condition it would be able to attack the middle lamella ahead of it, and also synthesise cellulose and other wall materials. The thin wall areas were later considered to be pit fields by Wardrop[124] in his early study of enlarging xylem elements in *Pinus radiata*, *Eucalyptus elaeophora* and *Ulmus* sp. Sterling and Spit[117] supported this interpretation but demonstrated that the tips of growing cells are not naked. Preston and Ripley[95] were unable to find mosaic growth areas in cambial cell walls, and at the same time confirmed electron microscopically that cellulose microfibrils are oriented at a large angle to the long axis of these cells.

Roelofsen and Houwink[53, 109] demonstrated that microfibrils in the older, outer layers of the walls of growing cells were oriented more nearly parallel to the axis of extension, while microfibrils of the younger, inner layers were oriented, and hence had been deposited, transversely to this axis. In the intermediate layers a range of orientations was found, such, that as one passed from younger to older microfibrils so the orientation became more and more aligned with the growing axis of the cell. They considered this reorientation was brought about passively and reflected longitudinal movement of wall components during growth. This 'multi-net' hypothesis of microfibril rearrangement was applied by Wardrop[125] to his observations on growing walls of developing *Pinus radiata* tracheids. He found that the microfibrils of the outer layers of the primary wall were disperse near the cell centre, but oriented longitudinally near the cell tips. The inner layers were oriented transversely to the long axis of the tracheid throughout its length. The major reorientation of microfibrils at the cell tips is good evidence that most growth occurs in the tip region.

Doubts arose about the universal applicability of the multi-net hypothesis as a consequence of studies on the structure of the walls of enlarging collenchyma cells[22, 128]. These were found to be made up of lamellae of transversely oriented microfibrils alternating with lamellae in which they are longitudinally oriented. This system persisted through the entire wall thickness and no re-orientation of the type required by multi-net growth could be found. Wardrop[128] considered that in the hydrated state, the microfibrils and the lamellae they comprise were sufficiently widely separated for them to be able to slide past one another. Chafe and Wardrop[28] later found that epidermal cells are even more complex, with their tangential walls having the crossed poly-lamellate structure found in collenchyma, and radial walls having mainly transversely oriented microfibrils. From this it became apparent that some modification of the multi-net hypothesis would be necessary in explaining growth in these cell types.

Roland *et al.*[111] examined elongating parenchyma of *Phaseolus aureus* and *Pisum sativum* roots and collenchyma of *Apium graveolens*. Here again the bulk of the primary wall was well ordered with no progressive evolution from a transverse microfibril orientation near the plasmalemma to a scattered or longitudinal orientation near the middle lamella. It was proposed that growth occurs by an ordered sliding mechanism in which microfibrils retain, during growth, more or less the orientation in which they were deposited. Thus transversely oriented microfibrils become more widely spaced as the wall grows longitudinally, while longitudinally oriented microfibrils slide past one another. Itoh[59, 60] supported this view on the basis of his freeze-etching studies of primary walls. He concludes that microfibrils in these walls are in fact highly ordered and that random textures observed in replicas are an artefact caused by drying and extraction procedures prior to replication.

Wardrop *et al.* recently reconsidered the multi-net hypothesis using a variety of techniques including freeze-etching[130]. They reaffirmed that in *Avena* sub-epidermal parenchyma cells the microfibrils of the wall layers nearest the plasmalemma are transversely oriented, while those of the outer layers are disperse. In *Apium* collenchyma it was found, as before, that microfibril lamellae are alternately transversely and longitudinally oriented, but it was also found that some layers were present which did not fit this pattern. They conclude that a type of multi-net growth is still occurring in these cells.

Boyd and Foster[16] have queried one of the basic assumptions of the multi-net hypothesis; that microfibrils are free to slide over one another during growth. In micrographs of algal cell walls they observed a tendency for microfibrils to be firmly bonded to one another in certain regions. By separation of unbonded regions during expansion of the wall a trellis-like arrangement would be produced. They argue that in extreme cases the separations may be great enough for several underlying layers to be visible and that this would be responsible for the felt-like wall structure recorded by early electron microscopists studying this type of material.

To what extent may these observations, made on a variety of cell types, be relevant to the walls of enlarging xylem cell precursors? Preston and Wardrop[93] found that cellulose comprises about eight percent only of the volume of a cambial cell wall. Recent work by Roland[110] and Catesson[21] has demonstrated that there are major differences in composition and structure between tangential and radial walls in cambial cells. In dicotyledons it was found that radial walls contained mostly pectin and hemicellulose, with little cellulose, while tangential walls contained a higher proportion of cellulose. Thus, the radial wall, which will undergo greatest expansion, is unlikely to contain a sufficient density of microfibrils for bonding between them to impose any restraints on growth or to produce a trellis-like structure. It is far more likely that the microfibrils which are present are sufficiently separated for passive reorientation to occur in the manner proposed in the multi-net hypothesis. This fits well with Wardrop's[124, 125] early observations on enlarging fibre and tracheid walls.

Since only a small proportion of the primary wall consists of cellulose, most growth must occur by movement and synthesis of wall matrix substances. While the nature of these substances is known[30, 90], their interaction with one another and with the cellulose microfibrils is not understood. Keegstra *et al.*[62] have devised a model for the arrangement of primary wall components based on the composition of the walls of sycamore callus cells while Preston has presented a model of possible bonding schemes between the components of the primary wall (Introduction, Fig. 4). A complete understanding of the mechanism of cell wall growth will require a full picture of the structure of the wall. The electron microscope has made an important contribution in terms of the microfibrillar component and its behaviour.

PIT FORMATION

Pits in mature secondary xylem elements vary in shape, size, structure, abundance and arrangement from one cell type to another, and between similar cell types in different species. Consequently, they have long been used as a diagnostic feature in wood identification. Pits perform several functions. Firstly, the persistence of areas of wall free from secondary thickening permits continued entry of nutrients and wall precursors into the cell. At maturity, in conducting elements, the pits become apertures through which water may pass from one cell to the next in its progress up the stem. Pits between vessel elements and parenchyma cells are the points at which tylosis formation begins in many angiosperms, and in older xylem they allow the movement of heartwood substances from the rays into the axial elements of the stem during heartwood formation.

Pit development begins with the commencement of cell enlargement in the cambium, and is completed at the end of differentiation. In the case of tracheid and vessel pits, the final stage is, for the cell, a post-mortem change, since it is the autolytic enzymes released by cell death which perforate the pit membranes and turn the cells into functional water conducting elements (see chapter 1).

Early stages in pit formation: The role of plasmodesmata

Plasmodesmatal connections are known to persist in the walls between a variety of cell types. A recent work[46] was devoted entirely to these structures and their role in differentiation. Juniper[61] and Gunning[46] are among those who have indicated that plasmodesmata play an important part in determining the fate of a meristematic cell derivative.

Early workers regarded plasmodesmata-containing pit fields in the walls of cambial cells as the sites of future pits, and assumed as a corollary that all pits arose from such pit fields. Electron microscopy of developing fibres in angiosperm wood revealed that numerous plasmodesmata are present in the pit membranes[127]. It was not unreasonable to suppose, therefore, that plasmodesmata are required to ensure accurate alignment of pit pairs between adjacent cells, and by their presence, to prevent secondary wall deposition on the pit membrane[61]. Cronshaw and Wardrop[34], however, in carrying out the first electron microscope examination of sectioned differentiating xylem elements noted that plasmodesmata were absent from cell walls at sites of future pit development in tracheids of *Pinus radiata*. Barnett and Harris[11] carried out an extensive examination of cambium cells and enlarging xylary derivatives in the same species and confirmed the absence of plasmodesmata from pit fields in the radial walls of these cells (it is in these walls that bordered pits are eventually formed).

In an early study, Kerr and Bailey[64] reported that the radial walls of cambial cells in *Pinus strobus* have a beaded appearance when viewed in tangential longitudinal section with a light microscope. They referred to the constrictions in the wall, between the beads, as plasmodesmata-containing pit fields. Their micrographs[64] do not show plasmodesmata, however. In the above-mentioned examination of *Pinus radiata*[11], a series of tangential, longitudinal sections through the cambial zone revealed a region just interior to differentiating phloem cells in which no beading is present in the radial walls. Such a region is illustrated in Fig. 2.27. It is believed that these are the meristematic cells. Immediately interior to this section the cell walls take on a beaded appearance suggesting that the pit fields appear with the onset of cell enlargement (Fig. 2.28). Figs. 2.29 and 2.30 illustrate sections further into the stem in which pits are further advanced in their development. It is felt that the beaded walls illustrated and described by Kerr and Bailey[64] in fact belong to cells in which enlargement has commenced.

This view is supported by examination of radial longitudinal sections of the zone of cell enlargement using polarisation optics (Fig. 2.31). The beads appear as birefringent rims enclosing thinner, non-crystalline circular zones corresponding to the constrictions in the beaded walls seen in tangential longitudinal sections. This birefringence appears long before secondary wall formation has begun. Barnett and Harris[11] proposed that the rim is produced by a localised rearrangement of wall microfibrils during cell growth. This could be achieved, for example, by the injection of a loosening agent into the primary wall/middle lamella by one of the cells sharing the common wall. In the wall of an enlarging cell subject to turgor pressure this

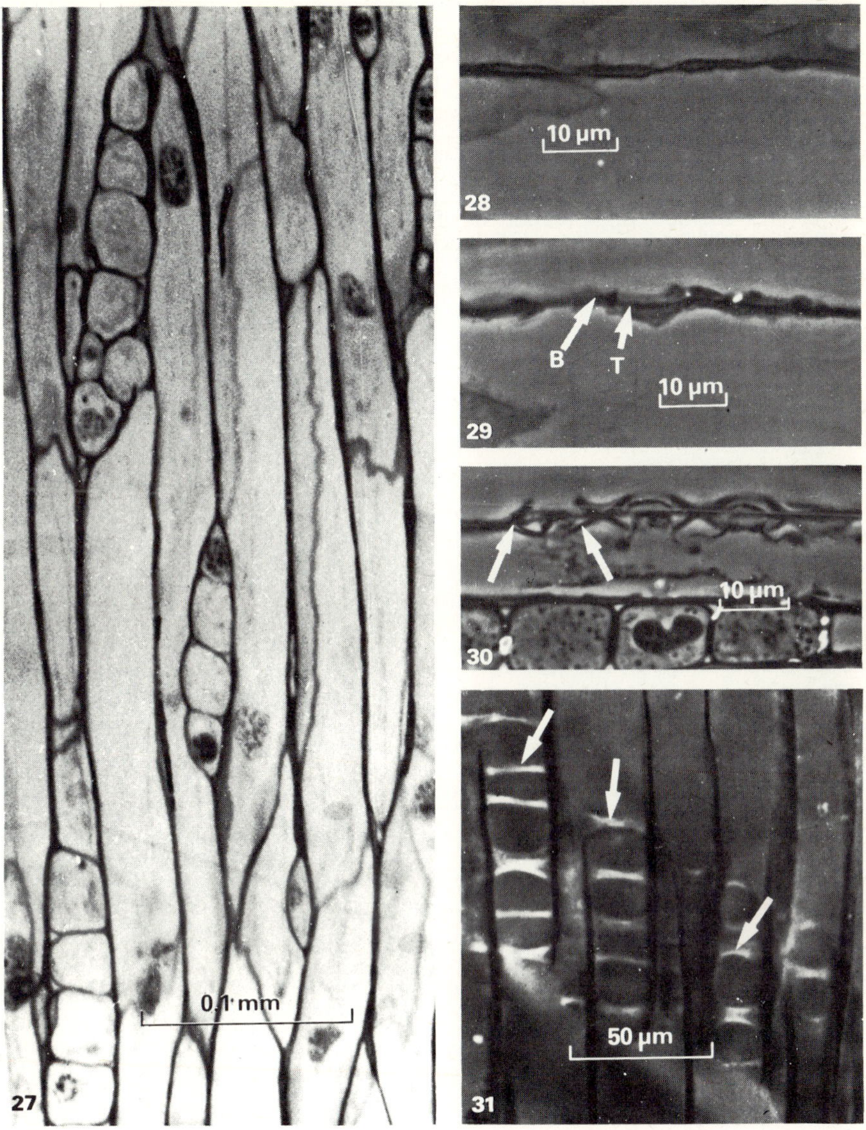

Figure 2.27: *Tangential longitudinal section (TLS) through the cambium of* P. radiata. *Note the absence of beading in the walls. (Light micrograph—bright field)*

Figure 2.28: *TLS slightly further into the stem than Figure 2.27. Beading is now visible in this enlarging radial wall. (Light micrograph—phase contrast)*

Figure 2.29: *TLS slightly further into the stem than Figure 2.28. The first signs of torus (T) and border (B) development may be seen in this radial wall. (Light micrograph—phase contrast)*

Figure 2.30: *TLS yet further into the stem. The pit borders in this wall are now clearly visible (arrows). (Light micrograph—phase contrast)*

Figure 2.31: *Radial longitudinal section through the zone of xylem cell enlargement. Older cells are on the left of the micrograph. Note the birefringent rims enclosing the pit fields (arrows). Secondary wall deposition has not yet commenced in these cells. (Light micrograph—polarisation)*

would create a point of weakness where stretching would occur. Displacement of microfibrils during this process would tend to align them around the periphery of the affected zone, producing the conditions necessary for birefringence to be observed. The 'piling up' of microfibrillar material at the periphery in this way would also lead to a swollen rim being formed, giving rise to the beaded appearance described by Kerr and Bailey[64]. The rim produced, protrudes into each of the cells sharing the common wall, and therefore localises the site for pit formation unequivocally for both cells. Creation of a rim around the pit field would also prevent material from an adjacent loosened area of wall from entering the existing field.

Such a rim of oriented microfibrils around the pit field has been described before by Wardrop[124], and by Imamura and Harada[55], although the latter consider it to be deposited on the primary wall to mark out the pit field. This would require complex organisation of the cellulose synthetic mechanism, since two adjacent cells expanding in opposite directions would need to co-ordinate microfibril production at the surface of primary walls which are probably moving in different directions. The idea that a single 'shot' of wall loosening agent could produce this effect as suggested above offers a simpler explanation, and has also been put forward by Bartnicki-Garcia[13] to explain local growth in fungal hyphae.

The existence of a rim around the pit field helps to explain the form and structure of the pit border in terms of a hypothesis proposed by Young and Service[140]. They suggest that a circular obstacle in the path of an advancing front of microfibril synthesis would divert the synthetic mechanism around the pit field in such a way that it would produce an overarching border. The rim around the pit field, possibly formed as described above, could well fulfil this function.

While these ideas may be applied to *P. radiata* and, probably, other conifers, it cannot be overlooked that the presence of plasmodesmata in the membranes of developing fibre pits in some angiosperms has been convincingly demonstrated. Examination of *A. hippocastanum* reveals that plasmodesmata are common in the membranes of developing fibre pits (Fig. 2.32). Their presence in the radial walls of cambial cells in this species has been described above (Fig. 2.5). On the other hand, however, plasmodesmata do not seem to be present in the pit membranes of developing vessel elements, either where these form part of intervessel pits, or vessel to parenchyma pits (Figs. 2.33 and 2.34). Micrographs published by other authors[15, 24, 139] indicate that plasmodesmata may be typically absent from the developing pit membranes of vessels. As yet, however, only a few species have been examined ultrastructurally, and no systematic study has been made of the development of vessel pitting. Caution must therefore be used in generalising about the role of plasmodesmata in development of vessel pitting.

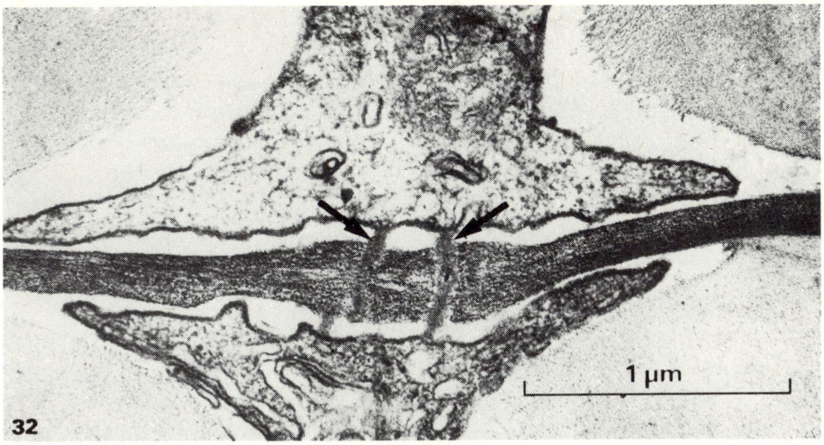

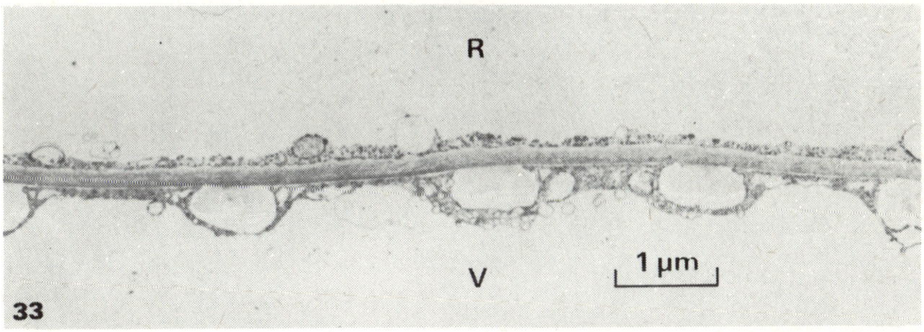

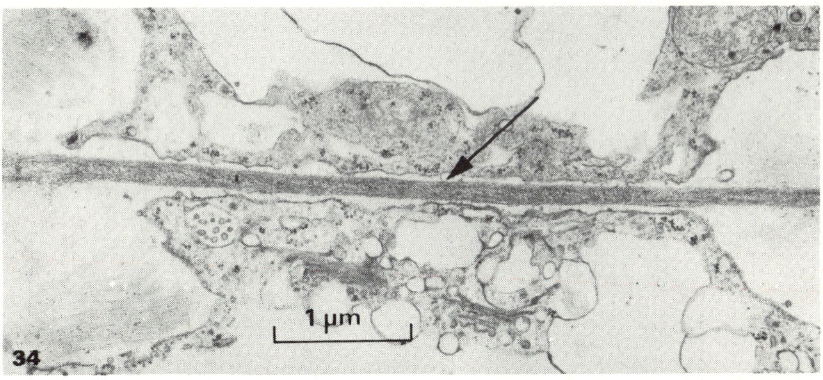

Figure 2.32: *Developing fibre-fibre pit in* A. hippocastanum. *Plasmodesmata are always found in the membranes of these pits (arrows).*

Figure 2.33: *Vessel-ray parenchyma pitting in* A. hippocastanum. *Note the thickenings on the vessel (V) wall and the absence of plasmodesmata from the membranes separating it from the ray cell (R).*

Figure 2.34: *Developing inter-vessel pit in* A. hippocastanum. *The pit membrane (arrow) is entirely free of plasmodesmata.*

Clearly, plasmodesmata are not essential for pit development in some cases. Further, the idea that their presence acts to inhibit deposition of secondary wall over the pit membrane does not seem to hold in all cases since, as described above, thickenings may develop on the primary walls of ray cells without regard for the location of plasmodesmata (Figs. 2.14 and 2.15). This is apparently also the case with spiral thickening in primary xylem elements (Fig. 1.13).

Although it is not possible to draw definite conclusions from examination of such a small number of species, it is interesting that developing pits in cells which ultimately have a water conducting function lack plasmodesmata in their membranes, while the pit membranes of cells with a purely structural function possess them. Gunning[45] has proposed that the density of plasmodesmata in walls of meristematic cells may be a factor in determining the fate of their derivatives. This suggestion might be applied to secondary xylem, albeit cautiously, in an attempt to explain how the fate of a cambial derivative is determined. The presence of plasmodesmata in developing pit membranes might increase the efficiency with which nutrients and wall precursors pass into the developing cell. This in turn would reduce the area required for this transport, i.e. reduce the number of pits needed in the wall, and hence the number of points of weakness in the structure, thereby increasing the suitability of the particular cell for its structural role. Large messenger molecules might also be enabled to enter the cell via plasmodesmata, permitting some external control over the fate of the cell. Absence of, or the presence of small numbers only of plasmodesmata in the pit membranes of a cambial derivative would isolate the cell from such external control and in some way lead to the formation of a water conducting element.

The whole question of the role of plasmodesmata in the walls of differentiating xylem elements obviously requires much more attention. We need to know why they appear in the pit membranes of some differentiating cambial derivatives and not others. An understanding of their pattern of formation may provide a key to the problem of how different cell types develop from apparently similar cambial cells.

Later stages in pit development

The bordered pits of conifers have received much more attention from electron microscopists than have the pits of angiosperms. Consequently their structure is well documented[14, 30, 68, 90]. The pit membrane undergoes considerable change once its location has been determined, the central region becoming thickened to form a torus which in the mature state is suspended by a meshwork of microfibrils, the margo (Fig. 2.35). These margo microfibrils were considered by Brown and Baker[17] to be remnants of the original primary wall. Thomas[120], however, has shown that the margo comprises the randomly arranged microfibrils of the primary wall over which radially oriented (with respect to the centre of the pit membrane) microfibrils have been deposited. Parham and Baird[85] supported this latter view.

The torus is apparently formed by the deposition of matrix materials and circularly oriented microfibrils over the central region of the pit field. The microfibrils are certainly deposited after those of the margo since they obscure the latter where they pass through the torus (Fig. 2.35)[40, 85, 119]. Deposition of osmiophilic material into the almost mature tori of pits in *Pinus radiata* has been illustrated by Barnett[10]. This material enters the torus via channels which are superficially similar to plasmodesmata, but which in fact never pass right through the membrane, ending instead in a median nodule. What may represent the scars of these channels are visible

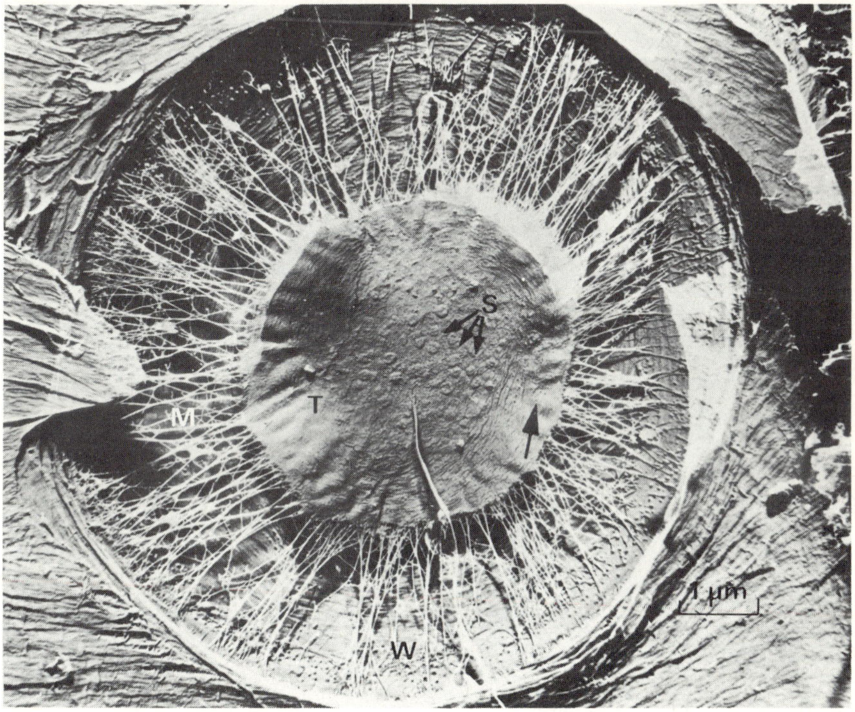

Figure 2.35: *Replica of a bordered pit margo and torus in* P. radiata. *The margo (M) consists of a 'cobweb' of cellulose microfibrils from which the thickened torus (T) is suspended. Note the circular orientation of microfibrils in the peripheral region of the torus, and scars (S), probably representing the positions of channels by which impregnating material entered the maturing torus. The interior of the pit border underlying the margo shows the circular orientation of the S1 microfibrils in this region (arrow). Note the warts (W) on this inner surface of the border. (Micrograph by courtesy of Dr J. A. Kininmonth, New Zealand Forest Service).*

at the surface of the torus in Fig. 2.35. The tendency to interpret these structures as plasmodesmata has in the past lent weight to the view that these structures are involved in pit development in tracheids. Material impregnating the torus in this way has been cited as protecting the structure from attack by autolytic enzymes at cell death. These certainly remove primary wall matrix substances from the margo leaving only the microfibrillar web. The role of lignification in protecting the secondary wall and wall thickenings from dissolution by these enzymes has been mentioned elsewhere (chapters 1 and 4). Bauch *et al.*, however, have shown that the torus is not lignified, but rather that it is impregnated with polyphenols[14]. O'Brien[83] suggested that the resistance to attack by enzymes which is possessed by the torus is a result of its polyuronide content. Fengel[37] on the other hand, proposed that the torus persists simply because it is thicker than the margo and that dissolution is halted before the torus has been significantly attacked. Imamura *et al.*[56] examined the effect of hemicellulase and pectinase on pit membranes and found that while the

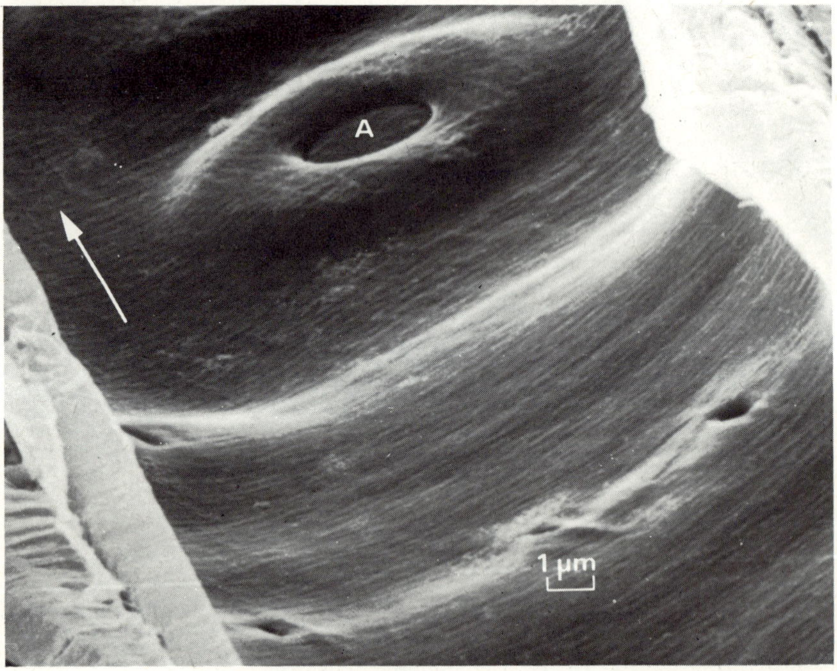

Figure 2.36: *Scanning electron micrograph showing the interior of a tracheid of* P. radiata. *The orientation of the S3 microfibrils may be seen to be almost perpendicular to the main axis of the cell (arrow indicates the cell's long axis). Very little deviation of microfibril orientation is necessary in this layer for the pit aperture (A) to be by-passed. The smaller apertures visible in this micrograph represent half bordered pits to ray tracheids.*

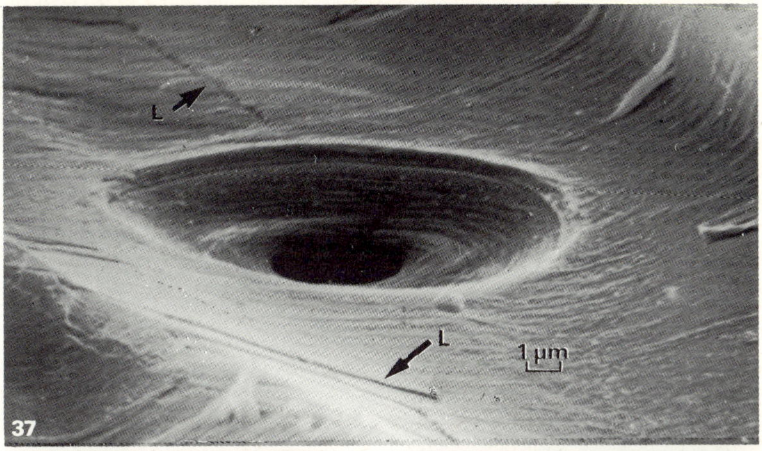

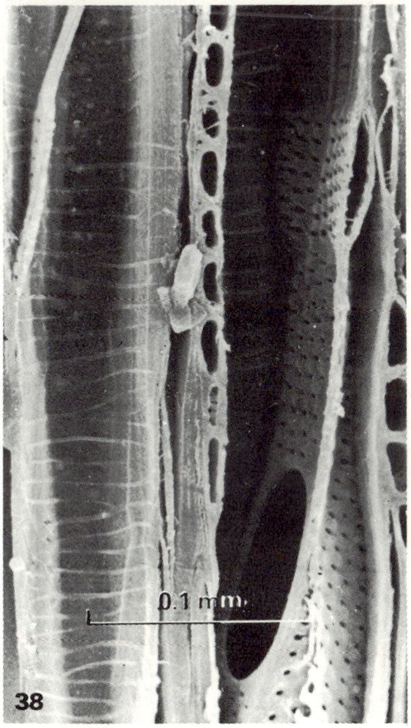

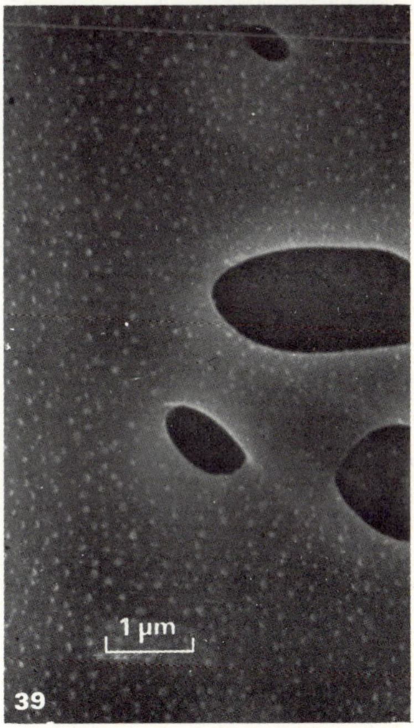

Figure 2.37: *Scanning electron micrograph illustrating a region in which two tracheids have been separated during specimen preparation along the S1 layer of one of them. The circular orientation of the S1 microfibrils lining the pit border may be seen. Here again, the major microfibril orientation is at right angles to the cell's long axis. The lamellar organisation of the S1 microfibrils may be seen where such a lamella has been torn during preparation (arrow labelled L).*

Figure 2.38: *Scanning electron micrograph of a tangential longitudinal face showing vessels in* A. hippocastanum. *Note the helical thickenings on the vessel walls.*

Figure 2.39: *Scanning electron micrograph of the lumen surface of a tracheid in* P. radiata. *Note that the surface carries numerous warts.*

matrix material of the developing pit margo was degraded by hemicellulase, the torus was unaffected. The torus was found to be susceptible to pectinase, however. From this experiment, the authors concluded that the embedding substances are indeed responsible for the protection of the torus.

The pit membrane of angiosperm vessels is not elaborated to the same extent as that of conifer tracheids. Instead it appears to consist of the original middle lamella and primary wall only. No rearrangement of wall material to form pit fields enclosed by a circular rim has been observed. At maturity, the autolytic enzymes produced, remove wall matrix substances throughout the entire membrane. Czaninski[35] has carried out an electron microscope cytochemical study of walls and pit membranes in lignified parenchyma and vessels in a number of angiosperms, and shown that while pit membranes between parenchyma cells are lignified, and therefore resistant to attack, vessel pit membranes are unlignified and susceptible to dissolution.

Imamura and Harada[55] consider that the structure of the pit membrane is completed prior to the onset of border formation. The border is then built up by deposition of S1 microfibrils until the aperture is reduced to its final diameter. The appearance of the inner border surface with its circularly oriented microfibrils supports this view (Figs. 2.35 and 2.37). This orientation is a measure of the amount of deviation undergone by the synthetic system, when it encounters the rim of the pit field (see above). As Preston has pointed out, the contribution made to the border by the other secondary wall layers is uncertain[90]. That the S3 layer plays a part is, however clear when the pit is viewed from the cell lumen (Fig. 2.36). Here there is a less marked deviation of microfibrils to by pass the by now small aperture.

SECONDARY WALL DEPOSITION

The secondary wall of xylem elements is made up of three major components; cellulose, hemicellulose and lignin. For the same reason that it has formed the focus of attention in growing walls, *viz*. it occurs in the form of microfibrils which are readily resolvable, cellulose has been the secondary wall component most extensively investigated by electron microscopy. Hemicelluloses and lignin do not form such convenient structures in xylem and may be studied microscopically only by use of stains, whose action and specificity may be uncertain, or by selective removal of components to reveal their location in the wall. This being the case, and lignification forming the subject of a separate chapter in this book (chapter 4), it is intended in this section to concentrate on the cellulose component of the wall.

The arrangement of cellulose in the secondary wall

The distribution of components in the secondary wall has been reviewed by Côté[30] and Preston[90] (see also Introduction, p. 6, and Chapter 4, p. 120). It is known that the cellulose in the secondary walls of fibres and tracheids is organised into three layers known as S1, S2 and S3 (see Figs. 3 and 4.5). The first-formed, outermost layer is the S1, the last formed, innermost layer (adjacent the lumen), the S3. These two layers comprise several lamellae of oriented microfibrils arranged in shallow helices around the cell (Figs. 2.36 and 2.37). Between these, the thickest layer, the S2, is made up of microfibrils oriented more nearly parallel to the long axis of the cell, forming steep helices. The widely held view that the S2 layer is made up of lamellae in the same way as the S1 and S3 layers has been questioned by Chafe[23] who suggested that the effect of lamellation seen in electron microscope preparations might be an artefact of preparation techniques or an optical effect resulting from the angle made by microfibrils to the section. Kerr and Goring[63], however, have produced a model for the S2 layer in which both cellulose and hemicelluloses are arranged in tangential lamellae.

This basic three-layered wall pattern is subject to a number of variations. These range from differences in the relative proportions of the three layers; to the presence of intermediate layers with microfibrillar orientations between those of the major layers; to complete absence of a layer (the S3 in compression wood) or the modification or addition of extra layers (e.g. the gelatinous layer in tension wood fibres).

Less is known of the structure of vessel walls. Wardrop[126] has shown that layering is present and detectable with the polarisation microscope, and that less specialised vessels show more layering than more specialised ones. There is, however, only slight variation in the microfibrillar orientation between the layers, which Preston[88, 90] showed formed a rather flat spiral around the cell. He also demonstrated that the inclination of the spiral changes from point to point within the cell and has pointed out the difficulty of ascertaining the main microfibril orientations owing to the large number of pits in the wall which cause numerous local deviations.

With regard to xylem parenchyma cells, a varied picture is emerging. Chafe and Chauret[27] have reviewed the early work on wall structure in these cells and found the walls of axial parenchyma in trembling aspen to be more complex than had been expected. Layering of microfibrils, similar to that observed in fibres and tracheids is present in these cells, with groups of two or three layers separated by a concentric isotropic layer similar to primary wall. The effect is of a cell with two or three concentric secondary walls separated by layers of primary wall. Such layering was found in five out of six angiosperm woods examined, but was not found in any of seven gymnosperms. In a study of parenchyma walls in *Cryptomeria japonica* Chafe[25] demonstrated that they have a typical crossed polylamellate structure.

Ray parenchyma cells in all species examined by Chafe[24, 25] and Chafe and Chauret[27] were found to have a crossed polylamellate structure in their walls, which persisted through localised wall thickenings.

Thus in all these types of cell the cellulose of secondary walls is in layers of oriented microfibrils. The extent to which these layers consist of lamellations is not clear in some cases, but what is clear is that the mechanism by which the cellulose component of the wall is produced must be capable both of producing microfibrils of uniform size, laying them down parallel to one another either in lamellae or in thicker layers, and of orienting the microfibrils in such a way that the strength properties of the wall are optimised for the cell's function at maturity.

Two other features of the secondary xylem cell wall which ought to be mentioned are helical thickenings and warts. Helical thickenings (Fig. 2.38) are deposited by some species as the last-formed wall layer in tracheary elements. Hodge and Wardrop[50] demonstrated that these thickenings comprise cellulose microfibrils running parallel to the thickening. Preston[90] suggested that they may represent an attempt by the cell to continue wall formation when the cell is near death.

Warts (Fig. 2.35 and 2.39) are aptly named protruberances on the inner surface of the wall. Their origin is uncertain, Wardrop and Davis[129] and Liese and Ledbetter[69] considered them to arise by drying down of cytoplasm at the end of differentiation. Cronshaw[33], however, suggested they are actively deposited by the cell.

Microfibril synthesis and deposition

The mechanism by which cellulose microfibrils are synthesised and oriented remains one of the great biological mysteries. The nature of the precursor molecule(s), the enzyme system, and the mechanism controlling orientation are still subjects of speculation and controversy. Most of the positive information we possess has come from the study of lower organisms such as bacteria or algae, or from higher plant protoplasts in culture. Little evidence about the process has come from the study of higher plants *in vivo*.

A number of nineteenth-century workers reported that plant cell walls apparently increased in thickness by the addition of granular cytoplasm to their inner surfaces. This granular material was found by Straburger[118] to give a positive reaction when tested for cellulose.

No really marked advance was made from this position until the application of electron microscopy to the problem. Observations on *Valonia*[94] and cambium[95] led Preston and co-workers to believe that cellulose microfibrils were produced at the protoplasmic surface by regions which they called islands of synthesis. The most significant observation, however, was that made by Frei and Preston[38] who found granular material attached to the inner surface of developing walls in *Chaetomorpha melagonium*. These granules were organised into rows, and from some, microfibrils appeared to emanate. This observation led Preston[89] to formulate his by now well known 'ordered granule hypothesis' for cellulose biosynthesis. A model was proposed to explain how microfibrils might be simultaneously synthesised and oriented, while permitting such observed algal wall phenomena as the presence of two major cellulose orientations at right angles to one another with the occasional appearance of a lamella containing microfibrils with an orientation of 45° to the major ones, and interweaving of microfibrils between lamellae.

While there is no firm evidence to suggest that a three-dimensional, cubically packed enzyme granule array as described in the model is present covering the surface of the plasmalemma of cells synthesising cellulose walls, there is increasing and strong support for the involvement of granules in both synthesis and assembly of microfibrils. Most of the evidence has come from the application of the freeze-etching technique. Granule systems of varying complexity from individual granules to arrangements of granule 'bands' to strings of granules which give rise to microfibrils, have been observed at the plasmalemma surface of algal cells known to be synthesising a cellulose wall[12, 73, 86, 107, 108]. Similar observations have been made in some higher plant cell protoplasts. Willison and Grout[134] observed single globules at the ends of developing microfibrils in isolated tobacco protoplasts, while Robenek and Peveling[102] found two particulate systems at the plasmalemma surface of protoplasts of *Skimmia japonica*. One set was formed into hexagonal arrays such as are commonly observed at the plasmalemma of freeze-etched yeast cells, while the other set comprised rows of fourteen particles.

Such wide variability in granule arrangement from one plant to another makes it impossible at the moment to determine how the granules act to synthesise microfibrils, although there is an increasing conviction among workers that this is what they do. This view has been strengthened by the work of Bartnicki-Garcia's group, who have been able to extract a granular component from *Mucor rouxii* and use this to synthesise chitin microfibrils *in vitro*[112].

The view that individual granules both synthesise cellulose and assemble the microfibril is an essential feature of the ordered granule hypothesis. In this way the crystallisation process may be envisaged as being directed to form the naturally occurring crystalline form of cellulose known as cellulose I. Under *in vitro* conditions, without the intervention of a biological system, cellulose crystallises into a different form known as cellulose II, readily distinguishable from cellulose I by X-ray crystallography[90]. Despite this, a number of workers have suggested that the cellulose microfibril is formed by crystallisation of cellulose molecules some time after their synthesis. Grout[44] proposed that cellulose precursors are assembled in the plasmalemma until a critical concentration is reached at which cellulose is synthesised and the microfibril lifted out of the plasmalemma after crystallisation. Colvin and Leppard[29] observed large diameter 'nascent' microfibrils in fresh walls of *Acetobacter xylinum* and *Acetobacter acetigenus*. These nascent microfibrils they consider to have arisen following association of intermediate polyglucan molecules released from the cytoplasm. The more compact, typical appearance of microfibrils seen in electron microscope preparations is considered to arise by compaction (crystallisation?) of the nascent microfibril during drying. Burgess and Linstead[18] are the most recent authors to have invoked a mechanism involving crystallisation of previously formed molecules.

These workers do not consider the problem of how this might give rise to cellulose I when crystallisation normally gives the cellulose II form. Preston[90] has discussed cases in which cellulose II has been claimed as a naturally occurring form in cell walls and has shown them to be spurious. The difference between the two forms is in the arrangement of the crystal lattice and it may well be that in order for the cellulose I type lattice to form the individual chains must be presented to one another in a way that does not occur in a purely physical crystallisation. The arrangement of enzyme molecules within a synthetic granule might therefore reasonably be considered as providing a system in which this might be achieved.

However the microfibrils are assembled, there remains the problem of how they are oriented. The often-repeated observation that cells synthesising oriented lamellae of cellulose microfibrils possess microtubules just beneath the surface of the plasmalemma, whose orientation parallels that of the last-formed microfibrils, has led many authors to infer that microtubules are the controlling agency in microfibril orientation (see review by Robinson[104] for references). A much cited piece of evidence in support of this idea depends on the action of colchicine on cells forming walls. Colchicine causes the dismantling of microtubules, and cells treated in this way produce walls in which cellulose microfibrils are randomly arranged. This experiment has been repeated many times on different cell types (for some recent examples see references 51, 71, 105, 116).

Preston[90] considers that the observed parallelism between microfibrils and microtubules might be explained by the action of a third factor which orientates both structures. In his view, the enzyme granules responsible for cellulose synthesis also control orientation. As these would be embedded in the plasmalemma, this third factor might well be the plasmalemma itself. Since microtubules frequently appear to be joined to the plasmalemma by bridges (Fig. 2.43), such an idea seems feasible. Further, colchicine has been shown by Wunderlich *et al.*[138] to affect the mobility of the plasmalemma in addition to destroying microtubules. They conclude that '. . . one cannot attribute colchicine-induced alterations of membrane phenomena uniquely to microtubule involvement.'

Robinson *et al.*[105] supported a role for microtubules in orienting the plasmalemma and through it the microfibrils in *Oocystis*, but later[103] found that in *Glaucocystis*, the microtubules were separated from the plasmalemma by a system of lacunae which would restrict such an effect. It was also found[106] that a variety of known microtubule inhibitors were inconsistent in their effects on cellulose orientation, some causing the cell of *Oocystis* to produce a wall made up of randomly arranged microfibrils, while others did not affect wall production in this way. Other evidence suggesting that walls made up of oriented microfibrils may be formed in the absence of microtubules has come from the studies of Nelmes *et al.*[82]. They found that in apple trees suffering from a condition known as rubbery wood, in which cell walls remain unlignified, microtubules are absent from cells in which normally-oriented microfibrils are being formed. Montezinos and Brown[73] claim that microtubules present in *Oocystis apiculata* disappear as the oriented secondary wall forms and suggest that microtubules function to maintain the cell's elliptical shape until the wall is sufficiently advanced to take over this function. At the same time, these authors observed granule bands comprising three rows of particles, with 30–40 particles per row at the termini of grooves in the plasmalemma containing developing microfibrils. Further, they found that the granule bands were disrupted if the cells were treated with colchicine. Thus, the superficially attractive hypothesis that microtubules are responsible for orientation of cellulose microfibrils is still far from being proved.*

In most cells in which the presence of granules which might be responsible for cellulose synthesis has been reported, the granules are present in a monolayer, and in small groups as bands, strings or even single granules. Thus the observed pattern does not fit that required by Preston's hypothesis. There is an increasing tendency to consider that individual, or small groups of granules assemble the microfibril as they move along the plasmalemma, in a way analogous to a spider spinning a web[26, 48, 73, 133]. Heath[48] has attempted to reconcile the granule hypothesis with the microtubule orientation hypothesis, suggesting that microtubules, by their attachment to the plasmalemma form tracks along which synthesising granules move. Willison and Brown[133] observed dark bands in longitudinal views of microtubules which they postulated might represent attachments to the plasmalemma and tracks along which such movement might take place.

* See also Chapter 1 for discussion of the role of microtubules in primary xylem differentiation (p. 25).

In Chafe's[26] view, orientation of microfibrils might be achieved by single synthesising-particles moving over the cell surface along geodesic lines. Once the orientation of the microfibril has been determined, right at the beginning of its formation, this orientation would be presumably maintained by the relative rigidity of the crystalline microfibril structure. Near the tip of the cell, the geodesic line would flow around the cell and thereby reverse its direction, giving a microfibril the opposite spirality to that which it originally possessed. In this way the production of alternate lamellae oriented at right angles to one another such as are found in algal cell walls and parenchyma cell walls may be envisaged. There are, however, several difficulties inherent in this hypothesis. Chafe[26] himself points out that the formation of thick layers of microfibril lamellae having similar orientations, for example the S2 layer in secondary xylem cells, is difficult to explain in terms of reversal of geodesic lines. If orientation of microfibrils, once initiated is perpetuated in this kind of way, then it must be supposed that in a case such as the S2 layer, a change from the geodesic pattern occurs near the cell tips, or that no reversal takes place, a granule's useful synthetic life ending when it reaches the tip.

Thus none of the current ideas on the mechanism of microfibril orientation is without its problems. Confusion still exists about the nature of the synthetic system and it may be that this will not be resolved until the system is isolated and made to perform *in vitro*, like the chitin synthesising granules mentioned above[112].

THE CYTOLOGY OF SECONDARY XYLEM CELLS DURING THE PHASE OF SECONDARY WALL FORMATION

The main activity of differentiating xylem elements following growth, is the formation of secondary wall. One would therefore expect to see ample evidence of cytoplasmic activity associated with this process in electron microscope preparations of tissue at this stage. The uncertainty which exists about the cellulose biosynthetic system, despite extensive research, has been indicated above, and it must be said that little conclusive evidence about the role of electron microscopically visible cytoplasmic structures in wall formation has appeared from the study of secondary xylem elements. These have been examined in a small number of gymnosperms[10, 34, 77, 78, 79] and angiosperms[31, 101, 127, 137]. Robards[99] and Murmanis[77] both reported the presence of particles outside the plasmalemma and implicate these in cellulose biosynthesis. Murmanis and Sachs[78] observed fibrillar material in 'Golgi-derived vesicles' which appeared to be undergoing incorporation into the wall. They considered this material to be cellulose, but as their fibrillar material was well stained, and cellulose is inert to the stains normally used in electron microscopy, their inference may be incorrect.

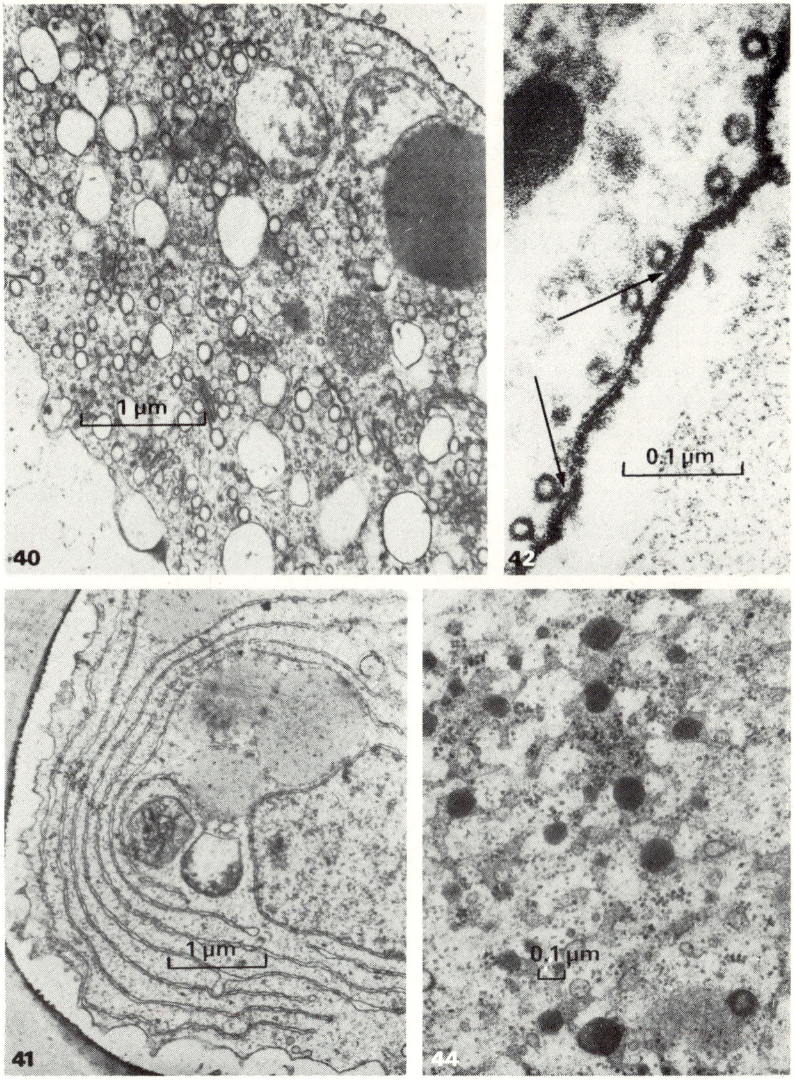

Figure 2.40: *Part of the cytoplasm of a developing fibre in* A. hippocastanum. *Numerous vesicle-producing dictyosomes are present.*

Figure 2.41: *Endoplasmic reticulum in a developing fibre of* A. hippocastanum *seen in TS. The reticulum is organised into lamellae parallel to the cell surface.*

Figure 2.42: *Section taken parallel to the surface of a developing fibre in* A. hippocastanum. *The organisation of ER into anastomosing tubules, with swollen regions containing osmiophilic material, is apparent.*

Figure 2.43: *Transverse section of microtubules in a developing fibre of* A. hippocastanum. *The microtubules appear to be joined to the plasmalemma by bridges (arrows).*

Examination of fibres in *Aesculus hippocastanum* has revealed a number of differences between the cytoplasm of cells undergoing wall formation and that of cambium cells. The observations are in many ways similar to those made on *Acer rubrum* by Cronshaw[31, 32] and confirm many of his findings. Among these are that free ribosomes disappear to be replaced by polysomes, microtubules undergo a great increase in numbers and ER becomes more prominent.

The cells are still highly vacuolate with cytoplasm confined to a generally thin parietal layer. In places, however, there are concentrations of cytoplasm which contain large numbers of dictyosomes actively producing vesicles with electron lucent contents (Fig. 2.40). Other regions are rich in endoplasmic reticulum, of which the smooth form appears to predominate (Fig. 2.41). The ER is arranged in lamellae parallel to the cell surface. In sections cut more nearly tangential to the lamellae it is seen that the elements of the reticulum are organised into anastomosing tubules (Fig. 2.42). These tubules frequently contain osmiophilic bodies. Such osmiophilic bodies are, in addition to the dictyosome derived vesicles, abundant in cells undergoing secondary wall formation, and it is possible that these have their origin in the ER.

Always to be found in cells at this stage are numerous microtubules. In TS (Fig. 2.43) they often possess what are generally thought to be bridges to the plasmalemma. In oblique sections (that is those taken at an angle of less than 90° to the long axis of the cell) and sections cut tangential to the cell surface, the microtubules are seen to be oriented parallel to the microfibril orientation in the secondary wall (Figs. 2.44 and 2.45). This orientation is indicated by the presence of stain between the microfibrils which are thereby negatively stained. In oblique sections, rows of particles may be seen exterior to the plasmalemma, but interior to the wall, and aligned parallel to both microtubules and microfibrils (Fig. 2.46). In other regions strings of particles occur. Where they are in contact with the plasmalemma they parallel the microfibrils of the wall, but where they lie in the space between the plasmalemma and the wall they have a more random orientation (Fig. 2.47).

Most of these observations are what would be expected on the basis of what has been described by other authors working on differentiating xylem. It is, perhaps, reasonably safe to suppose that ER and dictyosome derived vesicles are contributing materials to the developing wall, but how they are assembled, collected into the vesicles, and even what they are, has still to be demonstrated. It may be that the strings of particles observed at the plasmalemma surface are similar to the granule strings of Peng and Jaffe[86], but their presence in an electron micrograph in no way proves that this is so. It is possible that *in vivo* the particle strings more perfectly reflect the orientation of cellulose in the wall, and that separation of the plasmalemma from the wall during fixation and embedding has disrupted their arrangement. It is known that in cells prepared for electron microscopy by freeze-etching, without prior treatment with cryoprotectants, the plasmalemma is closely adpressed to the wall[132].

An important question concerns the role of microtubules in these cells. The question of their involvement in microfibril orientation remains, as indicated above, a matter of dispute. In the secondary xylem cells of *Aesculus* during the main phase of secondary wall formation, they are so abundant as to occupy between a quarter and a third of the inner surface area of the plasmalemma. They form a cage-like structure such that material passing out of the cytoplasm through the plasmalemma must also pass through the gaps between the 'bars' of this cage. It is probable that

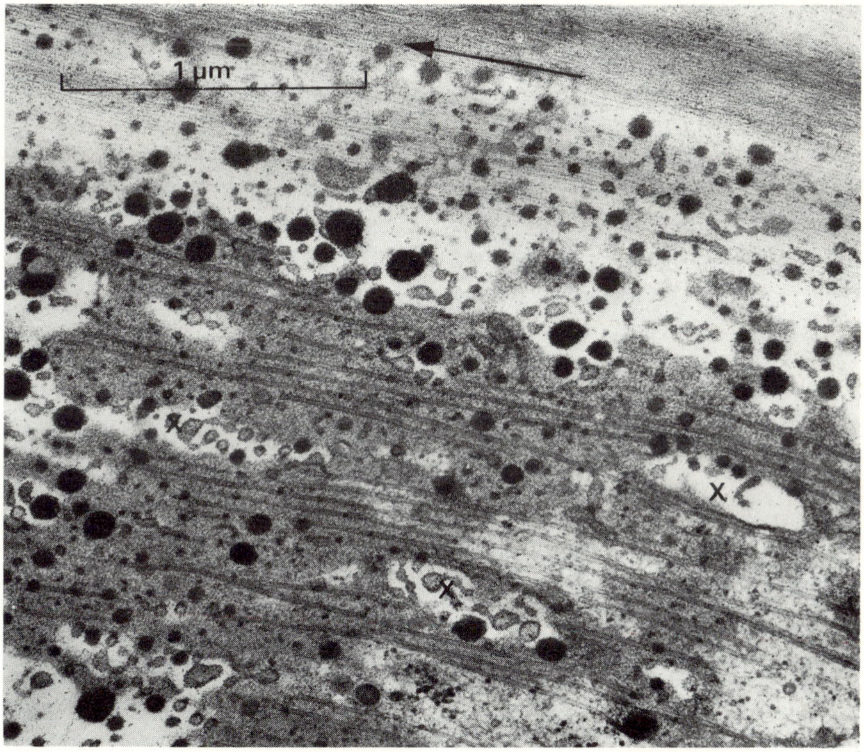

Figure 2.44: *Section cut parallel to the surface of a differentiating fibre in* A. hippocastanum. *Included in the thickness of the section are the plasmalemma, which appears as a grey smear, the underlying microtubules, and the space between the plasmalemma and the wall, probably produced during specimen preparation. The microtubules lie parallel to the long axis of the cell (arrow) and show a tendency to deviate around infoldings of the plasmalemma. These appear as unstained areas (X). A large amount of vesicular material and osmiophilic globular material (similar to that in the ER in Figure 2.42) is present in the space between the plasmalemma and the wall.*

both the packages of material passing out of the cell, and the microtubules are sufficiently flexible to permit even large vesicles to pass through. In the tangential view of the cell shown in Fig. 2.48 it may be seen that microtubules curve, and move out of parallel in some regions, possibly to accommodate plasmalemma infoldings. As such disturbances in the plasmalemma may be artefacts, as indicated by the above-mentioned freeze-etching study[132], it is possible that in the natural, living condition the microtubules are more regularly spaced than is apparent from the micrographs shown here.

An alternative analogy for the arrangement of microtubules in these cells (in which they are oriented almost parallel to the cell's long axis), is with the wooden stringers in a canoe or aircraft fuselage, over which fabric is stretched. Perhaps it is not beyond the bounds of possibility that they serve to keep the plasmalemma stretched and in contact with the developing wall, thereby ensuring coherence in wall formation? Increasingly, there is a tendency among authors to consider that microtubules serve to maintain cell shape until the wall is sufficiently well formed to take this function on itself. This is the view of Montezinos and Brown[73] who found that in *Oocystis apiculata*, microtubules were present in naked cells but disappeared as cell wall formation proceeded, an observation which had been recorded earlier by Preston and Goodman[92]. Marchant and Hines[71]

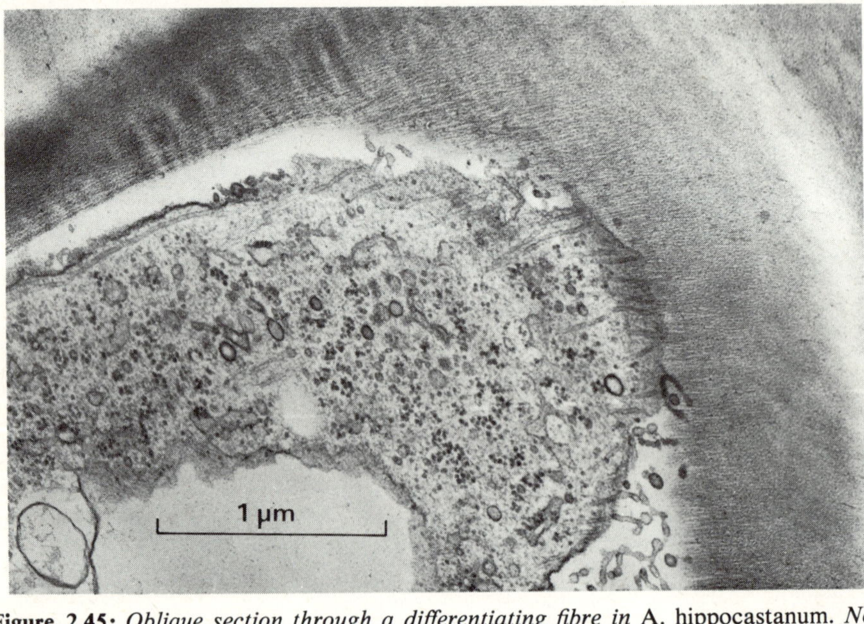

Figure 2.45: *Oblique section through a differentiating fibre in* A. hippocastanum. *Note that the microtubules appear at all points to parallel the innermost wall microfibril orientation as indicated by the negative staining of these structures.*

have also implicated microtubules in maintenance of shape in naked cells, but have extended this role to include control of microfibril orientation. If the role of microtubules is, in fact, to 'stretch' the plasmalemma, colchicine treatment, in destroying these microtubules would remove this framework supporting the plasmalemma. In naked cells this would lead to the observed tendency for them to remain spherical when they would otherwise become ovoid or more elongate[71]. In cells already well advanced in wall formation, destabilisation of the plasmalemma might lead to its separation from the wall, as in fixed material, and disruption of the ordered synthesising mechanism. The result would be the production of randomly arranged microfibrils as is indeed so often observed.

The answer to these problems is likely to remain a matter of controversy until we have a better understanding of the biochemistry of cellulose biosynthesis, the nature of the plasmalemma, and the function(s) of microtubules.

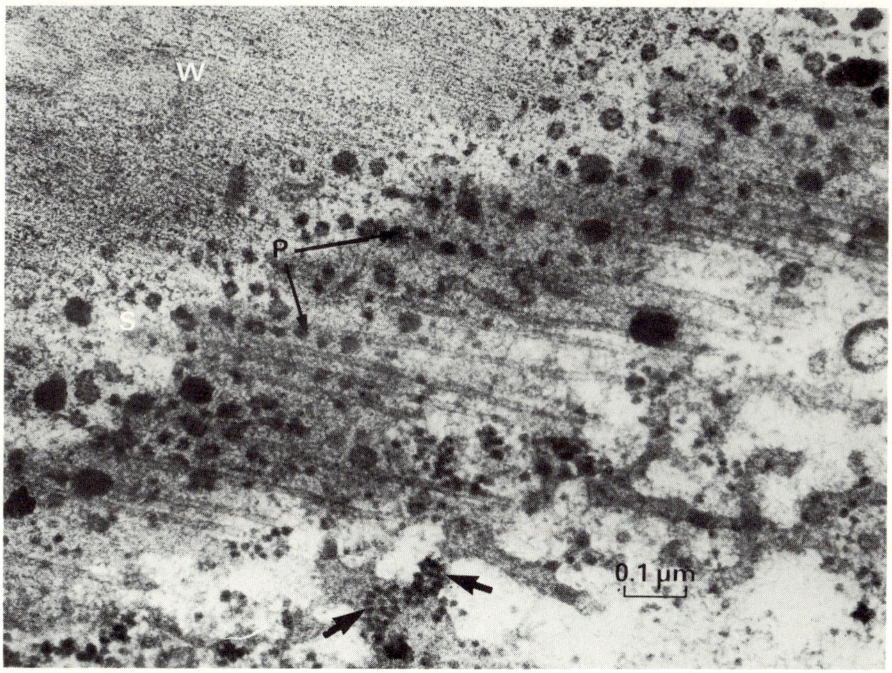

Figure 2.46: *Oblique section through a differentiating fibre in* A. hippocastanum. *As in Figure 2.44, several layers are present: the cell wall (W), the space between the wall and the plasmalemma (S), the plasmalemma (as a grey smear), the underlying microtubules, and the cytoplasm containing ER and polysomes (arrows). In places, vesicles appear in rows oriented parallel to microtubules and microfibrils of the wall (P). That these are external to the plasmalemma is indicated by their continuation into the space between the plasmalemma and the wall.*

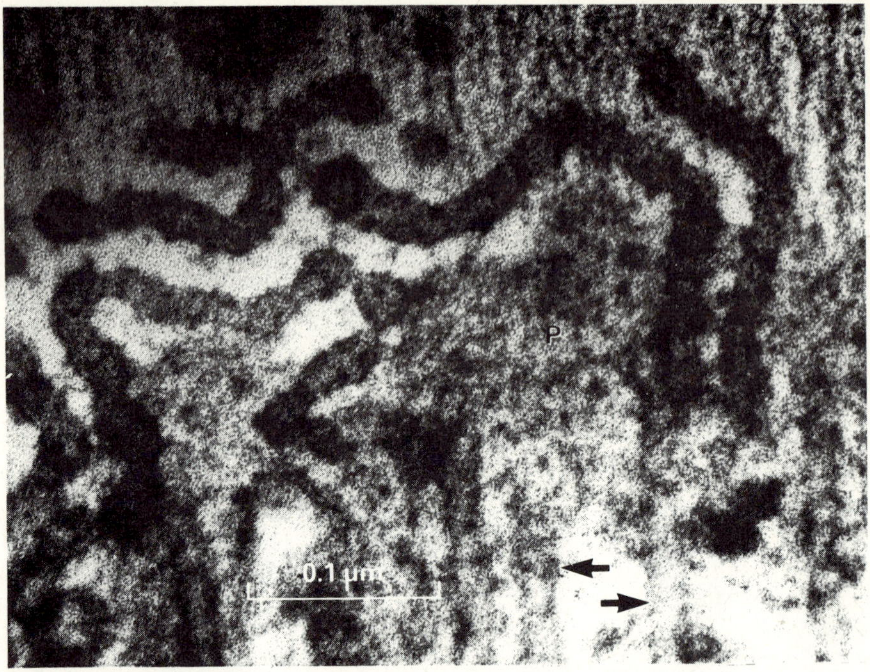

Figure 2.47: *Detail of the plasmalemma, wall, and the space between them in a differentiating fibre of* A. hippocastanum. *Chains of particles, apparently attached to the plasmalemma where it lies in section (P), and here paralleling the microtubules (arrows) and microfibrils, become more randomly oriented where they enter the space between the plasmalemma and the wall. This randomness may arise as a consequence of plasmolysis during fixation.*

THE FINAL STAGES

Wodzicki and Humphreys[135, 136] appear to be the only workers to have examined the structure of the senescent cytoplasm in maturing secondary xylem elements. Using scanning and transmission electron microscopy, they examined the final stages of differentiation in tracheids of *Pinus echinata*. They observed the formation of cytoplasmic spherules and the breakdown of cytoplasmic bridges which normally traverse the vacuole. They propose that organelles are subsequently digested in the vacuole. Some doubt must exist, however, as to whether the changes they describe are real, or represent damage caused during preparation. The samples they used were cut to 1 mm cubes prior to fixation, and since differentiating tracheids in pines are usually at least this long, it is probable that few intact tracheids would remain in their tissue blocks after such a trimming. Applied to larger pieces of tissue, and therefore cells which would certainly be intact at the time of fixation, their techniques might usefully be applied to all stages of differentiation.

CONCLUSION

Electron microscopy has made a significant and positive contribution to our knowledge of the process of xylem differentiation. Unfortunately, much of this information has been derived from studies of tissues other than xylem, notably in the case of cellulose biosynthesis. The difficulties to be overcome in preparing developing xylem for electron microscopy have been a major factor in persuading workers to turn to more tractable tissues. Such an approach does, fortunately, often yield results which can be applied to cell types other than those under investigation, although ultimately, they will have to be tested on xylem itself.

Application of electron microscopy alone will, however, provide no complete solutions. We need to know more of the chemical nature of the cell wall and the relationships between its components, we must understand the biochemical processes involved in differentiation, and how these are controlled by genetic, hormonal and environmental factors. Only by integration of such information can we hope to achieve a full understanding of how cells differentiate.

REFERENCES

[1] Bailey, I. W. Phenomena of cell division in the cambium of arborescent gymnosperms and their cytological significance. *Proc. Nat. Acad. Sci.* **5,** 283-285 (1919)

[2] Bailey, I. W. The cambium and its derivative tissues. II. Size variations of cambial initials in gymnosperms and angiosperms. *Amer. J. Bot.* **7,** 355-367 (1920)

[3] Bailey, I. W. The cambium and its derivative tissues. IV. The increase in girth of the cambium. *Amer. J. Bot.* **10,** 499-509 (1923)

[4] Bailey, I. W. The cambium and its derivative tissues. V. A reconnaissance of the vacuome in living cells. *Z. Zellforsch. Mikroskop. Anat.* **10,** 651-682 (1930)

[5] Bannan, M. W. The frequency of anticlinal divisions in fusiform cambial cells of *Chamaecyparis. Amer. J. Bot.* **37,** 511-519 (1950)

[6] Bannan, M. W. Some aspects of the elongation of fusiform cambial cells in *Thuja occidentalis* L. *Canad. J. Bot.* **34,** 175-196 (1956)

[7] Barnett, J. R. Winter activity in the cambium of *Pinus radiata. N.Z.J. For. Sci.* **1,** 208-222 (1971)

[8] Barnett, J. R. Seasonal variation in the ultrastructure of the cambium in New Zealand grown *Pinus radiata* D. Don. *Ann. Bot.* **37,** 1005-1011 (1973)

[9] Barnett, J. R. Seasonal variation of organelle numbers in sections of fusiform cambium cells of *Pinus radiata* D. Don. *N.Z. J. Bot.* **13,** 325-332 (1975)

[10] Barnett, J. R. Tracheid differentiation in *Pinus radiata. Wood Sci. Technol.* **11,** 83-92 (1977)

[11] Barnett, J. R., Harris, J. M. Early stages of bordered pit formation in radiata pine. *Wood Sci. Technol.* **9,** 233-241 (1975)

[12] Barnett, J. R., Preston, R. D. Arrays of granules associated with the plasmalemma in swarmers of *Cladophora. Ann. Bot.* **34,** 1011-1017 (1970)

[13] Bartnicki-Garcia, S. Fundamental aspects of hyphal morphogenesis. *Symp. Soc. Gen. Microbiol.* **XXIII,** 245-267 (1973)

[14] Bauch, J., Liese, W., Schulz, F. Über die Entwicklung und stoffliche

Zusammensetzung der Hoftüpfel membranen von Langstracheiden in Coniferen. *Holzforsch.* **22,** 144-153 (1968)

15 Bonner, L. D., Thomas, R. J. The ultrastructure of intercellular passageways in vessels of yellow poplar (*Liriodendron tulipifera* L.). Part 1: Vessel pitting. *Wood Sci. Technol.* **6,** 196-203 (1972)

16 Boyd, J. D., Foster, R. C. Microfibrils in primary and secondary wall growth develop trellis configurations. *Canad. J. Bot.* **53,** 2687-2701 (1975)

17 Brown, F. L., Baker, H. M. Scanning electron microscopy of mature Douglas-fir earlywood and inter-tracheid pitting. *Wood and Fiber* **2,** 52-64 (1970)

18 Burgess, J., Linstead, P. J. Structure and association of wall fibrils produced by regenerating tobacco protoplasts. *Planta* **146,** 203-210 (1979)

19 Catesson, A. M. Origine, fonctionnement et variations cytologique saisonnières du cambium de l'*Acer pseudoplatanus* L. (*Acéracées*) *Ann. Sc. Nat. Bot. 12e serie* **5,** 229-498 (1964)

20 Catesson, A. M. Cambial Cells. In *Dynamic Aspects of Plant Ultrastructure* A. W. Robards, (ed.) (London: McGraw Hill, 1974)

21 Catesson, A. M. The Vascular Cambium. *Proc. IUFRO Workshop on Control of Shoot Growth in Trees.* In press

22 Chafe, S. C. The fine structure of the collenchyma cell wall. *Planta* **90,** 12-21 (1970)

23 Chafe, S. C. On the lamellate structure of the S2 layer. *Protoplasma* **79,** 145-158 (1974)

24 Chafe, S. C. Cell wall formation and protective layer development in the xylem parenchyma of trembling aspen. *Protoplasma* **80,** 335-354 (1974)

25 Chafe, S. C. Cell wall structure in the xylem parenchyma of *Cryptomeria. Protoplasma* **81,** 63-76 (1974)

26 Chafe, S. C. On the mechanism of cell wall microfibrillar orientation. *Wood Sci. Technol.* **12,** 203-217 (1978)

27 Chafe, S. C., Chauret, G. Cell wall structure in the xylem parenchyma of trembling aspen. *Protoplasma* **80,** 129-147 (1974)

28 Chafe, S. C., Wardrop, A. B. Fine structural observations on the epidermis. 1. The epidermal cell wall. *Planta* **107,** 269-278 (1972)

29 Colvin, J. R., Leppard, G. G. The biosynthesis of cellulose by *Acetobacter xylinum* and *Acetobacter acetigenus. Canad. J. Microbiol.* **23,** 701-709 (1977)

30 Côté, W. A. Wood ultrastructure in relation to chemical composition. In *Recent Advances in Phytochemistry: The Structure, Biosynthesis and Degradation of Wood.* F. A. Loewus, V. C. Runeckles eds. (New York: Plenum 1977)

31 Cronshaw, J. Cytoplasmic fine structure and cell wall development in differentiating xylem elements. In *Cellular Ultrastructure of Woody Plants* W. A. Côté, (ed.) (Syracuse New York: Syracusé University Press 1965)

32 Cronshaw, J. The organisation of the cytoplasmic components during the phase of cell wall thickening in differentiating cambial derivatives of *Acer rubrum. Canad. J. Bot.* **43,** 1401-1407 (1965)

33 Cronshaw, J. The formation of the wart structure in tracheids of *Pinus radiata. Protoplasma* **60,** 233-242 (1965)

34 Cronshaw, J., Wardrop, A. B. The organisation of cytoplasm in differentiating xylem. *Aust. J. Bot.* **12,** 15-23 (1963)

35 Czaninski, Y. Cytochimie ultrastructurale des parois du xylème secondaire. *Biol. Cell.* **35,** 97-102 (1979)

36 Evert, R. F., Deshpande, B. P. An ultrastructural study of cell division in the cambium. *Amer. J. Bot.* **57,** 942-961 (1970)

37 Fengel, D. Structure and function of the membrane in softwood bordered pits. *Holzforsch.* **26,** 1-9 (1972)

[38] Frei, E., Preston, R. D. Cell wall organisation and wall growth in the filamentous green algae *Cladophora* and *Chaetomorpha*. *Proc. Roy. Soc.* **B154,** 70-94 and **B155,** 55-77 (1961)

[39] Frey-Wyssling, A. Growth of plant cell walls. *Symp. Soc. Expt. Biol.* **6,** 320-328 (1952)

[40] Frey-Wyssling, A., Bosshard, H. H., Mühlethaler, K. Die submikroskopische Entwicklung der Hoftüpfel. *Planta* **47,** 115-126 (1956)

[41] Frey-Wyssling, A., Mühlethaler, K., Wyckoff, R. W. G. Mikrofibrillenbau der pflanzlichen Zellwände. *Experimentia* **4,** 475-476 (1948)

[42] Frey-Wyssling, A., Stecher, H. Das Flächenwachstum der pflanzlichen Zellwände. *Experimentia* **7,** 420-421 (1951)

[43] Goosen de Roo, L., van Spronsen, P. C. Electron microscopy of the active cambial zone of *Fraxinus excelsior* L. *IAWA Bull.* 59-64 (1978/4)

[44] Grout, B. W. W. Cellulose microfibril deposition at the plasmalemma surface of regenerating tobacco mesophyll protoplasts: A deep etch study. *Planta* **123,** 275-282 (1975)

[45] Gunning, B. E. S. Age-related and origin-related control of the number of plasmodesmata in cell walls of developing *Azolla* roots. *Planta* **143,** 181-190 (1978)

[46] Gunning, B. E. S., Robards, A. W. Intercellular Communication in Plants: Studies on Plasmodesmata. (Berlin, Heidelberg, New York: Springer-Verlag 1976)

[47] Hartig, R. Ueber den Drehwuchs der Kiefer. *Sitzungsber. Bayer. Akad. Wiss. Math. Phys.* **25,** 199-217. Quoted from Bailey.[3]

[48] Heath, I. B. A unified hypothesis for the role of membrane bound enzyme complexes and microtubules in plant cell wall synthesis. *J. Theor. Biol.* **48,** 445-449 (1974)

[49] Hejnowicz, Z. Tensional stress in the cambium and its developmental significance. *Amer. J. Bot.* **67,** 1-5 (1980)

[50] Hodge, A. J., Wardrop, A. B. An electron microscopic investigation of the cell wall organisation of conifer tracheids and conifer cambium. *Aust. J. Bot.* **3,** 265-269 (1950)

[51] Hogetsu, T., Shibaoka, H. Effects of colchicine on cell shape and on microfibril arrangement in the cell wall of *Closterium acerosum*. *Planta* **140,** 15-18 (1978)

[52] Hohl, H. R. Über die submikroskopische Struktur normaler und hyperplastischer Gewebe von *Datura stramonium* L. 1. Normalgewebe. *Ber. Schweiz. Bot. Ges.* **70,** 395-439 (1960)

[53] Houwink, A. L., Roelofsen, P. A. Fibrillar architecture of growing plant cell walls. *Acta Bot. Neerl.* **3,** 385-395 (1954)

[54] IAWA. *Multilingual glossary of terms used in wood anatomy*. (Verlagsanstalt Buchdrucherei Konkordia Winterthur 1964)

[55] Imamura, Y., Harada, H. Electron microscopic study on the development of the bordered pit in coniferous tracheids. *Wood Sci. Technol.* **7,** 189-205 (1973)

[56] Imamura, Y., Harada, H., Saiki, H. Embedding substances of pit membranes in softwood tracheids and their degradation by enzymes. *Wood Sci. Technol.* **8,** 243-254 (1974)

[57] Iterson, G. van. The formation of the cell wall. *Proc. 6th Int. Bot. Cong. Amsterdam. Vol. II*, 291-293 (1935)

[58] Itoh, T. On the ultrastructure of dormant and active cambium of conifers. *Wood Res.* (*Bull. of the Wood Res. Inst. Kyoto Univ.*) **51,** 33-45 (1971)

[59] Itoh, T. Cell wall organisation of cortical parenchyma of angiosperms observed by the freeze-etching technique. *Bot. Mag. Tokyo* **88,** 145-156 (1975)

[60] Itoh, T. Studies on the structure and growth of primary walls of woody plants. *Wood Res.* (*Bull. of the Wood Res. Inst. Kyoto Univ.*) **65,** 54-110 (1979)

[61] Juniper, B. E. Some speculation on the possible roles of the plasmodesmata in the control of differentiation. *J. Theor. Biol.* **66,** 583-592 (1977)

[62] Keegstra, K., Talmadge, K. W., Bauer, W. D., Albersheim, P. The structure of plant cell walls III. A model of the walls of suspension-cultured sycamore cells based on the interconnections of the macromolecular components. *Plant Physiol.* **51,** 188-196 (1973)

[63] Kerr, A. J., Goring, D. A. Lamellation of hemicellulose in the fibre wall of birch wood. *Wood Sci.* **9,** 136-139 (1977)

[64] Kerr, T., Bailey, I. W. The cambium and its derivative tissues. X. Structure, optical properties and chemical composition of the so-called middle lamella. *J. Arn. Arb.* **15,** 327-349 (1934)

[65] Kidwai, P., Robards, A. W. On the ultrastructure of resting cambium of *Fagus sylvatica* L. *Planta* **89,** 361-368 (1969)

[66] Klinken, J. Ueber das gleitende Wachstum der Initialen im Kambium der Koniferen und der Markstrahlenverlauf in ihrer sekundären Rinde. *Bibl. Bot.* **84,** 1-37 (1914)

[67] Krabbe, G. Das gleitende Wachstum bei der Gewebebildung der Gefässpflanzen. Berlin 1886. Quoted from Scott, D. H. Review of Krabbe. *Ann. Bot.* **2,** 127-136 (1888-1889)

[68] Liese, W. The fine structure of bordered pits in softwoods. In *Cellular Ultrastructure of Woody Plants* W. A. Côté (ed.) (Syracuse, New York: Syracuse University Press 1965)

[69] Liese, W., Ledbetter, M. C. Occurrence of a warty layer in vascular cells of plants. *Nature* **192,** 201 (1963)

[70] Mahmood, A. Cell grouping and primary wall generations in the cambial zone, xylem and phloem in *Pinus. Aust. J. Bot.* **16,** 177-196 (1968)

[71] Marchant, H. J., Hines, E. R. The role of microtubules and cell wall deposition in elongation of regenerating protoplasts of *Mougeotia. Planta* **146,** 41-48 (1979)

[72] Mia, A. J. Fine structure of active, dormant and aging cambial cells in *Tilia americana. Wood Sci.* **3,** 34-42 (1970)

[73] Montezinos, D., Brown, R. M. Surface architecture of the plant cell: Biogenesis of the cell wall, with special emphasis on the role of the plasma membrane in cellulose biosynthesis. *J. Supramol. Struct.* **5,** 277-290 (1976)

[74] Mueller, S. C., Brown, R. M., Scott, T. K. Cellulosic microfibrils: Nascent stages of synthesis in a higher plant cell. *Science* **194,** 949-951 (1976)

[75] Murmanis, L. Locating the initial in the vascular cambium of *Pinus strobus* L. by electron microscopy. *Wood Sci. Technol.* **4,** 1-14 (1970)

[76] Murmanis, L. Structural changes in the vascular cambium.of *Pinus strobus* L. during an annual cycle. *Ann. Bot.* **35,** 133-141 (1971)

[77] Murmanis, L. Particles and microtubules in vascular cells of *Pinus strobus* L. during cell wall formation. *New Phytol.* **70,** 1089-1093 (1971)

[78] Murmanis, L., Sachs, I. B. Seasonal development of secondary xylem in *Pinus strobus* L. *Wood Sci. Technol.* **3,** 177-193 (1969)

[79] Murmanis, L., Sachs, I. B. Cell wall formation in secondary xylem of *Pinus strobus* L. *Wood Sci. Technol.* **7,** 173-188 (1973)

[80] Nečesaný, V. Die Vitalitätsveränderung der Parenchymzellen als physiologische Grundlage der Kernholzbildung. *Holzforschung und Holzverwertung* **18,** 61-65 (1966)

[81] Neeff, F. Über Zellumlagerung: Ein Beitrag zur experimentellen. Anatomie. *Z. Bot.* **6,** 465-547 (1914)

[82] Nelmes, B. J., Preston, R. D., Ashworth, D. A possible function of microtubules suggested by their abnormal distribution in rubbery wood. *J. Cell Sci.* **13,** 741-751 (1973)

[83] O'Brien, T. P. Further observations of hydrolysis of the cell wall in the xylem. *Protoplasma* **69,** 1-14 (1970)
[84] Parameswaran, N., Bauch, J. On the origin of phenolic compounds in the wood rays of *Abies alba. Wood Sci. Technol.* **9,** 165-173 (1975)
[85] Parham, R. A., Baird, W. M. The bordered pit membrane in differentiating balsam fir. *Wood and Fiber* **5,** 80-86 (1973)
[86] Peng, H. B., Jaffe, L. F. Cell wall formation in *Pelvetia* embryos. A freeze-fracture study. *Planta* **133,** 57-71 (1976)
[87] Philipson, W. R., Ward, J. M., Butterfield, B. G. *The Vascular Cambium.* (London: Chapman and Hall 1971)
[88] Preston, R. D. Wall structure and growth. I. Spring vessels in some ring porous dicotyledons. *Ann. Bot.* **3,** 507-530 (1939)
[89] Preston, R. D. Structural and mechanical aspects of plant cell walls with particular reference to synthesis and growth. In *The Formation of Wood in Forest Trees* M. H. Zimmermann (ed.). (London: Academic Press 1964)
[90] Preston, R. D. *Physical Biology of Plant Cell Walls.* (London: Chapman and Hall 1974)
[91] Preston, R. D., Nicolai, E., Reed, R., Millard, A. An electron microscope study of cellulose in the wall of *Valonia ventricosa. Nature* **162,** 665-667 (1948)
[92] Preston, R. D., Goodman, G. N. Some aspects of cellulose microfibril biosynthesis. *Proc. Roy. Mic. Soc.* **88,** 513-527 (1968)
[93] Preston, R. D., Wardrop, A. B. The submicroscopic organisation of the walls of conifer cambium. *Biochim. Biophys. Acta.* **3,** 549-559 (1949)
[94] Preston, R. D., Nicolai, E., Kuyper, B. Electron microscopic investigations of the walls of green algae. III. The wall/cytoplasm relationship of freeze-dried *Valonia macrophysa. J. Exper. Bot.* **4,** 40-51 (1953)
[95] Preston, R. D., Ripley, G. W. An electron microscopic investigation of the walls of conifer cambium. *J. Exper. Bot.* **5,** 410-413 (1954)
[96] Priestley, J. H. Studies in the physiology of cambial activity. 1. Contrasted types of cambial activity. *New Phytol.* **29,** 56-73 (1930)
[97] Priestley, J. H. Studies in the physiology of cambial activity. 2. The concept of sliding growth. *New Phytol.* **29,** 96-140 (1930)
[98] Robards, A. W. On the ultrastructure of differentiating secondary xylem in willow. *Protoplasma* **65,** 449-462 (1968)
[99] Robards, A. W. Particles associated with developing plant cell walls. *Planta* **88,** 376-379 (1969)
[100] Robards, A. W., Kidwai, P. A comparative study of the ultrastructure of resting and active cambium of *Salix fragilis* L. *Planta* **84,** 239-249 (1968)
[101] Robards, A. W., Kidwai, P. Microtubules and microfibrils in xylem fibres during secondary wall formation. *Cytobiologie* **6,** 1-21 (1972)
[102] Robenek, H., Peveling, E. Ultrastructure of the cell wall regeneration of isolated protoplasts of *Skimmia japonica* Thunb. *Planta* **136,** 135-145 (1977)
[103] Robinson, D. G. Structure, synthesis and orientation of microfibrils. IV. Microtubules and microfibrils in *Glaucocystis. Cytobiologie* **15,** 475-484 (1977)
[104] Robinson, D. G. Plant cell wall synthesis. *Adv. in Bot. Res.* **5,** 89-151 (1978)
[105] Robinson, D. G., Grimm, I., Sachs, H. Colchicine and microfibril orientation. *Protoplasma* **89,** 375-380 (1976)
[106] Robinson, D. G., Herzog, W. Structure, synthesis and orientation of microfibrils. III. A survey of the action of microtubule inhibitors on microtubules and microfibril orientation in *Oocystis solitaria. Cytobiologie* **15,** 463-474 (1977)
[107] Robinson, D. G., Preston, R. D. Plasmalemma structure in relation to microfibril biosynthesis in *Oocystis. Planta* **104,** 234-246 (1972)

[108] Robinson, D. G., White, R. K., Preston, R. D. Fine structure of swarmers of *Cladophora* and *Chaetomorpha*. III. Wall synthesis and development. *Planta* **107,** 131-144 (1972)

[109] Roelofsen, P. A., Houwink, A. L. Architecture and growth of the primary wall in some plant hairs and in the *Phycomyces* sporangiophore. *Acta Bot. Neerl.* **2,** 218-225 (1953)

[110] Roland, J. C. Early differences between radial walls and tangential walls of actively growing cambial zone. *I.A.W.A. Bull.* 7-10 (1978/1)

[111] Roland, J. C., Vian, B., Reis, D. Observations with cytochemistry and ultracryotomy on the fine structure of the expanding walls in actively elongating plant cells. *J. Cell Sci.* **19,** 239-259 (1975)

[112] Ruiz-Herrera, J., Sing, V. O., van der Woude, W. J., Bartnicki-Garcia, S. Microfibril assembly by granules of chitin synthetase. *Proc. Nat. Acad. Sci.* **72,** 2706-2710 (1975)

[113] Sinnott, E. W., Bloch, R. Changes in intercellular relationships during the growth and differentiation of living plant tissues. *Amer. J. Bot.* **26,** 625-634 (1939)

[114] Srivastava, L. M. On the fine structure of the cambium of *Fraxinus americana* L. *J. Cell Biol.* **31,** 79-93 (1966)

[115] Srivastava, L. M., O'Brien, T. P. On the ultrastructure of the cambium and its vascular derivatives. 1. Cambium of *Pinus strobus* L. *Protoplasma* **61,** 257-276 (1966)

[116] Srivastava, L. M., Sawhney, V. K., Bonnettemaker, M. Cell growth, wall deposition, and correlated fine structure of colchicine-treated lettuce hypocotyl cells. *Canad. J. Bot.* **55,** 902-917 (1977)

[117] Sterling, C., Spit, B. J. Microfibrillar arrangement in developing fibres of asparagus. *Amer. J. Bot.* **44,** 851-859 (1957)

[118] Strasburger, E. Ueber dem Bau und das Wachstum der Zellhaute. Gustav Fischer. (Jena. 1882)

[119] Thomas, R. J. The development and ultrastructure of the bordered pit membrane in the southern yellow pines. *Holzforsch.* **22,** 38-44 (1968)

[120] Thomas, R. J. Origin of bordered pit margo microfibrils. *Wood and Fiber* **2,** 285-288 (1970)

[121] Timell, T. E. Ultrastructure of the dormant and active cambial zone and the dormant phloem associated with formation of normal and compression woods in *Picea abies* L. Karst. *Technical Publication No. 96*. State University of New York, College of Environmental Science and Forestry (1973)

[122] Timell, T. E. Formation of compression wood in Balsam fir (*Abies balsamea*). 1. Ultrastructure of the active cambial zone and its enlarging derivatives. *Holzforsch.* **33,** 137-143 (1979)

[123] Tsuda, M. The ultrastructure of the vascular cambium and its derivatives in coniferous species. 1. Cambial cells. *Bull. Tokyo Univ. For.* **67,** 158-226 (1975)

[124] Wardrop, A. B. The mechanism of surface growth involved in the differentiation of fibres and tracheids. *Aust. J. Bot.* **2,** 165-175 (1954)

[125] Wardrop, A. B. The organisation of the primary wall in differentiating conifer tracheids. *Aust. J. Bot.* **6,** 299-305 (1958)

[126] Wardrop, A. B. The structure and formation of the cell wall in xylem. In *The Formation of Wood in Forest Trees*. M. H. Zimmermann (ed.). (London: Academic Press 1964)

[127] Wardrop, A. B. Cellular differentiation in xylem. In *Cellular Ultrastructure of Woody Plants*. W. A. Côté (ed.). (Syracuse, New York: Syracuse University Press 1965)

[128] Wardrop, A. B. The structure of the cell wall in lignified collenchyma of Eryngium spp. (Umbelliferae). *Aust. J. Bot.* **17,** 229-240 (1969)

[129] Wardrop, A. B., Davies, G. W. Wart structure of gymnosperm tracheids. *Nature* **194,** 497 (1962)
[130] Wardrop, A. B., Wolters-Arts, M., Sassen, M. M. A. Changes in microfibril orientation in the walls of elongating plant cells. *Acta Bot. Neerl.* **28,** 313-333 (1979)
[131] Wenham, M. W., Cusick, F. The growth of secondary wood fibres. *New Phytol.* **74,** 247-261 (1975)
[132] Willison, J. H. M. Plant cell wall disposition revealed by freeze-fractured plasmalemma not treated with glycerol. *Planta* **126,** 93-96 (1975)
[133] Willison, J. H. M., Brown, R. M. An examination of the developing cotton fibre: Wall and plasmalemma. *Protoplasma* **92,** 21-41 (1977)
[134] Willison, J. H. M., Grout, B. W. W. Further observations on cell wall formation around isolated protoplasts of tobacco and tomato. *Planta* **140,** 53-58 (1978)
[135] Wodzicki, T. J., Humphreys, W. J. Cytodifferentiation of maturing pine tracheids: The final stage. *Tissue and Cell* **4,** 525-528 (1972)
[136] Wodzicki, T. J., Humphreys, W. J. Maturing pine tracheids. Organisation of intravacuolar cytoplasm. *J. Cell Biol.* **56,** 263-265 (1973)
[137] Wooding, F. B. P., Northcote, D. H. The development of the secondary wall of the xylem in *Acer pseudoplatanus*. *J. Cell Biol.* **23,** 327-337 (1964)
[138] Wunderlich, F., Müller, R., Speth, V. Direct evidence for a colchicine-induced impairment in the mobility of membrane components. *Science* **182,** 1136-1138 (1973)
[139] Yang, K. C. The fine structure of pits in yellow birch (*Betula allaghaniensis* Britton). *I.A.W.A. Bull.* 71-77 (1978/4)
[140] Young, O. A., Service, M. J. Slow viscous flow and the organisation of the cell wall in conifer tracheids. *Wood Sci. Technol.* **5,** 1-5 (1971)

3. Perforation Plate Differentiation in the Vessels of Hardwoods

B A MEYLAN
Physics and Engineering Laboratory, DSIR, Lower Hutt, New Zealand

B G BUTTERFIELD
Botany Department, University of Canterbury, Christchurch, New Zealand

Vessels provide the major pathway for the ascent of sap in the secondary xylem of all dicotyledonous angiosperms except those belonging to the Winteraceae, Trochodendraceae, Tetracentraceae[35], Amborellaceae and *Sarcandra* of the Chloranthaceae[15]. They are also present in many arborescent monocotyledons and the Gnetales. Vessels are built up of a longitudinal series of individual cells, termed vessel members or elements, to form a conduit. They form a three-dimensional pathway for the movement of water in wood[5, 6, 7, 8, 60]. In ring-porous woods, vessels may extend the full length of the trunk but in diffuse-porous woods they are usually less than one metre long. Where vessels run in pairs and groups, vessel to vessel movement of water occurs through the pit-pairs. Within each vessel sap passes relatively freely between contiguous vessel elements through perforations in the common walls. A certain amount also diffuses through the intervessel pits providing conductional continuity in diffuse-porous woods with determinate vessels.

Perforations are normally aggregated near to the end of each vessel element to form a perforation plate. A simple perforation plate consists of a single large opening (Fig. 3.1) whereas a multiple perforation plate is an aggregation of perforations variously grouped into a scalariform, reticulate or foraminate arrangement. Typically a scalariform perforation plate is made up of a series of regular parallel openings (Fig. 3.2), a reticulate perforation plate is made up of an irregular or net-like arrangement of small perforations (Fig. 3.3) and a foraminate perforation plate is made up of a small number of large circular perforations. In reality few multiple perforation plates fit exactly into any precise category and are usually described in a way that best reflects their nearest relationship[28]. For example a scalariform perforation plate with a few branched dividing bars but otherwise regular linear perforations is described as a branched scalariform perforation plate. Perforation plates of complex intermediate forms have been observed in numerous woods using scanning electron

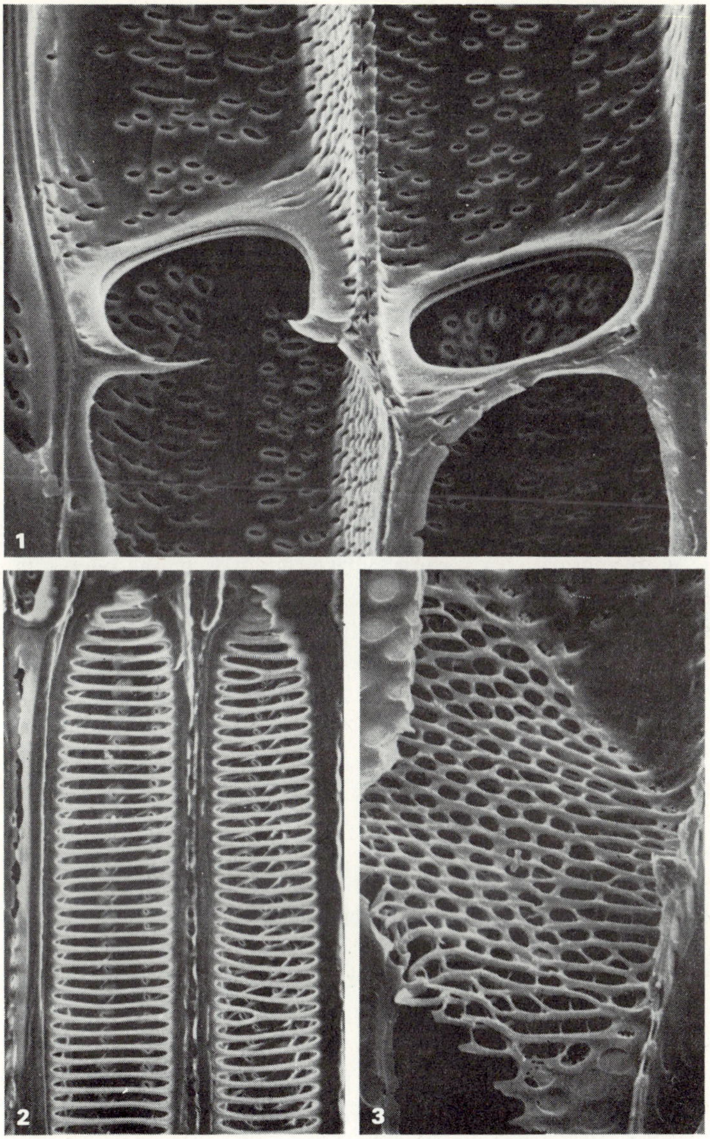

Key. *TF—Transverse face. TLF—Tangential longitudinal face. RLF—Radial longitudinal face.*

Figure 3.1: *Simple perforation plates between vessel elements in* Myrsine salicina *Hew. ex Hook.f (Myrsinaceae). Note the intact pit membranes separating the intervessel pits. (RLF ×800)*

Figure 3.2: *Scalariform perforation plates between vessel elements in* Carpodetus serratus *J. R. et G. Forst. (Escalloniaceae). (RLF ×500)*

Figure 3.3: *A reticulate perforation plate between vessel elements in* Coprosma tenuicaulis *Hook.f. (Rubiaceae). Most of the perforation plates in this genus are simple but multiple plates occur occasionally. (RLF ×1600)*

microscopy[40, 41, 54, 55]. Where a wood has more than one type of perforation plate, dimorphic or combination plates can develop between contiguous vessel elements[12, 16, 40, 41, 52, 55, 63]. Simple to scalariform combination perforation plates (Fig. 3.4), for example, can occur in woods that have mostly simple perforation plates together with a few scalariform perforation plates, and also in woods with mostly scalariform perforation plates but also having a few simple perforation plates. Simple to reticulate combination perforation plates (Fig. 3.5) have been recorded in a number of woods[16, 27, 43, 55] but true scalariform to reticulate perforation plates appear to be rare[13].

The origin and phylogenetic specialization of perforation plates have been the subject of numerous investigations[17] with the specialization of perforation plates in the vessel elements of dicotyledonous woods receiving special attention[2, 3, 16, 24, 25, 26, 31, 34, 62, 63]. It has generally been assumed that simple perforation plates have evolved from scalariform plates by the loss of the dividing bars and that scalariform perforation plates evolved from tracheids with scalariform pitting. More recently, it has been suggested by Parameswaran and Liese[55] that an alternative evolutionary line could be: tracheids with scalariform pitting to tracheids with net-like pitting to vessel elements with multiperforate plates to vessel elements with simple perforation plates. Such a sequence is consistent with various observations published elsewhere[41, 54]. The functional and adaptive significance of the various perforation plate types have been summarised by Carlquist[15] and Baas[1] and no attempt will be made here to review this aspect of vessel element morphology.

Perforation plates develop during the differentiation of a column of vessel elements behind the the cambium. Priestley, Scott and Malins[57] suggested that the perforations in the end walls of each vessel element develop during the expansion growth phase of the differentiating cell. They further suggested that the 'rims' around perforations are formed by the sudden contraction of the end wall material. This theory was subsequently discredited by Esau and Hewitt[19] who observed that the vessel element secondary wall was deposited on the primary wall in a specific pattern after the cells had reached their final size and prior to the removal of the end wall. The pits and areas to be perforated did not become overlaid with secondary walls.

Studies of perforation plate development using light, transmission electron[18, 46, 59, 61, 64] and scanning electron microscopy[9, 38, 53] have now confirmed that the secondary wall is laid down on all areas of the vessel element primary wall except those areas to be perforated, and leaving of course the pit apertures, prior to the removal of the end wall (Figs. 3.8 and 3.13). The secondary wall usually, but not always, overarches the area to be perforated. This area has traditionally been termed the 'end wall' but

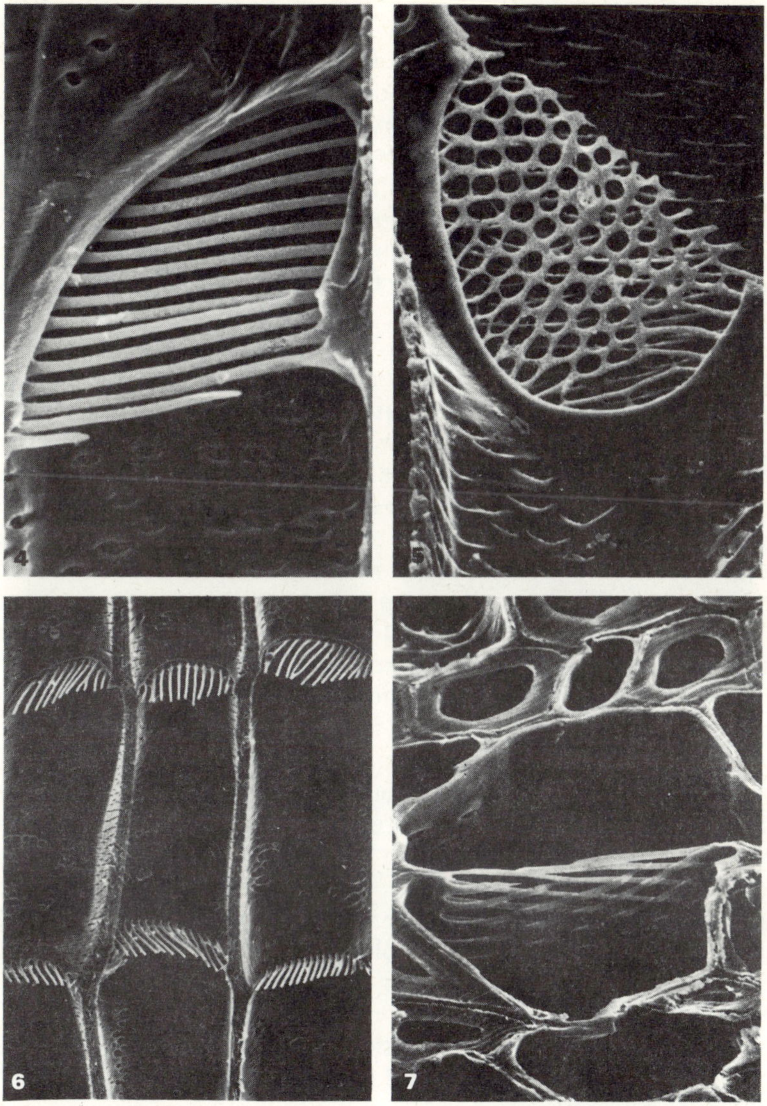

Figure 3.4: *A simple to scalariform combination perforation plate in* Brachyglottis repanda *J. R. and G. Forst (Compositae). The upper vessel element in this micrograph has formed a simple perforation while the lower vessel element has deposited its secondary wall to form a scalariform perforation plate. (RLF ×1700)*

Figure 3.5: *A simple to reticulate combination perforation plate in* Coprosma arborea *Kirk (Rubiaceae). (RLF ×1200)*

Figure 3.6: *Three vessels from a radial longitudinal file in* Vitex lucens *Kirk (Verbenaceae). The three vessel elements in the centre of the micrograph, probably derived from the same fusiform cambial initial, have all developed scalariform half plates at each end to form combination perforation plates with the simple half plates of the adjoining vessel elements. (RLF ×250)*

Figure 3.7: *A crossed perforation plate between two vessel elements in* Quintinia acutifolia *Kirk (Escalloniaceae). (TF ×1000)*

there seems little reason to perpetuate the use of this term as perforations are by no means confined to the ends of individual vessel elements. Nor does 'perforation plate membrane', as used by Thomas and Bonner[61] and others, seem a very satisfactory term. We prefer the alternative term 'perforation partition' as used by us[9, 13, 38] to designate the common wall area between two contiguous cells prior to its removal.

Esau and Charvat[18] note that investigators of xylem differentiation emphasize the common phenomenon that secondary walls appear opposite one another in adjacent tracheary cells. Our examination of tens of thousands of perforation plates in hundreds of different woods, has indicated that the deposition of secondary walls around the areas to be perforated is usually a well-coordinated process between contiguous cells where simple and normal scalariform perforations are involved. However, the deposition pattern of the secondary walls in contiguous cells *does* fail sometimes. Blind simple perforations resembling a large pit can develop in one cell where there is no matching opening in the secondary wall of the contiguous vessel element. Although we have never observed entirely blind perforation plates, blind openings can occur at the end of long scalariform plates, and in rare instances the total secondary wall deposition coordination appears to break down and 'crossed perforation plates' result (Fig. 3.7)[13, 41, 54]. This is perhaps not an unexpected phenomenon since a shift in the predominant microfibril direction is necessary in order that the dividing bars of two contiguous cells can match in a transverse pattern.

Simple to scalariform, simple to reticulate, and scalariform to reticulate combination perforation plates are further examples of differences in the pattern of secondary wall deposition between contiguous vessel elements (Figs. 3.4 and 3.5). An examination of two woods where the vessel elements lie in radial files in the secondary xylem and which are occasionally interconnected by simple to scalariform perforation plates[12] revealed that there is a tendency for all the vessel elements in one file to produce identical types of perforation plate (Fig. 3.6). This suggests that the pattern of secondary wall deposition in any one vessel element may be determined by genetic factors passed into all the derivatives of a particular fusiform cambial initial.

In combination perforation plates, however, it is generally the case that the wall area occupied by the perforations on both sides of the plate coincides almost exactly despite different patterns of wall thickening in the contiguous cells.

The exact process by which the perforation partition (made up of the two primary walls and intervening middle lamella) is removed remains unclear. Esau and Hewitt[19] noted that the perforation partition increases in thickness prior to its removal. This has since been confirmed by several authors but their explanations as to why this happens vary. Esau *et al.*[20]

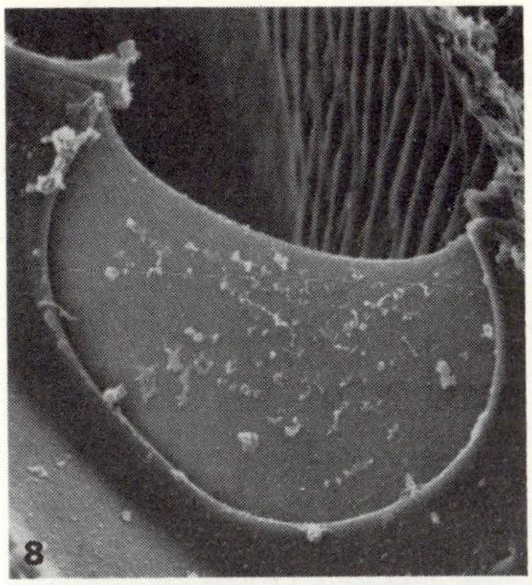

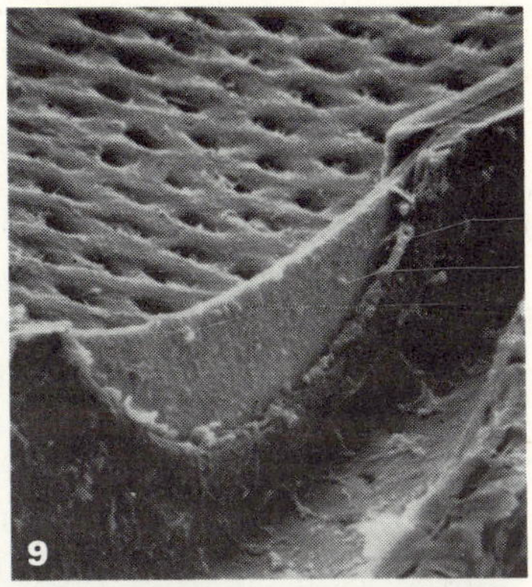

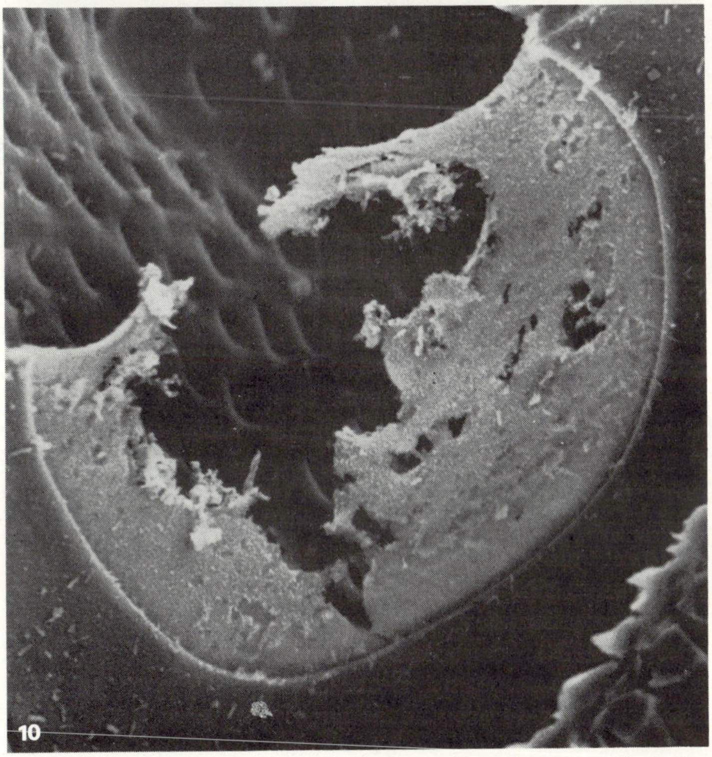

Figure 3.8: *An intact perforation plate partition between two differentiating vessel elements in* Knightia excelsa *R.Br. (Proteaceae). At this stage the secondary walls of the two adjoining vessel elements have been deposited but the breakdown of the partition is incomplete. The upper cut edge appears curved due to dehydration during specimen preparation. (RLF ×900)*

Figure 3.9: *Half a perforation plate between the differentiating vessel elements in* Knightia excelsa. *The perforation partition is still present and its continuity with the primary walls and middle lamella can be seen. Note the overarching borders of the secondary walls. (RLF ×1200)*

Figure 3.10: *A partly degraded perforation plate partition in* Knightia excelsa. *(×2000)*

suggested that a swelling of pectic substances, particularly those of the middle lamella, could be involved. Esau and Charvat[18] further stated that in *Phaseolus vulgaris* L. the swelling phenomenon is caused by the removal of some of the matrix materials by wall degrading enzymes and the consequent uptake of water by the remaining carbohydrates. The partition acquires a dense homogeneous structure except for the presence of surface vesicles and in unstained section it has a gelatinous, swollen appearance. Buvat[14], Sassen[59], Niedermeyer[47] and Murmanis[46] however, have all interpreted this swelling of the partition as a normal deposition of new wall material during differentiation. Our observations of perforation plate development in *Knightia excelsa* R.Br.[38] clearly indicate that, prior to its removal, the perforation partition is considerably thicker than neighbouring pit membranes whose basic composition should be essentially the same. We have found no evidence to suggest that this thickening is secondary in nature. In fact, although the perforation partition is thicker than the adjacent primary walls and middle lamellae from which it is suspended, it shows complete continuity with them (Fig. 3.9).

Recent observations in *Knightia excelsa* R.Br., *Aristotelia serrata* (J. R. and G. Forst), *Elaeocarpus dentatus* (J. R. and G. Forst) and *Nothofagus* sp. have confirmed earlier findings[9, 38] that prior to the removal of the perforation partition its surface has a granular appearance (Figs. 3.9 and 3.13). We interpreted this as the first indication of the partition breakdown process. Esau and Charvat[18] have suggested however, that the granular appearance may be caused by numerous vesicles such as those they observed on the lumen surfaces of the partitions in *Phaseolus vulgaris*. Murmanis[46] also reported vesicles in contact with the partitions in *Quercus rubra* L. some of which appear to fuse with the plasmalemma and which she suggests may also participate in the breakdown process.

Nearly all studies in this field have shown that the subsequent removal of the perforation partition proceeds very rapidly and few intermediate stages have been observed. Meylan and Butterfield[38] showed that after the partition becomes granular it develops numerous small holes before its final disappearance. This stage has now been confirmed by Murmanis[46] using transmission electron microscopy. She considered the breakdown process to be gradual because intermediate stages were frequently found. Figure 3.10 illustrates one of our recent observations of this breakdown stage in *Knightia excelsa* R.Br. Most authors claim that both the non-cellulosic and the cellulosic components of the perforation partition are removed enzymatically[33, 46, 47, 53, 58, 59], though the exact digestive process remains unknown.

Other authors including Priestley *et al.*[57], Frey-Wyssling[22] and Frey-Wyssling and Mühlethaler[23] have concluded that the breakdown of the perforation partition is largely mechanical. The non-cellulosic matrix is dissolved enzymatically, while the cellulosic microfibrils are pushed aside mechanically.

O'Brien and Thimann[50] showed that the destruction of the protoplast is accompanied by the loss of the non-cellulosic polysaccharides from the unlignified parts of the primary walls of tracheary elements in oat coleoptiles. The residue consists of a mat of fine fibrils which they termed a hydrolysed wall. Two types of material protected the walls from attack by hydrolases: lignin and an unidentified component of the middle lamella. Wall hydrolysis did not proceed beyond the middle lamella when a tracheary element abutted a living parenchyma cell. The middle lamella did disappear however if attacked from both sides. O'Brien[48] later extended this work to show that the pit membranes of intervessel pits between tracheary elements of willow are also a hydrolysed wall consisting of a mat of fine cellulose fibrils.

Other possible intermediate stages in thc breakdown of the perforation partition have been observed where the cellulosic microfibrils are still present after most of the non-cellulosic components have been removed. Such microfibrillar webs traversing the individual openings in scalariform perforation plates (Figs. 3.16 and 3.17) have been observed by several authors[4, 9, 13, 29, 30, 37, 38, 41, 51, 52, 54, 56, 61]. In general these webs are confined to the lateral extremities of openings close to the end of long scalariform plates though they may traverse the entire openings of the last few perforations (Fig. 3.18). Often such 'webbed' openings grade into normal scalariform pit-pairs at each end of the perforation plate.

The exact nature of these microfibrillar webs remains unresolved but they are almost universally assumed to be aggregations of cellulose microfibrils left after the digestion of the non-cellulosic components of the perforation partition.

Meyer and Muhammad[37] distinguished between reticulate and orthogonal webs. The former have a random arrangement of fibrils while the latter have major fibrils traversing the scalariform openings in a perpendicular direction leaving regular rectangular spaces. While webs of such different types can be demonstrated, it is likely that the variation is a consequence of the specimen preparation technique rather than a natural phenomenon.

Microfibrillar webs are absent from most of the scalariform perforation plate openings in mature wood. This has led to the suggestion that they may be removed by the transpiration stream[64]. Butterfield and Meylan[9] observed scalariform perforation plates with complete webs traversing every opening only in vessels close to the cambium in *Laurelia novae-zelandiae* A. Cunn, while Thomas and Bonner[61] observed webs only in the smaller openings (5 to 8 μm) and not in the larger openings (13 to 20 μm) of the plates in mature wood of *Liriodendron tulipifera*. Both of these observations are consistent with the theory of web removal by the transpiration stream. Nevertheless the removal by mechanical means of all surviving microfibrillar webs in the large openings must be viewed with some caution. The manner in which the sap flow starts between contiguous cells and its mass flow and velocity at any particular time in the degradation sequence must clearly be important factors. The total area perforated and the angle of the perforation plate are also of importance. These will vary from one wood to another and also between different vessels within the same material.

Ohtani and Ishida[53] found remnants of microfibrillar webs commonly occurring between the overarching borders of developing simple perforations. In our earlier studies of simple perforation partitions[38], however, we were unable to observe any microfibrillar webs associated with their breakdown, although we have now found them in *Knightia excelsa* R.Br. (Fig. 3.12) and in several other woods. Ohtani and Ishida[53] suggest that these remnants represent a stage when the transpiration stream has not effectively begun. It is of interest to note that in the stage shown in Fig. 3.10, (presumably an earlier one than that in Fig. 3.12), no microfibrillar component could be detected. Although this may appear to agree with Murmanis' findings[46], it is more probable that the microfibrils are present but concealed by matrix material.

Until recently it had been assumed that complete or almost complete microfibrillar webs do not survive as a stage in the breakdown of the large perforation partitions in simple perforation plates. Such a stage has now been observed[13] (Fig. 3.11) and so must be assumed to exist, even if only briefly, in the breakdown sequence, and perhaps only then in situations where the enzymatic processes have proceeded with the involvement of only minimal mechanical forces from the sap flow. It must be assumed that the much greater diameter of the simple perforation plate over the scalariform plate makes the likelihood of web survival in the former very small. Further evidence can be obtained from those examples of perforation plates of unusual form which have small half-formed bars projecting into the main opening. Micrographs of these have been published by Meylan and Butterfield[40] and Ohtani and Ishida[54]. In both cases microfibrillar webs have survived, but only in the small segment formed by the bar.

It can, therefore, reasonably be inferred that the microfibrillar webs represent a stage in the breakdown of the perforation partition, where the non-cellulosic, and also probably some of the cellulosic components have been digested. Presumably the sap flow has not yet attained sufficient force to remove the webs. Microfibrillar webs that seem to survive the full force of the transpiration stream, at least for a time, occur across narrow openings only, or in the corners of larger openings where they are well supported. They are also more likely to survive in the last few openings in

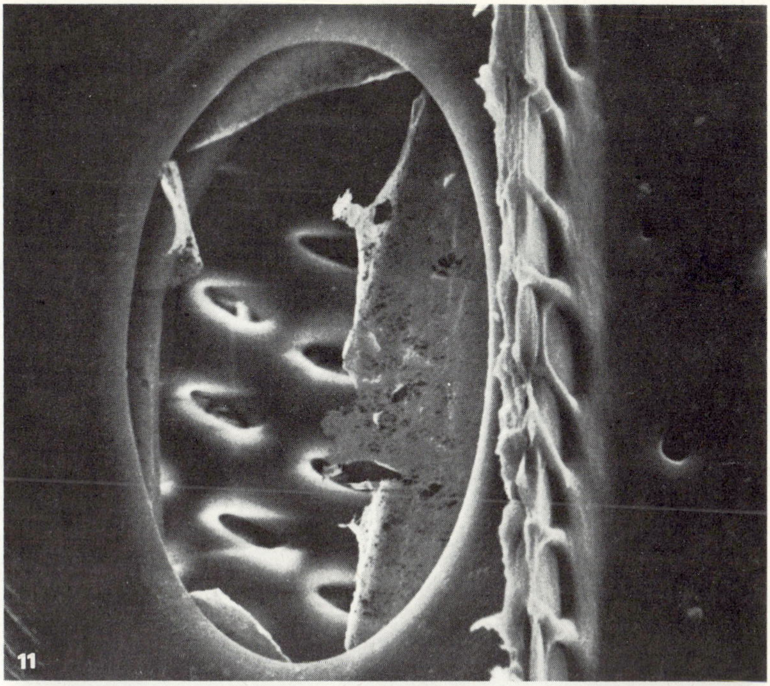

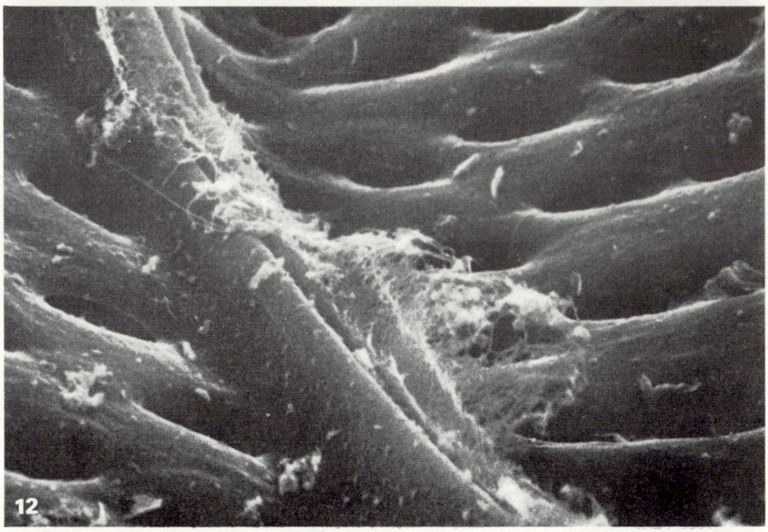

Figure 3.11: *A torn cellulose microfibrillar web traversing a simple perforation plate in* Knightia excelsa. *Most of the non-cellulosic and some of the cellulosic components of the partition have been enzymatically digested. It is not clear whether this membrane was ruptured by the onset of the transpiration stream or by the electron beam of the microscope. (RLF ×2700)*

Figure 3.12: *Remnants of the cellulose component of a perforation partition in* Knightia excelsa. *(RLF ×4000)*

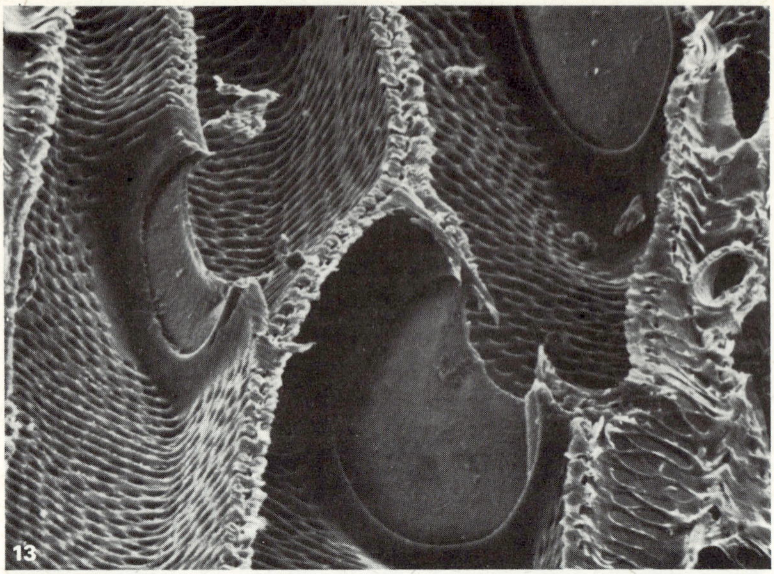

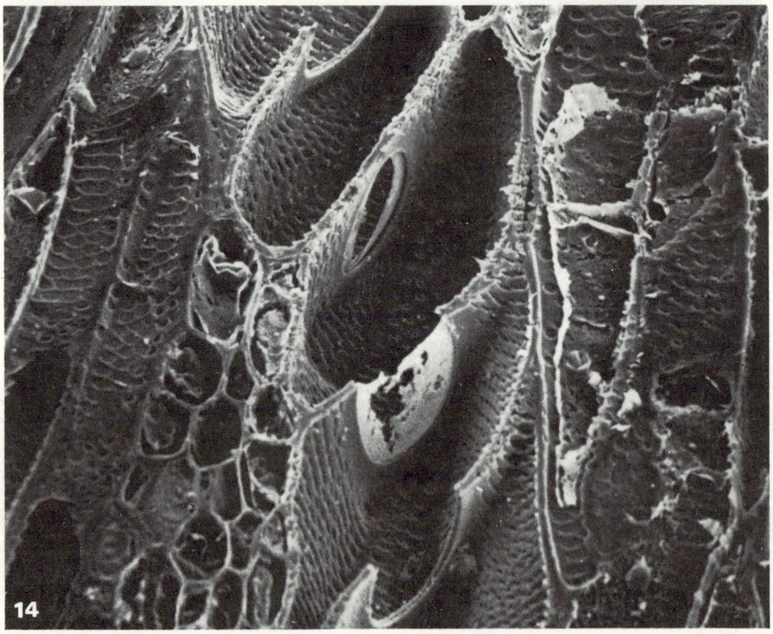

Figure 3.13: *Differentiating vessel elements in* Knightia excelsa. *The perforation partitions are still intact in this early stage close to the cambium. (TLF ×700)*

Figure 3.14: *A later stage of perforation plate development in* Knightia excelsa. *Hydrolysis of the partition is complete in two of the perforations but incomplete in two others. At what stage of differentiation the sap flow commences is not known for certain but it seems likely that it begins before some at least, of the partitions are completely removed. (TLF ×350)*

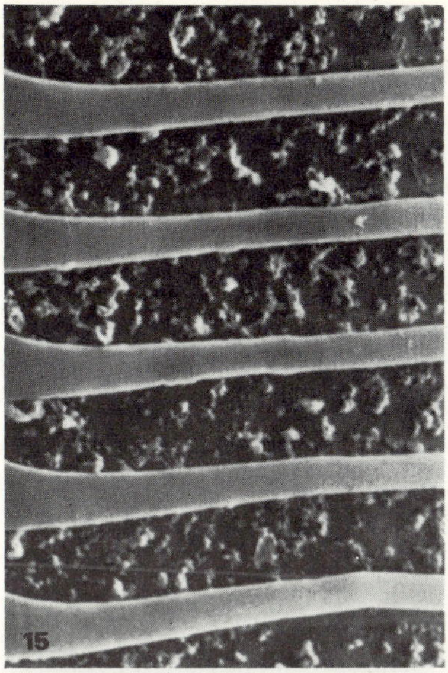

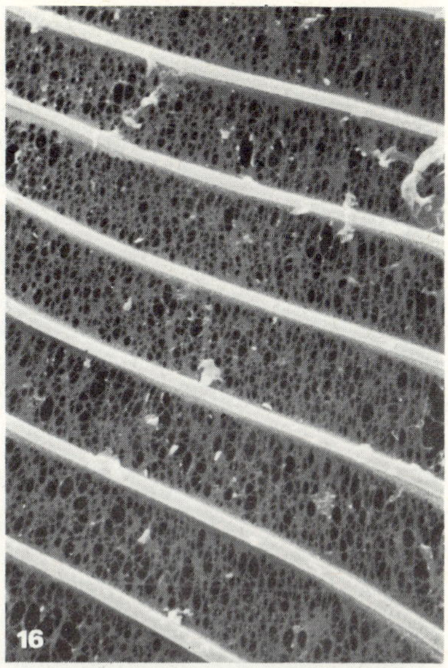

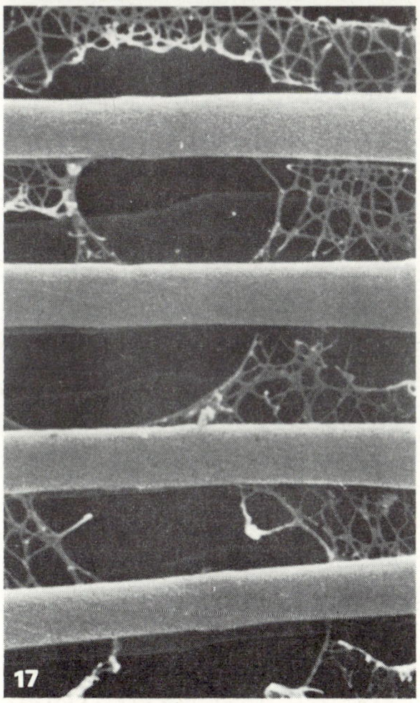

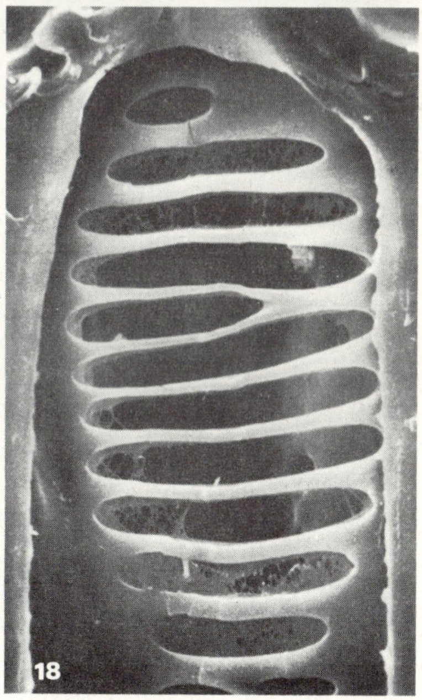

Figure 3.15: *An early stage in the breakdown of the perforation partitions in a scalariform plate of* Ascarina lucida *Hook.f. (Chloranthaceae). (TLF ×2600)*

Figure 3.16: *Microfibrillar webs traversing all the openings in a scalariform perforation plate of* Liquidamber styraciflua *L. (Hamamelidaceae). (RLF ×1650)*

Figure 3.17: *Remnants of former microfibrillar webs in a scalariform perforation plate of* Ascarina lucida. *(RLF ×3800)*

Figure 3.18: *Microfibrillar webs normally only survive in mature wood in the corner of the individual opening or the last few openings in a scalariform perforation plate as in this plate from* Dracophyllum longifolium *(J. R. et G. Forst). R.Br. (Epacridaceae). (RLF ×2700)*

very long scalariform plates where the force of the sap flow is expected to be less.

A slightly different concept of perforation plate development has been proposed by Sassen[59]. Using thin section microscopy he deduced that the perforation partition separates from the vessel wall initially around its periphery. He went on to suggest that a weak mechanical force, such as the onset of sap flow, could displace the entire end wall. The presence of active enzymes in the end wall is then assumed to complete the dissolution of the partition after it has become detached from the perforation. Such a theory has not been substantiated by any other study. In our own extensive work on perforation plate development in several woods we have, however, very occasionally noted a perforation partition that has been displaced intact, but this always appears to be a mechanical effect due to the action of the razor blade used in preparing the samples for examination. We have never seen any evidence to suggest that intact perforation partitions can be displaced and later dissolved away.

Pit membranes are apparently not removed during vessel differentiation (Fig. 3.1). Yata *et al.*[64] were of the opinion that pits and perforation partitions were both attacked enzymatically and lost their non-cellulosic component and possibly part of their cellulosic component. They explained the more or less total loss of the perforation partitions as a consequence of the transpiration stream. O'Brien[48, 49] has also suggested that the primary parts of both the side and end walls (perforation partitions) are hydrolysed in a similar manner but that in the side walls the cellulose component persists. Scanning electron microscope observations of vessel element pit membranes do sometimes suggest a loose microfibrillar texture (Fig. 3.19)[10]. It is also quite common to see part digested intervessel pit membranes or pit membranes consisting of microfibrillar webs and numerous openings (Fig. 3.20) similar to the margo region of conifer pit membranes. It has commonly been assumed that such observations are specimen preparation induced artifacts but this may not be so. The removal of the non-cellulosic components of pit membranes must be a necessary requirement for intervessel sap flow. This is certainly so in the tracheids of vessel-less dicotyledons (Fig. 3.21). Recent studies on the length of vessels in diffuse porous woods makes such a flow essential if long distance transport is to occur.

It has been known for many years that the margo region of the intertracheid conifer pit membrane consists solely of cellulose microfibrils. O'Brien and Thimann[50] have suggested that this region is also a hydrolysed wall, the polyuronide enrichment of the torus saving it from similar attack. In hardwoods the pit membranes and the perforation partitions are the only areas of secondary xylem tracheary elements to remain unlignified, hence both are vulnerable to enzymatic attack[48].

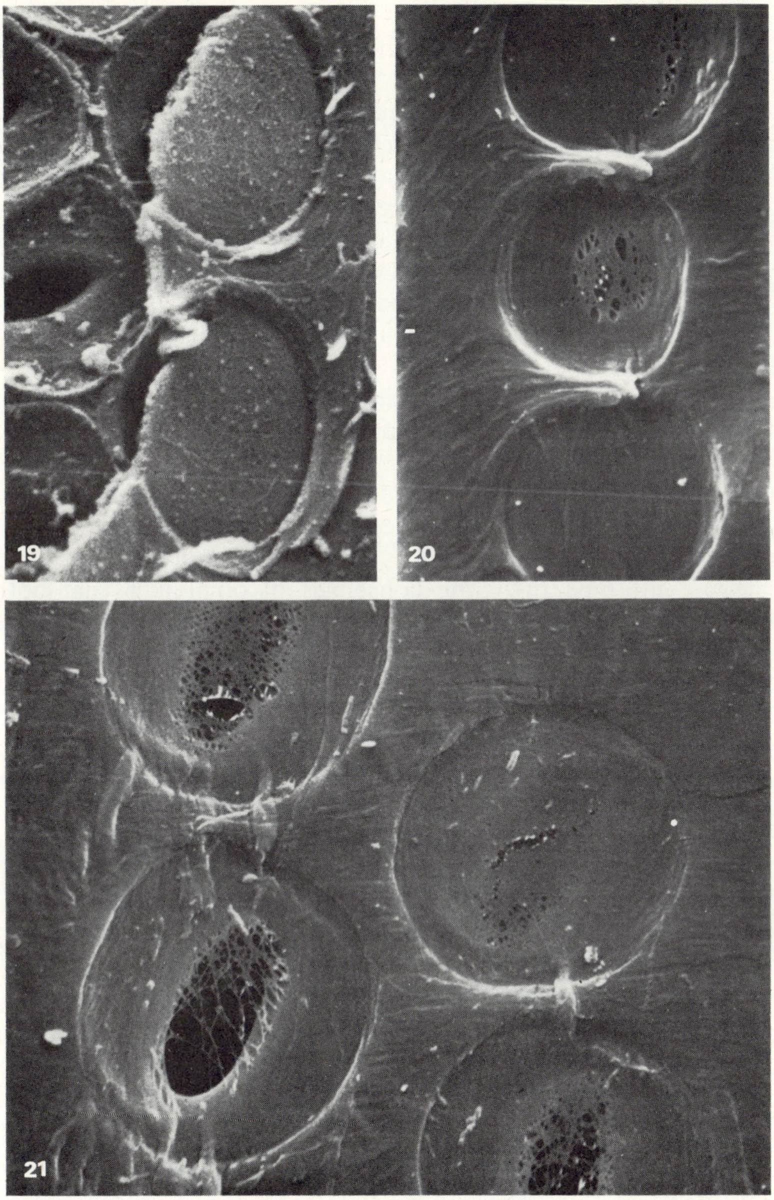

Figure 3.19: *Intervessel pit membranes exposed by wall fracture in* Knightia excelsa. *Note the loose microfibrillar texture of the membranes. (×6700)*

Figure 3.20: *Hydrolysed vessel pit membranes exposed by wall fracture in* Nothofagus menziesii *(Hook.f.) Oerst. (Fagaceae). (×5400)*

Figure 3.21: *Hydrolysed intertracheid pit membranes in the vessel-less wood of* Pseudowintera axillaris *(J. R. et G. Forst). Dandy (Winteraceae). Such hydrolysis is presumably necessary in order to allow water movement between tracheids that do not have specialised margo regions to their pit membranes. (RLF ×4400)*

Prior to the degradation of the perforation partition and the hydrolysis of the vessel pit membranes the cell walls of axial paratracheal and ray parenchyma cells often become modified by the deposition of an additional wall layer usually termed the 'protective layer' (Fig. 3.22). This layer resembles the primary wall but is laid down on the lumen side of the secondary wall towards the end of cell differentiation. It is usually laid down over the entire lumen surface of the wall including the pits but is usually thicker on the walls adjacent to the tracheary element. Its role in tylose development is well documented (Fig. 3.23)[21, 36, 44, 45]. The deposition of this layer may also function in isolating the living parenchyma cell from hydrolysing agents.

Vestures around perforation borders present an interesting morphological addition to the perforation plates in some woods. They have been observed around both scalariform[11] and simple[32, 54] perforation plates (Fig. 3.24). They resemble pit vestures in all respects. Their position on the pit borders suggests that they are deposited prior to the removal of the perforation partition.

At the present time then, the consensus of opinion is that both the cellulosic and non-cellulosic components of the perforation plate partition are enzymatically removed. The fact that microfibrillar webs are commonly observed in scalariform plates and occasionally in simple plates suggests that the rate of removal of the cellulose component is much slower. This is in accord with the known stability of cellulose to such attack. The situation is clearly complicated by mechanical effects because the manner in which the flow of sap builds up is not clear. Presumably the flow pattern in the vessel changes at different points along its length depending on the exact pathway taken by the sap and will start relatively slowly at any particular vessel element. Thus the mechanical force associated with the sap stream will vary considerably both in time and within the vessel. The many intermediate breakdown stages illustrated here and elsewhere are the result of a delicate balance between the digestion processes and the onset of the mechanical force. It is perhaps not surprising that there is some conflict among different workers. In more mature elements the lack of any web remnants suggests these have been removed by the full force of the transpiration stream.

As suggested by O'Brien[48] a similar situation may well apply to vessel pits. Hydrolysis of their membranes removes most of the non-cellulosic components but being smaller than perforation partitions they are much less fragile. Also the mechanical effect of the sap flow is likely to be practically non-existent. The breakdown of vessel to vessel pits is nearly always observed to be much more common than that of other vessel pits presumably because the enzymatic attack occurs here on both sides of the pit membrane.

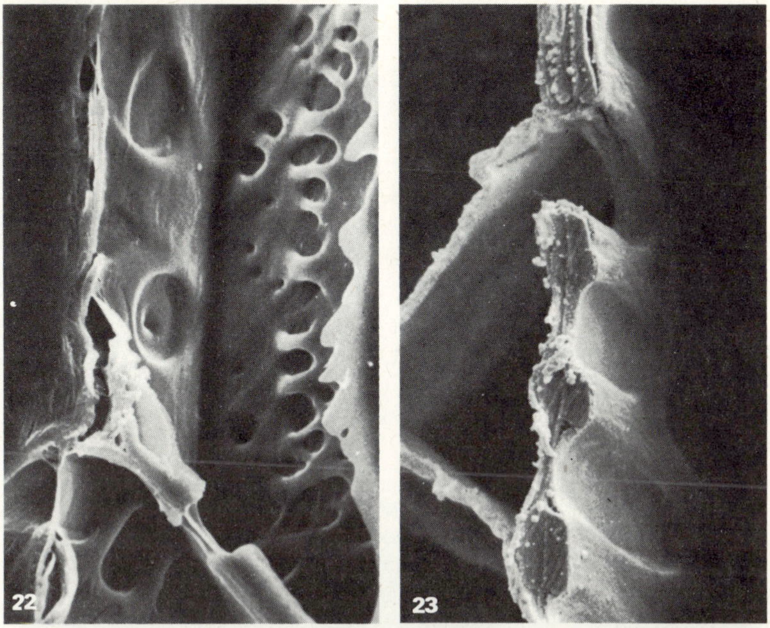

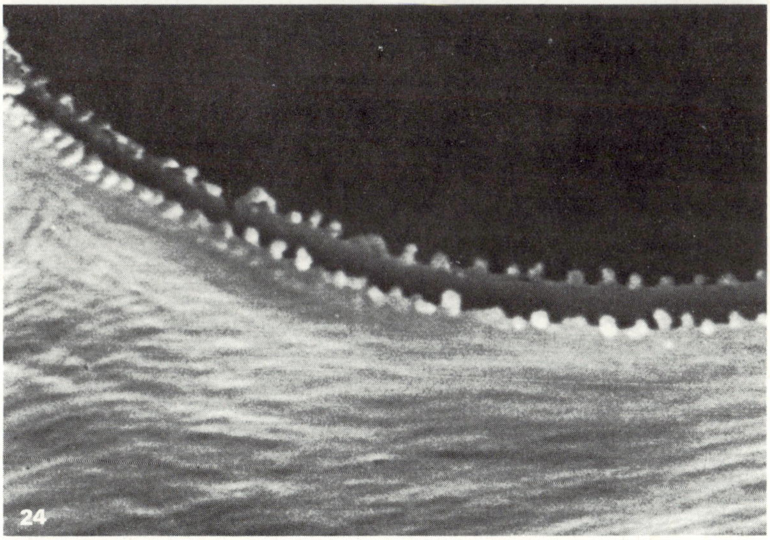

Figure 3.22: *The protective layer overlying the normal secondary wall and pits in a vasicentric axial parenchyma cell of* Paratrophis microphylla *(Raoul) Ckn. (Moraceae). (RLF ×1700)*

Figure 3.23: *The protective layer in a parenchyma cell of* Metrosideros fulgens *Sol. ex Gaertn. (Myrtaceae) showing its role in tylose development. (RLF ×3000)*

Figure 3.24: *Vestures lining the border of a simple perforation plate in* Coprosma repens *A. Rich. (Rubiaceae). (×5400)*

ACKNOWLEDGEMENTS

Some of our figures have been used previously. Permission to reproduce them is as acknowledged from the editors of the following books and journals.

New Zealand DSIR Bulletin 222 Structure of New Zealand Woods by B. A. Meylan and B. G. Butterfield 1978, for Figs. 3.2 and 3.3.

International Association of Wood Anatomist's Bulletin for Figs. 3.5 and 3.19.

Australian Journal of Botany, for Fig. 3.9.

Three Dimensional Structure of Wood: An ultrastructural approach by B. G. Butterfield and B. A. Meylan. (London: Chapman & Hall 1980) for Figs. 3.4, 3.7 and 3.11.

REFERENCES

[1] Baas, P. Some functional and adaptive aspects of vessel member morphology. In: Baas, P., Bolton, A. J., Catling, D. M. editors. Wood structure in biological and technological research. *Leiden Botanical Series* **3,** 157-181 (1976)

[2] Bailey, I. W. The development of vessels in angiosperms and its significance in morphological research. *Amer. J. Bot.* **31,** 421-428 (1944)

[3] Bailey, I. W., Howard, R. A. The comparative morphology of Icacinaceae. II. Vessels. *J. Arnold Arbor.* **22,** 171-187 (1941)

[4] Bonner, L. D. Ultrastructure and polymer composition of intercellular passageways in in yellow poplar. *Ph.D. dissertation*, North Carolina State University N.C. (1970)

[5] Bosshard, H. H., Kucera, L. J. The three dimensional structure analysis of wood. I. The network of the vessel, system in *Fagus sylvatica*. *Holz als Roh-und Werkstoff* **31,** 437-445 (1973)

[6] Braun, H. J. Die Vernatzung der Gefässe bei *Populus*. *Seitschr. f. Bot.* **47,** 421-435 (1959)

[7] Braun, H. J. Die Organisation des Stammes von Bäumen und Sträuchern. *Wiss. Verlagsgesellschaft, Stuttgart* (1963)

[8] Burgraaf, P. D. Some observations on the course of the vessels in the wood of *Fraxinus excelsior* L. *Acta Bot. Neerl.* **21,** 32-47 (1972)

[9] Butterfield, B. G., Meylan, B. A. Scalariform perforation plate development in *Laurelia novae-zelandiae* A. Cunn. A scanning electron microscope study. *Aust. J. Bot.* **20,** 252-259 (1972)

[10] Butterfield, B. G., Meylan, B. A. Intervessel pit membranes in *Knightia excelsa* R.Br. *I.A.W.A. Bull.* **4,** 3-9 (1972)

[11] Butterfield, B. G., Meylan, B. A. Vestured scalariform perforation plate openings in *Neomyrtus pedunculata, Aust. J. Bot.* **22,** 425-427 (1974)

[12] Butterfield, B. G., Meylan, B. A. Simple to scalariform combination perforation plates in *Vitex lucens* Kirk and *Brachyglottis repanda* J. R. and G. Forst. *I.A.W.A. Bull.* **3,** 39-42 (1975)

[13] Butterfield, B. G., Meylan, B. A. *The three dimensional structure of wood: An ultrastructural approach.* (London: Chapman & Hall 1980)

[14] Buvat, R. Observations infrastructurales sur les parois transversales des elements de vaisseaux (métaxylème de *Cucurbita pepo*) avant leur perforation. *C.r. hebd. Séanc. Acad. Sci. Paris* **258,** 6210-6212 (1964)

[15] Carlquist, S. *Ecological strategies of xylem evolution.* (Berkeley: Univ. Calif. Press 1975)

[16] Chalk, L. Multiperforate plates in vessels with special reference to the Bignoniaceae. *Forestry* **7,** 16-25 (1933)

[17] Cheadle, V. I. Independent origin of vessels in the monocotyledons and dicotyledons. *Phytomorph.* **3,** 23-44 (1953)

[18] Esau, K., Charvat, I. On vessel member differentiation in the bean (*Phaseolus vulgaris* L.) *Ann. Bot.* **42,** 665-677 (1978)

[19] Esau, K., Hewitt, W. M. B. Structure of end wall in differentiating vessels. *Hillgardia* **13,** 229-244 (1940)

[20] Esau, K., Cheadle, V. I., Risley, B. B. A view of ultrastructure of *Cucurbita* xylem. *Bot. Gaz.* **124,** 311-316 (1963)

[21] Foster, R. C. Fine structure of tyloses in three species of Myrtaceae. *Aust. J. Bot.* **15,** 25-34 (1967)

[22] Frey-Wyssling, A. *Die pflanzliche Zellwand.* (Berlin: Springer-Verlag 1959)

[23] Frey-Wyssling, A., Muhlethaler, K. *Ultrastructural plant cytology.* (New York: Elsevier 1965)

[24] Frost, F. H. Specialisation in secondary xylem of dicotyledons. I. Origin of vessel. *Bot. Gaz.* **89,** 67-94 (1930)

[25] Frost, F. H. Specialisation in secondary xylem of dicotyledons. II. Evolution of end wall of vessel segment. *Bot. Gaz.* **90,** 67-94 (1930)

[26] Frost, F. H. Specialisation in secondary xylem of dicotyledons. III. Specialisation of lateral wall of vessel segment. *Bot. Gaz.* **91,** 88-96 (1931)

[27] Gottwald, H., Parameswaran, N. Vielfache Gefa durchbrechungen in der Familie *Dipterocarpaceae. Z. Bot.* **52,** 321-334 (1964)

[28] Gray, R. L., de Zeeuw, C. H. Terminology for multiperforate plates in vessel elements. *I.A.W.A. Bull.* **2,** 22-27 (1974)

[29] Ishida, S. SEM photographs presented at the 19th meeting of the Japan Wood Research Society (in Japanese) Hokkaido Univ. (1969)

[30] Ishida, S. Observation of wood structure by SEM. *Wood Industry* **21,** 560-564 (1970)

[31] Jeffrey, E. C. *Anatomy of woody plants.* (Illinois: Univ. Chicago Press 1917)

[32] Kucera, L., Meylan, B. A., Butterfield, B. G. Vestured simple perforation plates. *I.A.W.A. Bull.* **1,** 3-6 (1977)

[33] Kuster, E. *Die Pflanzelle.* Gustaf Fischer Verlag, Jena. (1956)

[34] MacDuffie, R. C. The vessels in seed plants. *Bot. Gaz.* **71,** 438-445 (1921)

[35] Metcalfe, C. R., Chalk, L. *Anatomy of the dicotyledons.* Volumes 1 and 2. (Oxford: Clarendon Press 1950)

[36] Meyer, R. W., Côté, W. A. Formation of the protective layer and its role in tylosis development. *Wood Sci. and Tech.* **2,** 84-94 (1968)

[37] Meyer, R. W., Muhammad, A. F. Scalariform perforation plate fine structure. *Wood and Fiber* **3,** 139-145 (1971)

[38] Meylan, B. A., Butterfield, B. G. Perforation plate development in *Knightia excelsa* R.Br. A scanning electron microscope study. *Aust. J. Bot.* **20,** 79-86 (1972)

[39] Meylan, B. A. Butterfield, B. G. Scalariform perforation plates: Observations using scanning electron microscopy. *Wood and Fiber* **4,** 225-233 (1972)

[40] Meylan, B. A., Butterfield, B. G. Unusual perforation plates: Observations using scanning electron microscopy. *Micron* **4,** 47-59 (1973)

[41] Meylan, B. A., Butterfield, B. G. Occurrence of simple, multiple and combination perforation plates in the vessels of New Zealand woods. *N.Z. J. Bot.* **13,** 1-18 (1975)

[42] Meylan, B. A., Butterfield, B. G. *The structure of New Zealand woods*, DSIR Bull. 222, NZ Govt. Printer (1978)

[43] Muller-Stoll, W., Suss, H. Uber aussergewöhnlich gestaltete vielfache Gefassdurchbrechungen bie *Aeschynomene virginica* (L) B.S.P. (*Papili naceae*). *Ber dt. bot. Ges.* **82,** 613-619 (1969)

[44] Murmanis, L. Formation of tyloses in felled *Quercus rubra* L. *Wood Sci. Technol.* **9,** 3-14 (1975)
[45] Murmanis, L. The protective layer in xylem parenchyma cells of *Quercus rubra* L. *App. Pol. Symp.* **28,** 1283-1292 (1976)
[46] Murmanis, L. Breakdown of end walls in differentiating vessels of secondary xylem in *Quercus rubra* L. *Ann. Bot.* **42,** 679-682 (1978)
[47] Niedermeyer, W. Auflösung der Endwänden in differenzierenden Gefässzellen. *Ber. dt. bot. Ges.* **86,** 529-536 (1974)
[48] O'Brien, T. P. Further observations on hydrolysis of the cell wall in the xylem. *Protoplasma* **69,** 1-14 (1970)
[49] O'Brien, T. P. Primary vascular tissues. In: Robards, A. W., ed. *Dynamic aspects of plant ultrastructure.* (London: McGraw-Hill 1974)
[50] O'Brien, T. D., Thimann, K. V. Observations on the fine structure of the oat coleoptile. III. Correlated light and electron microscopy of vascular tissues. *Protoplasma* **63,** 443-478 (1967)
[51] Ohtani, J. Representation summary of the 20th meeting of the Japan Wood Research Society (in Japanese). Hokkaido Univ. (1970)
[52] Ohtani, J., Ishida, S. An observation of the sculptures of the vessel wall of *Fagus crenata* Bl. using scanning electron microscopy. *Res. Bull. College Exp. For. Hokkaido Univ.* **33,** 115-126 (1973)
[53] Ohtani, J., Ishida, S. An observation on perforation plate differentiation in *Fagus crenata* Bl. using scanning electron microscopy. *Res. Bull. College Exp. For. Hokkaido Univ.* **33,** 115-131 (1976)
[54] Ohtani, J., Ishida, S. An observation on the perforation plates in Japanese dicotyledonous woods using scanning electron microscopy. *Res. Bull. College Exp. For. Hokkaido Univ.* **35,** 65-116 (1978)
[55] Parameswaran, N., Liese, W. Scanning electron microscopy of multiperforate perforation plates. *Holzforsch.* **27,** 181-186 (1973)
[56] Parham, R. A. On the substructure of scalariform perforation plates. *Wood and Fiber* **4,** 342-346 (1973)
[57] Priestley, J. H., Scott, L. I., Malins, M. E. Vessel development in angiosperms. *Proc. Leeds Phil. Soc.* **3,** 42-54 (1935)
[58] Roelofsen, P. A. *The plant cell wall.* (Berlin: Borntraeger 1959)
[59] Sassen, M. A. A. Breakdown of the plant cell wall during the cell-fusion process. *Acta Bot. Neerl* **14,** 165-196 (1965)
[60] Skene, D. S. A three-dimensional reconstruction of the wood of *Eucalyptus maculata. Holzforschung* **23,** 33-37 (1969)
[61] Thomas, R. J., Bonner, L. D. The ultrastructure of intercellular passageways in vessels of yellow poplar (*Liriodendron tulipifera* L.) II. Scalariform perforation plates. *Wood Sci.* **6,** 193-199 (1974)
[62] Thompson, W. Independent evolution of vessels in Gnetales and angiosperms. *Bot. Gaz.* **65,** 83-90 (1918)
[63] Thompson, W. P. The relationship of the different types of angiospermic vessels. *Ann. Bot.* **37,** 183-192 (1923)
[64] Yata, S., Itoh, T., Kishima, T. Formation of perforation plates and bordered pits in differentiating vessel elements. *Wood Res.* **50,** 1-11 (1970)

4. Lignification and Xylogenesis

A B WARDROP
Department of Botany, La Trobe University, Bundoora, Victoria, Australia 3083

INTRODUCTION

In the cell walls of higher plants (Tracheophyta) four groups of components are recognised, although not all of these are present in all cells. There is an amorphous matrix of non-cellulosic polysaccharides in which are embedded partially crystalline fibrils* of cellulose constituting a structural framework of the wall. The matrix and framework are formed apparently in immediate succession and subsequently the chemically different (phenolic) component lignin, is deposited in association with the matrix components between the cellulose fibrils. The lignin is thus designated an encrusting component. The lignin may vary in its chemical nature. A fourth group of (adcrusting) components may occur in some cells and are added to the inner or outer surface of the wall. The deposition of some or all of these wall components is involved in the process of cellular differentiation.

Differentiation of cells involves a sequence of events which may be strictly consecutive or which overlap with each other. To describe this sequence of events the term 'cytodifferentiation sequence' has been used[100]. The cytodifferentiation sequence of a lignified cell such as a fibre or tracheid of secondary xylem involves the sequence (1) cell division and primary wall formation; (2) cell enlargement; (3) secondary wall formation and (4) lignification. In the differentiation of other lignified cells, however, the sequence may be different in that formation of the secondary wall may not take place. Lignification usually begins at the end of the phase of cell growth in the sequence. The process may proceed uniformly in all regions of the differentiating cell wall or the lignin may be deposited selectively in particular structures.

THE CHEMICAL NATURE OF LIGNIN

Lignin constitutes approximately 20 per cent of the cell walls of the secondary xylem of angiosperms and 35 per cent of that of gymnosperms. It is a complex molecule of molecular weight of ca. 11000 and is formed from the cinnamyl alcohols: coniferyl alcohol, synapyl alcohol and 4-coumaryl alcohol (Fig. 4.1).

The first structural formulation of the way in which the above

* In this discussion the term 'cellulose fibril' is used, eschewing the question of 'microfibrils' and 'elementary fibrils' of cellulose.

compounds were built into the lignin molecule resulted from the pioneering chemical investigations of Freudenberg and coworkers on the lignin of *Picea*[34].

More recently the structure of lignin from the xylem of *Fagus* has been investigated and a modification of the Freudenberg formula is shown in Fig. 4.2[89]. This structure has been the subject of some criticism[81].

Various attempts have been made to classify lignins according to their origin and composition. Thus, Gibbs[44] recognises two main classes, viz 'guaiacyl lignins' and 'guaiacyl-syringyl lignins'. Kawamura and Higuchi[64] have attempted a more detailed classification using the infra-red spectra of the lignins as criteria. Lignin is absent in the *Musci* although in some of the giant mosses such as *Dawsonia* a phenolic substance has been reported. The nature of this substance has been discussed by Gross[50]. As a generalisation the lignin from more primitive evolutionary groups, in the sense of existing groups similar or allied to those which have been detected in the earlier fossil records, contain more guaiacyl residues in the xylem than do plants of more recent origin.

In general, guaiacyl type of lignin is present in the Gymnospermae, Pteridophyta and Cycadales and the guaiacyl-syringyl type is present in the Angiospermae and Gnetales. Within these broad designations there are exceptions, and many variations in the composition of the lignin present. Thus in the Graminaceae, p-coumaryl as well as guaiacyl and syringyl units are present and curiously, lignin containing similar units is found in the Selaginellales. The question of the variation of lignin structure in relation to various taxonomic groups of plants has been the subject of extensive reviews[44, 50, 104] and can be explored to some degree by simple histochemical reactions (see below).

The composition of lignin can vary, not only between plant groups, but between tissues of the same plant or even between cells of the same tissues.

(1)
(2)
(3)

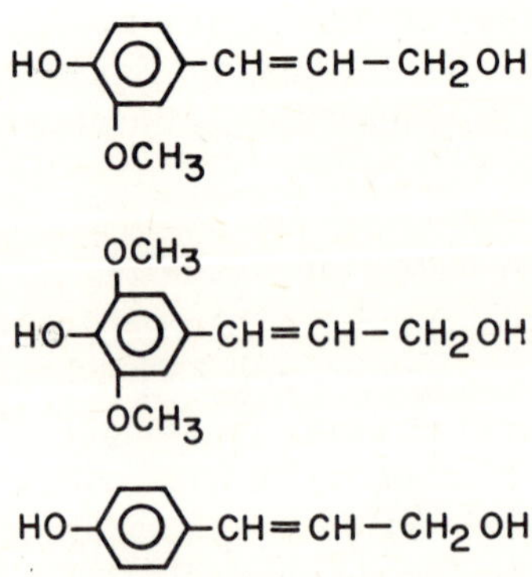

Figure 4.1: *The phenylpropane derived alcohols from which it is thought lignin is derived. (1) Coniferyl alcohol (2) synapyl alcohol (3) 4-coumaryl alcohol. cf. Fig. 4.2.*

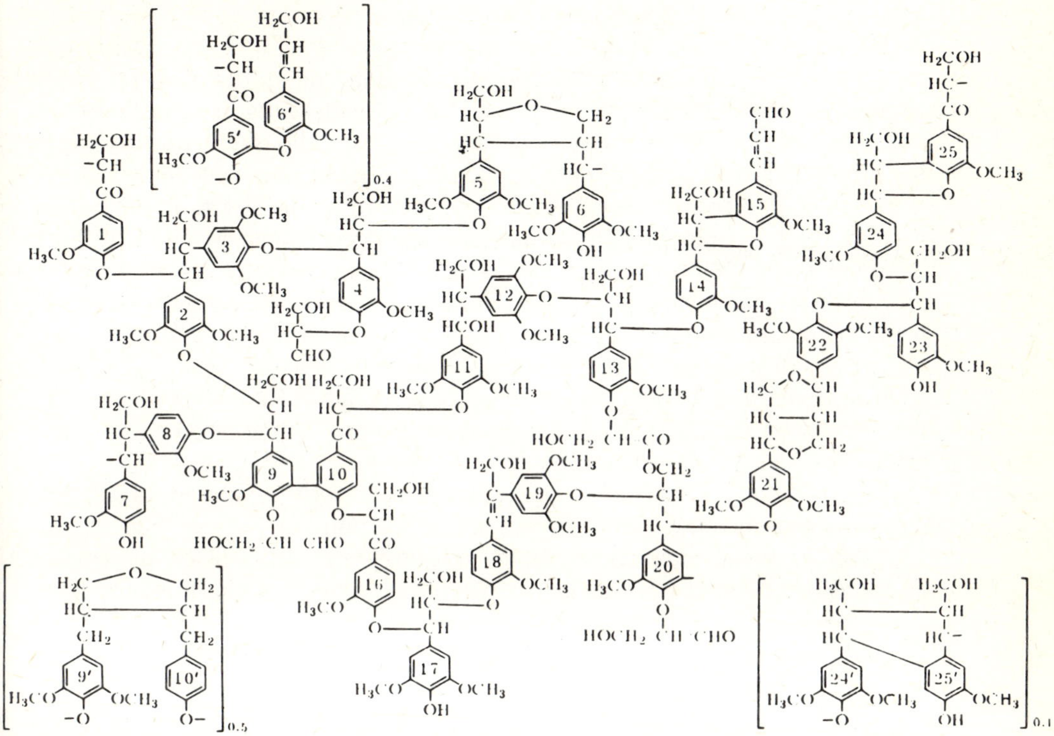

Figure 4.2: *The structure of lignin in* Fagus *(reproduced with permission from Nimz*[89]*).*

Thus, there is both histochemical and chemical evidence that the lignin of the primary xylem in angiosperms resembles that of gymnosperms, whereas the lignin of the secondary xylem has the composition of that generally present in angiosperms[11, 117].

In the secondary xylem of *Betula* the lignin of the vessel walls is of the guaiacyl type, whereas that of adjacent fibres is that of the syringyl type[30].

Lignin is usually confined to the walls of cells involved in transpiration, such as tracheids and vessels or cells with an assumed mechanical function, for example xylem fibres and sclerenchyma fibres[80, 117]. In some instances it also occurs in parenchyma[27], collenchyma[23, 120], sclereids[91, 121], and has been reported in sieve elements[70].

THE RELATION OF LIGNIN TO OTHER CELL WALL COMPONENTS

The relation between framework and matrix components of the cell wall has been widely investigated and has been reviewed by Albersheim[3]. The model proposed as a result of studies on suspension cultured sycamore cells showed the cellulose fibrils closely linked to xyloglucan components through arabinogalactan components to rhamnogalacturanans. Although it would not be justified to attempt to extrapolate the model of Albersheim, which was proposed for primary walls, to all cell walls, it is perhaps the most detailed model yet proposed. The nature of the chemical linkage of lignin to other wall components is difficult to determine, especially in view of its molecular complexity (Fig. 4.2). In attempts to determine the nature of the linkage, lignin-carbohydrate complexes have been isolated by both chemical and enzymic procedures. Hartley[56] successfully isolated carbohydrate esters of ferulic acid—which is an intermediate possibly involved in the biochemical pathway of lignification—from the cell wall of *Lolium multiflora*. In a further study[57], carbohydrate esters of diferulic acid were isolated. Evidence for the nature of linkage of lignin to carbohydrates has been reviewed by Lai and Sarkanen[71], and either the ether type of lignin linkage or the ester lignin linkage appears the most likely (Fig. 4.3).

More recently Eriksson and Lindgren[26] have isolated a lignin-hemicellulose fraction from *Picea* xylem containing 20 per cent hemicelluloses which was susceptible to enzymic and acid hydrolytic degradation. From these studies it was concluded that the linkages were

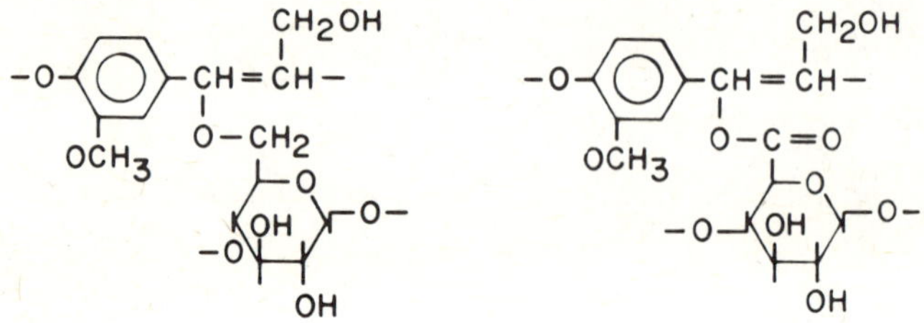

Figure 4.3: *Possible chemical linkages between wall components and lignin by either (a) ether or (b) ester linkages.*

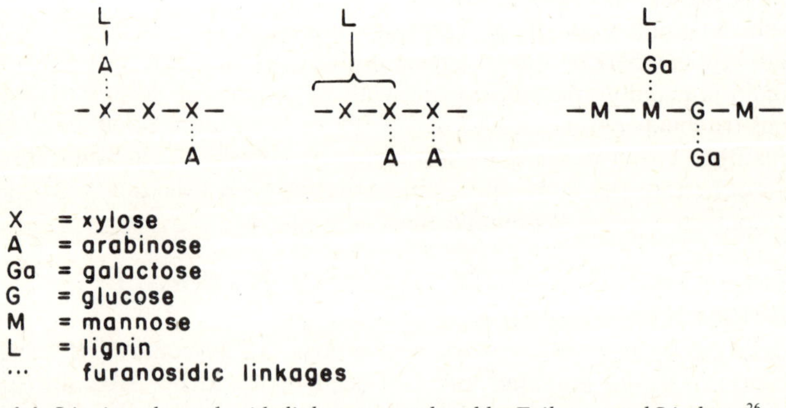

Figure 4.4: *Lignin-polysaccharide linkages postulated by Eriksson and Lindgren*[26].

mainly of the α-ether type and involved linkages to carbohydrates of the type shown in Fig. 4.4.

In a further study, Eriksson *et al.*[25] have isolated another lignin-carbohydrate complex. The results indicate that lignin is bound to all types of sugar units in the hemicellulose. Side chain linkages predominate in arabino-4-0-methylglucuronoxylan. The linkages to 4-0-methylglucuronic acid units are ester bonds, probably benzyl ester bonds, whereas those to other sugar units in the hemicellulose are ether bonds, probably benzyl ether bonds. At present it is a matter of speculation as to how the existence of linkages of the above type may be related to the Albersheim model.

HISTOCHEMICAL DETECTION OF LIGNIN

Because of its phenolic nature as distinct from the polysaccharide components of the matrix and framework components of the cell wall the presence of lignin can be detected by its optical properties, solubility properties and colour reactions. In some tissues, however, some substances such as ferulic acid may give some of the reactions of lignin so that in studying its distribution it is advisable where possible to use several different procedures on the same specimen. Perhaps the best method of detection of lignin is by its ultraviolet absorption. The method involves an examination of untreated or embedded material in an ultraviolet microscope at a wavelength of 260 nm. Since this method is capable of quantification by spectrophotometry it has been used to considerable effect in determining the distribution of lignin in the cell wall[29, 73, 117]. The only objection to the method would appear to be that when embedding procedures are used some lignin or lignin-associated compounds are removed by the reagents replacing water in the tissues. In the near ultraviolet lignin gives a green-blue fluorescent radiation and both transmission fluorescence and epifluorescence techniques have been employed. The disadvantage of the method is that the auto-fluorescence is weak and its intensity decays with time, so that standardised procedures must be established before quantitative conclusions can be drawn. However, the decay curve of the fluorescent radiation can be used to characterise the fluorescence of some wall components[45].

Procedures using combinations of dyes such as safranin/light green (lignin: red, cellulose: green) should be used only under standardised conditions and compared with results obtained by other procedures. An extremely useful and simple procedure was described by O'Brien *et al.*[90] using toluidine blue. This procedure is very rapid and stains lignified tissues green while unlignified tissues are blue-violet. The precise nature of the staining mechanism is not known.

A reaction which appears to have a chemical basis is the phloroglucinol-hydrochloric acid reaction. The phloroglucinol reacts with the coniferyl aldehyde and cinnamaldehyde groups in the lignin to yield a cationic chromatophore[1].

With chlorine water, followed by sodium sulphite, lignin gives a red colour with angiosperm (predominantly syringyl residues) and a brown colour with gymnosperm lignin (predominantly guaiacyl residues) (Unit 15 Fig. 4.2). The correlation of these colour reactions has been confirmed by the chemical investigations of Bland[11].

Analogous to the above reaction is the Maule reaction involving the treatment with aqueous potassium permanganate, followed by hydrochloric acid and ammonia[44]. At the electron microscope level potassium permanganate has been widely used as an electron stain. Its use, however, has been criticised by Feder and O'Brien[28] since starch, mucilage and some vacuolar deposits are electron opaque after staining with the reagent and of course, most of the internal membrane system of the cell is stained as well. However, Hepler *et al.*[59] have countered this criticism by pointing out that in the cell wall the regions stained by potassium permanganate are those which show lignin fluorescence and further, that in primary xylem permanganate stained only the regions where the secondary wall is attached so that it may be assumed that it is not reacting with other wall components.

The nature of the reaction with potassium permanganate has not been fully elucidated, but it appears to be reduced to manganese dioxide[19]. It has been suggested that the reaction involves the oxidation of the syringyl aldehyde or coniferyl aldehyde groups, or that intermediate cyclic manganese esters may be formed which decompose and deposit manganese dioxide[14].

There are many other reagents which have been used to detect lignin[51] but here discussion has been confined to those which have been most widely used.

THE DISTRIBUTION OF LIGNIN IN THE CELL WALL OF CONIFER TRACHEIDS

The fibrils of the cellulose framework in the cell walls of gymnosperm tracheids and angiosperm fibres are arranged in essentially a similar way. In the mature cell there is the primary wall which is formed during the phase of surface growth of the differentiating cell in the cytodifferentiation sequence and this encloses a secondary wall consisting typically of three layers, each of which is lamellated. In the outer layer (S1) a crossed helical organization is present. In the second formed layer (S2) the helix is steeper than in S1 and the helical organization is uniform. The pitch of the helix is much steeper than that in either S1 or the third formed layer (S3) which appears to have an organization similar to that of the layer S1 (Fig. 4.5).

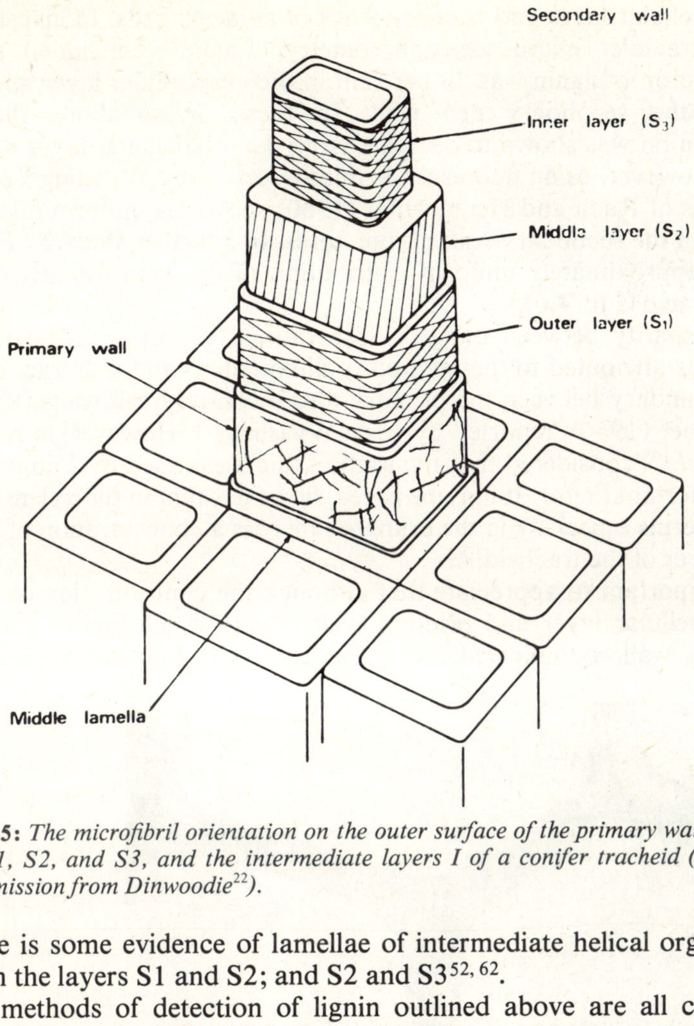

Figure 4.5: *The microfibril orientation on the outer surface of the primary wall P and the layers S1, S2, and S3, and the intermediate layers I of a conifer tracheid (reproduced with permission from Dinwoodie*[22]*).*

There is some evidence of lamellae of intermediate helical organization between the layers S1 and S2; and S2 and S3[52, 62].

The methods of detection of lignin outlined above are all capable of giving a qualitative indication of the distribution of lignin in the cell wall. However, for quantitative analysis, the method of ultraviolet microspectrophotometry is most reliable, especially when used in association with the methods of thin sectioning developed for electron microscopy.

An early investigation of the distribution of lignin in the cell wall of tracheids was that of Ritter[99]. From the appearance of the residue (Klason lignin) left after treatment with 72 per cent sulphuric acid he concluded that most of the lignin was located in the intercellular layer (probably the primary wall and the intercellular layer). A quantitative analysis of a fraction of the cell wall, isolated by microdissection gave a value for the lignin content of 71 per cent[6]. It is probable that this fraction consisted of the intercellular layer and primary walls of adjacent cells. In investigations using ultraviolet microspectrophotometry, Lange[73] concluded that the concentration of lignin was 73 per cent in the intercellular layer and 16 per cent in the secondary cell wall. In these investigations the lignin concentration was shown to decrease from the intercellular layer to the cell lumen. However, using fluorescence microscopy Frey-Wyssling[39] reporting the results of Ruch and Hentgartner (1960) showed a uniform distribution of lignin in the secondary wall of jute fibres. In a further study by Fergus *et al.*[29] an approximately uniform distribution of lignin in the cell wall was demonstrated (Fig. 4.6).

The disparity between Lange's observations and those of later investigators was attributed to the method of illumination and a diffraction effect at the boundary between the primary and secondary cell walls (Ruch and Hentgartner (1960), reported by Frey-Wyssling[39]). However, in retrospect Scott *et al.*[105] considered that the section thickness used by Lange was the more important factor. It may be noted, however, that in the xylem of some gymnosperms especially in the branches there is a concentration of lignin in the S3 layer of the tracheid[106].

It is important to appreciate that although the concentration of lignin in the intercellular layer and primary wall is very much higher than in the secondary wall, as indicated above; because of the greater volume of the

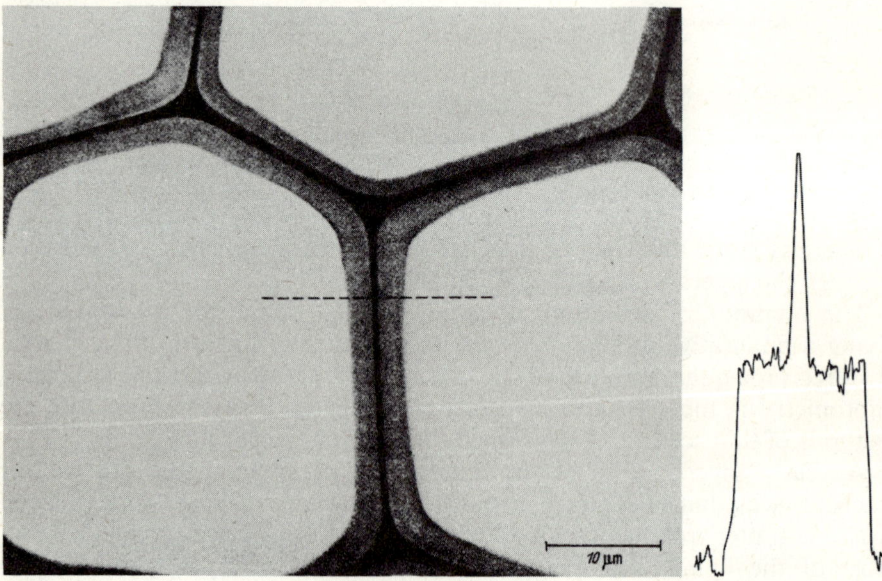

Figure 4.6: *An ultraviolet (240 nm) photomicrograph of a transverse section of* Picea mariana *Mill. At the right: a densitometer trace through the region of the dotted line (reproduced with permission from Fergus* et al.[29]*).*

secondary wall, most of the lignin is present in this structure[10]. Thus, for *Picea* the concentration of lignin in the intercellular layer was approximately four fold that in the secondary wall but for early wood, 72 per cent of the lignin was located in the secondary wall and only 28 per cent in the intercellular layer and primary wall. For late wood the values were 82 per cent and 18 per cent respectively[29]. It may also be noted that the concentration of lignin in the intercellular layer is greater in the radial layer compared with that in the tangential layer and the greatest absorption is at the cell corners[118].

Using the same technique, Fergus and Goring[31] studied the lignin distribution in *Betula papyrifera*. The pattern of distribution was similar to that in conifer tracheids and it was estimated that 77 per cent of the lignin in the fibres was located in the secondary cell wall.

The tracheids of reaction xylem of gymnosperms are characterized by a cell wall organization different from that shown in Fig. 4.5 and the distribution of lignin is also different. The mature cell is rounded with intercellular spaces between the tracheids and the wall consists of a primary wall and S1 layer of the secondary wall in which the cellulose fibrils are wound in a flat helix. The S2 layer of the secondary wall is characterized by helical fissures making an angle of ca. 45° to the cell axis. The layer S3 is absent[122]. In a study by Fujita *et al.*[41] it was shown that the concentration of lignin in the intercellular layer and primary wall was high. The concentration was lower in the S1 layer and was at a maximum in the outer regions of the S2 layer. The fibres of reaction xylem of angiosperms are characterized by an unlignified layer (G layer) which is formed in addition to the layers S1-S3 (Fig. 4.5) or may replace the layer S3 or the layers S2 and S3. The distribution of lignin in such fibres has been studied qualitatively by ultraviolet microscopy[108], and by examination of lignin residues[18].

A microspectrophotometric study of the thick-walled fibres of bamboo has been carried out by Parameswaran and Liese[92]. The secondary wall consists of many layers in which the cellulose fibrils are alternatively nearly transverse or parallel to the cell axis. The layers with a transverse orientation are thinner and are more heavily lignified than the wide layers in which the cellulose fibrils are of nearly longitudinal orientation. The level of ultraviolet absorption in the narrower layers was approximately the same as that in the intercellular layer. Parameswaran[91] has also investigated the lignin distribution in the multi-layered walls of sclereids from tree barks, but in these cells there did not appear to be any correlation between the lignin level and the orientation of the cellulose fibrils.

At the ultrastructural level the first studies of lignin texture were made by Mühlethaler[84]. In these observations lignin was isolated from the cell wall by the removal of cellulose and matrix substances with sulphuric acid. The residual lignin showed a porous structure resulting from the removal of the cellulose fibrils from the wall. In later studies Côté *et al.*[18] examined sections of tracheids and fibres which had been treated with hydrofluoric acid. Since this treatment involves less swelling of the wall the form of the residual lignin is better preserved. More recently, Kerr and Goring[66] have studied the distribution of lignin in transverse sections of the S2 layer of tracheids of gymnosperms and the fibres of angiosperms after staining with potassium permanganate. The electron micrographs of the 10 nm sections were examined by a microdensitometer both parallel and perpendicular to the intercellular layer. The average distance between peaks in the densitometer traces was 8.4 nm perpendicular to the intercellular layer and 15.2 nm parallel to the intercellular layer. The unstained regions of the wall were 3-4 nm in width. It has been pointed out by Kerr and Goring[66] that the greater periodicity between the stained areas in the tangential direction could be explained by the deposition of lignin between groups of laterally associated cellulose fibrils constituting the lamellae. In contrast, in the radial direction the lignin is located between lamellae consisting of sheets only one cellulose fibril in thickness. Goring has extended this observation further to the examination of the radial distribution of lignin in the fibres of some angiosperms where the densitometer peaks range between 9.7 nm and 10.8 nm. Attempts to draw more detailed conclusions of lignin distribution and that of hemicelluloses based on the area of section not stained by permanganate have been made by Kerr and Goring[66]. However, it will be appreciated that such conclusions must be treated with reserve as it is not known to what extend the permanganate penetrates the peripheral region of cellulose fibrils and to what extent there is growth of the manganese dioxide deposits following the initial reduction of the permanganate.

THE PROCESS OF LIGNIFICATION

Because of its economic importance, the most intensive studies on lignification have been made on the secondary xylem of both angiosperms and gymnosperms. A pioneering study was that of Griffoen[46]. The sequence of lignin deposition can be followed by using any of the methods listed above, but because they do not involve chemical treatment of the tissue, fluorescence microscopy and microspectrophotometry are the most useful. Some changes in the cell wall reactions of differentiating xylem with progressive lignification are indicated in Table 4.1. The reagents listed reflect chemical changes accompanying lignin deposition but the decreasing staining reaction with Congo Red most probably represents a decrease in the accessibility to the dye micelles due to the progressive deposition of lignin in the matrix regions between cellulose fibrils.

In early literature lignification was described as being initiated in the

Table 4.1: *Changes in the staining reactions of cells at progressive stages of lignification*

Reagent or optical property	*Unlignified cell*	*Partially lignified cell*	*Lignified cell*
Phloroglucinol —hydrochloric acid	unstained —	pink +	red ++
Chlorine water —sodium sulphite*	unstained —	pink +	red ++
Toluidine blue	purple	intermediate	blue-green
Potassium permanganate	unstained	brown	brown
Congo red	red	pink	unstained
Ruthenium red	red	pink	unstained
Ultraviolet	negative	partial absorption	strong absorption
Auto-fluorescence Excitation filter BG3 Barrier filter 50/49 (Zeiss)	negative	very light green fluorescence	green fluorescence

(* For angiosperms—see text)

intercellular layer between xylem tracheids and subsequently being deposited in the primary wall and then extending centripetally into the differentiating secondary wall. In a study using ultraviolet microscopy it was shown that lignin is first deposited in the corners of the cells[117] and this was subsequently confirmed in the autoradiographic studies of Saleh *et al*[102]. After this initial deposition, lignification extended to the intercellular layer at the cell corners and then to the primary and secondary walls of the developing cell. So far as is known, surface growth of the cell does not continue after the initiation of lignification and the process appears to continue until the death of the cell. This is supported by the observation that if the ultraviolet absorption of the intercellular layer of the cell wall is measured in successive cells from the cambium, the absorption increases until the formation of the third layer of the secondary wall is completed[117]. This is illustrated in Fig. 4.7, which shows variation in the absorption of ultraviolet radiation at points in the cell walls and the intercellular layer between cells 38 and 39, and cells 44 and 45 counted from the cambium. These observations imply that lignin deposition continues in the intercellular layer although the cytoplasm may be spatially removed from it. It may be noted however that pits may provide a pathway to this region of the wall.

As shown in Table 4.1, the above changes in u.v. absorption are accompanied by a decreased capacity of the cell wall to stain with the ruthenium red. This dye is thought to react with carboxyl groups and suggests a possible linkage of lignin to matrix components through ester linkages (see above). However the stain is not specific[79]. As stated above the decreased staining with Congo Red results from changed physical properties of the wall. The deposition of lignin may also be observed by fluorescence microscopy[59, 120, 121].

Ultraviolet absorption studies of lignification have also been carried out on reaction xylem of *Cryptomeria japonica* by Fujita *et al*[41]. As stated above the greatest ultraviolet absorption was observed in the outer region of the layer S2 with a second major peak in the intercellular layer and probably the primary wall. Lignin deposition began in the intercellular layer and was detected in the S1 layer in the third cell from the cambium. As with normal xylem, there was a continued deposition of lignin throughout the life of the cell. However, Imagawa *et al*[61] considered that the initial deposition of lignin occurred in the intercellular layer and not in the corners of the cell of the primary wall as observed by Wardrop[117].

The pattern of lignification of secondary xylem outlined above has been supported by studies using radioactive precursors of lignin. Saleh *et al.*[102] treated cuttings of *Populus trichocarpa*, *Pseudotsuga menziesii* and *Triticum vulgaris* with ^{3}H, 2-3-5 ferulic acid and subsequently observed the distribution of labelled lignin by autoradiographic techniques. The synthesis of possible radioactive precursors of lignin was described by Latif *et al.*[72] It was observed by Saleh *et al.*[102] that in the secondary xylem, labelled lignin was first deposited in the primary wall of the differentiating cells at the cell corners and was followed by deposition in the intercellular layer between radial walls before being deposited between the tangential walls. However, no radioactivity was detected in the secondary walls. In a subsequent study, Fujita and Harada[43] found that ^{3}H ferulic acid (2-6 ditritiated ferulic acid) was incorporated into the differentiating secondary wall. In this investigation the incorporation of ^{3}H-L-phenylalanine (generally labelled) into the lignin of the wall was also studied but the results were less conclusive than those obtained using ferulic acid since phenylalanine is further removed from lignin in the biosynthetic pathway (see below). The failure of Saleh *et al.*[102] to detect radioactivity in the secondary wall of differentiating secondary xylem perhaps may be attributed to the lower concentration of lignin in the secondary wall as opposed to the primary wall and intercellular layers although the amount of lignin in the secondary wall is greater than in these regions.

In the study of lignification of reaction xylem Fujita *et al.*[41] refer to 'primary wall lignin' and 'secondary wall lignin', the former being deposited in the intercellular layer and primary wall and the latter after the initiation of the secondary wall. However, as shown in Fig. 4.7 it is recognized that

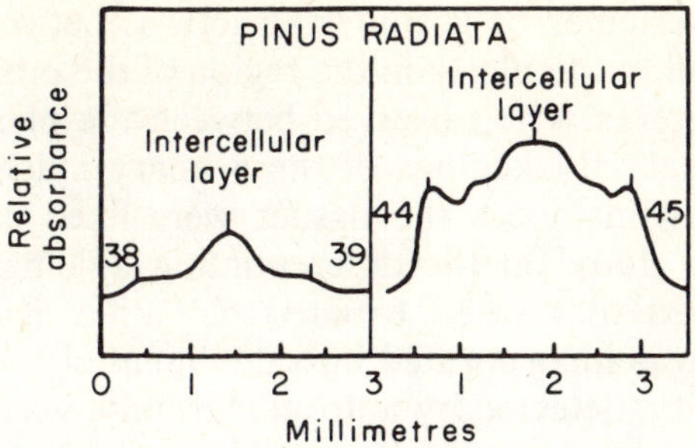

Figure 4.7: *Microphotometer traces through an ultra-violet photomicrograph showing greater absorption in the intercellular layer between tracheids of* Pinus radiata *of increasing maturity (after Wardrop and Bland*[123]*).*

the lignin is deposited continuously throughout the secondary wall, primary wall and intercellular layer. It will be clear that there must always be a temporal separation between the formation of an intercellular layer, the deposition of the matrix and framework components of the cell wall, and lignification. However, there is evidence that the chemical nature of the lignin deposited in the wall undergoes change during the process of lignification[123].

It was shown that in successive 32 μm thick tangential sections cut from the cambium, the form of the ultraviolet absorption curve progressively changed and only in sections 10–14 was the general form of the absorption curve lignin-like, although the curves differed from mature lignin in that there was still a definite long wavelength absorption band present. These observations were interpreted in terms of the hypothesis that lignin in the process of formation contains a higher proportion of conjugated units than the lignin of mature xylem. This would not indicate a fundamental difference in structure, but could be interpreted that at a lower stage of polymerisation the conjugated units could act as sites for the addition of further units during polymerisation. Another suggestion is that there are two types of precursor involved in lignification[8]. This suggestion was based on the observation that cells in which the whole wall showed positive staining by the Maule reaction, only the intercellular layer was stained when the phloroglucinol-hydrochloric acid reagent was used. However, it was pointed out by Bland[11] that although the phloroglucinol reaction is more sensitive than the Maule reaction there are few aldehyde groups present in lignin so that it may be that a positive phloroglucinol reaction is obtained only when a relatively high concentration of lignin is present. As stated in the preceding discussion this concentration is greatest in the intercellular layer.

Several attempts have been made to identify particular organelles or cytological phenomena with the process of lignification. Wardrop[119] observed the accumulation of vesicular or granular structures between the plasmalemma and the cell wall, particularly in the region of the pits of *Eucalyptus*. Similar granular structures were observed between the plasmalemma and the developing secondary thickenings of the primary xylem of *Coleus* by Hepler *et al.*[59] In both instances the tissues were fixed using potassium permanganate. In the study on the differentiating xylem of wheat roots Pickett-Heaps[94] showed that when tritiated cinnamic acid was administered to the roots it was incorporated into the lignin of the differentiating xylem. The radioactivity, detected by autoradiography, was observed to be associated with the vesicles derived from the Golgi apparatus and possibly from the endoplasmic reticulum[127]. Further, it was observed by Wardrop[121] that in the lignifying walls of *Pyrus communis*, vesicles with granular contents and staining in the same manner as the lignin in the cell wall appeared to undergo fusion with the plasmalemma, although the origin of the vesicles could not be determined (Fig. 4.8).

Working with the reaction xylem of *Cryptomeria japonica* Fujita *et al.*[42] have observed what appeared to be the Golgi vesicles fusing with the plasmalemma. Their presence was most obvious during the rapid thickening of the lignified layer S2 and these investigators postulate that lignin precursors are synthesized and stored in Golgi vesicles and that the precursors are incorporated into the wall by fusion of the vesicle membrane with the plasmalemma.

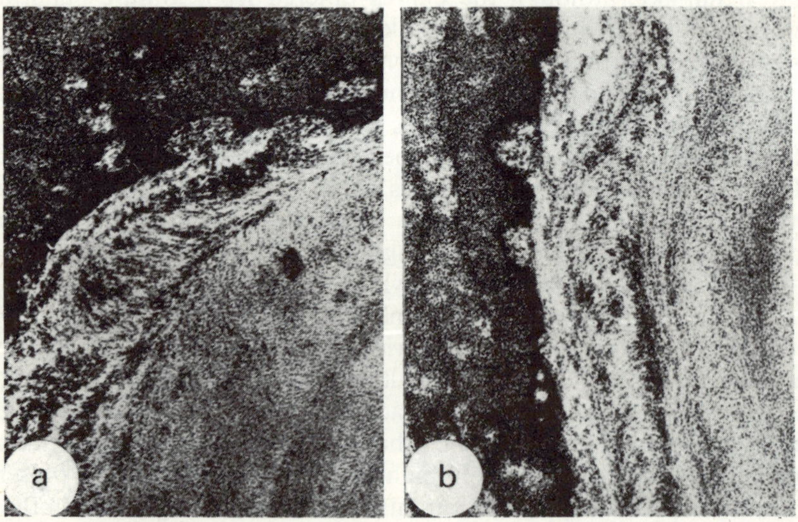

Figure 4.8: Pyrus communis. *Two sections cut approximately normal to the cell surface of a sclereid and showing what appears to be a process of the incorporation of the contents of vesicles into the cell walls. (×40,000) (reproduced with permission from Wardrop*[121]*).*

BIOCHEMISTRY OF LIGNIFICATION

Early investigations into the nature of lignification have been summarized by Griffoen[46]. The glucoside, coniferin, was detected in the cambial sap of conifers by Tiemann and Haarmann in 1874 and was recognised by Claesson in 1932 as a possible precursor of lignin. Claesson proposed the idea that precursors may arise in the meristematic zone and was subsequently transferred to the differentiating xylem. This idea was specifically formulated by Freudenberg *et al.*[35] In this work the presence of coniferin was demonstrated and ^{14}C-labelled coniferin was shown to be incorporated into the lignin of the differentiating xylem. In the same investigation consideration was given to the mechanism by which a possible precursor—glucoside might be elaborated and deposited as lignin in the cell wall. To detect β-glucosidase the glucoside indican was used. In the presence of β-glucosidase indoxyl is formed as a result of hydrolysis, and this compound in the presence of atmospheric oxygen is oxidised to indigotin (indigo blue). By this means it was shown that the β-glucosidase was apparently present in the walls of the lignifying tracheids. It was also shown[36, 60] that synthetic lignin could be made in vitro from coniferyl alcohol alone or mixtures of it with synapyl and p-coumaryl alcohols in the presence of fungal phenol oxidase which was later shown to be laccase. It was considered that this enzyme participated in the formation in an analogous way to form lignin in xylem. In later investigations it was shown, that depending on the ratio of precursors used, lignin-like substances produced in this manner could not be distinguished from lignin isolated from lignified tissues[33].

Freudenberg *et al.*[35] proposed an hypothesis of the mechanism of lignification according to which it was postulated that for stems with secondary xylem, lignin precursors were formed in the region of the cambium as glucosides such as coniferin and then diffused centripetally into the differentiating xylem where the sugar component was released from the glucoside by hydrolytic enzymes located in the wall of the lignifying cell. The aglycone (e.g. coniferyl alcohol) was considered to undergo oxidation by enzymes localized in the walls of the differentiating xylem and then through a series of steps to undergo a process of polymerisation which resulted in the deposition of lignin in the walls and intercellular substance of the cells of the xylem[33, 35].

Subsequently, the work of Harkin and Obst[55] showed that the oxidising enzyme present in differentiating xylem is a peroxidase. This was established using the reagent syringaldazine (Fig. 4.9(a)) which is oxidized to form the purple coloured tetramethoxyazo-p-methylene quinone (Fig. 4.9(b)) in the presence of peroxidase and hydrogen peroxide. In contrast, laccase produces this substance in the presence of atmospheric oxygen.

It will be appreciated that in the hypothesis of Freudenberg outlined

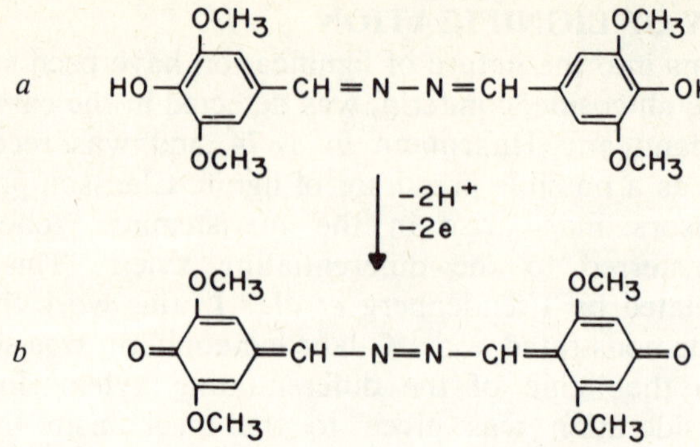

Figure 4.9: *Oxidation of syringaldazine (a) to tetramethoxyazo-p-methylene quinone (b) by peroxidase.*

above, the presence of β-glucosidase in the cell walls and its function of releasing for oxidation the phenylpropane derived aglycone are of central importance. To demonstrate the importance of β-glucosidase Freudenberg *et al.*[35] prepared ^{14}C-labelled L-coniferin and applied it to plants in the same manner as ^{14}C-labelled D-coniferin. The ^{14}C-labelled L-coniferin was not utilised to form lignin and could subsequently be extracted with water. These observations have been quoted as indicating the essential role of β-glucosidase in lignification[34, 103].

However, it will be understood that in coniferin synthesised from L-glucose the coniferyl alcohol is linked to the glucose by an α-linkage whereas in coniferin synthesised from D-glucose the coniferyl alcohol is linked by a β linkage. Thus the experiments do not indicate the essential role of β-glucosidase but only that α-L-glucosidase was apparently absent from the tissues. In a subsequent investigation, Wardrop[121] concluded that β-glucosidase is present mainly in the cytoplasm of the differentiating cells and not in the cell walls. In this study both indican and L-naphthyl-β-D-glucopyranoside were used to detect the presence of the enzyme. Manifestly further work is needed to demonstrate the role of β-glucosidase in lignification.

While at the time, the evidence in favour of Freudenberg's concept appeared strong, some reservations were expressed[117] on the basis of anatomical observations, such as the observed selective lignification of the secondary wall in primary xylem, selective lignification of particular layers of secondary walls in reaction xylem of angiosperms and also in lignified collenchyma[120]. These observations emphasized the possible intracellular origin of lignin precursors. This view was reinforced by the observation that the lignin in the secondary xylem of angiosperms is mainly of the syringyl type while that of the primary xylem is of the guaiacyl type[11, 123]. Further evidence suggesting intracellular origin of precursors was provided by the ultraviolet microspectrophotometric studies of Fergus and Goring[30]

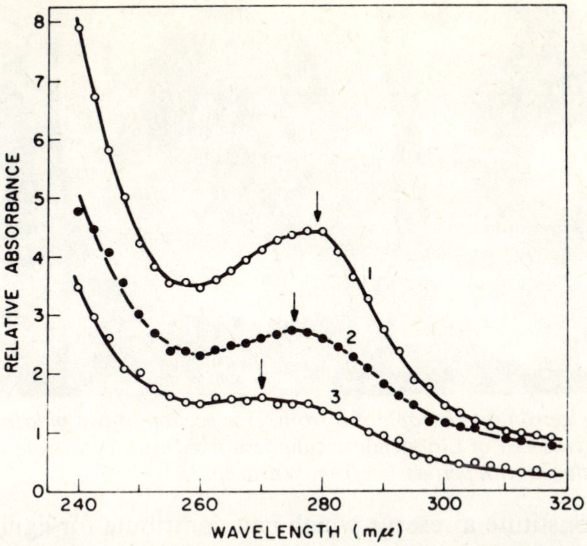

Figure 4.10: *The ultraviolet absorption curves in a transverse section of* Betula *for (1) the secondary wall of a vessel, (2) the cell corner region of the intercellular layer between fibres, and (3) the secondary wall of a fibre (reproduced with permission from Fergus and Goring[30]).*

in which it was shown that the lignin of the xylem vessels of *Betula* is mainly of the guaiacyl type, whereas that present in the fibres had a predominance of syringyl units. Furthermore, the intercellular substance between fibres and vessels contained both guaiacyl and syringyl units (Fig. 4.10).

These results are consistent with more recent chemical studies on tissues isolated by mechanical fractionation procedures[53, 54].

In a subsequent study[121], it was shown for sclerenchyma fibres in *Liriodendron tulipifera* and also for the sclereids of *Pyrus communis* that where two lignified cells were adjacent to each other the intercellular layer became lignified, but where the lignified fibre or sclereid was adjacent to a parenchyma cell lignification ceased at the primary cell wall (Fig. 4.11). These observations suggested further that the lignin precursors arise within the lignifying cell but that during differentiation there is a factor of cell recognition which governs whether the intercellular layer becomes lignified.

It should also be noted that following the formulation of the view that the glucosides of coniferyl, syringyl and p-coumaryl alcohols act as lignin precursors the analysis of the cambial sap of many species has shown one or more of these glucosides to be absent. This work has been reviewed by Sarkanen[103]. Furthermore, Hasegawa *et al.*[58] have demonstrated that ^{14}C-labelled glucose is incorporated into the lignin of tissue cultures of *Pinus strobus*. In view of this it would seem that glucosides detected in the

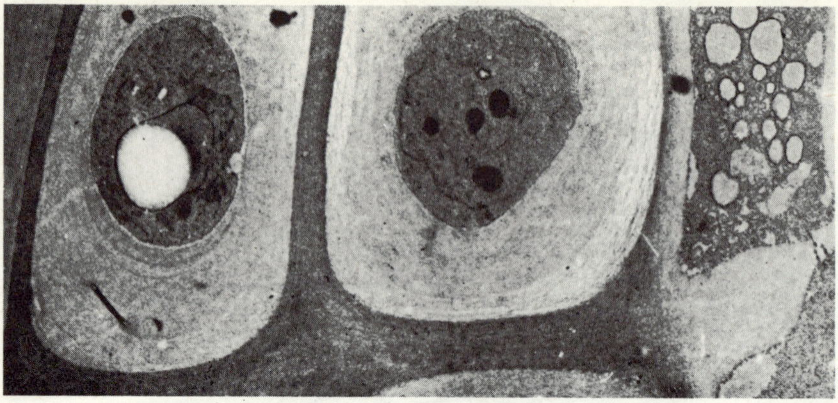

Figure 4.11: *An electron micrograph of a transverse section of two phloem fibres and an adjacent parenchyma cell of* Liriodendron tulipifera *fixed with potassium permanganate. (×52,000) (reproduced with permission from Wardrop*[121]*).*

cambial sap constitute a reserve which can contribute for lignin deposition in lignifying cells.

Evidence for the intracellular origin of lignin precursors was contained in the biochemical studies of Rubery and Northcote[101]. The significance of these observations can be seen in relation to the metabolic pathway leading to the formation of phenolic substances in plants. Thus, from the products of photosynthesis, phenylalanine and tyrosine are formed via the shikimic acid pathway (Fig. 4.12). The formation of cinnamic acid from phenylalanine involves its deamination by phenylalanine ammonia-lyase (PAL, Fig. 4.12). The steps following the formation of cinnamic acid to the formation of lignin precursors and ultimately of lignin, are also outlined (Fig. 4.12). In their investigations, Rubery and Northcote[101] dissected unlignified cambium and lignifying differentiating xylem from the stem of *Acer pseudoplatanus*. In lignifying differentiating xylem the PAL activity was high. Little of the enzyme was detected however, in the cambium zone. From Fig. 4.12 it can be seen that the enzyme is essential for the conversion of phenylalanine to the alcohol precursors of lignin.

While the general pathway of the formation of lignin precursors has been recognised for some time (see Neish[87]), more recent work has been directed towards identifying and elucidating the mode of action of the various enzymes which are involved. The biochemistry of these reactions has been reviewed by Grisebach[47] and by Gross[49, 50]. The main enzymes involved in the synthesis of coniferyl alcohol from the corresponding acid are indicated in Fig. 4.13.

Another field of biochemical interest concerns the roles of the peroxidases and the origin of hydrogen peroxide which is the substrate for the peroxidative oxidation of the cinnamyl alcohols. It has been recognised

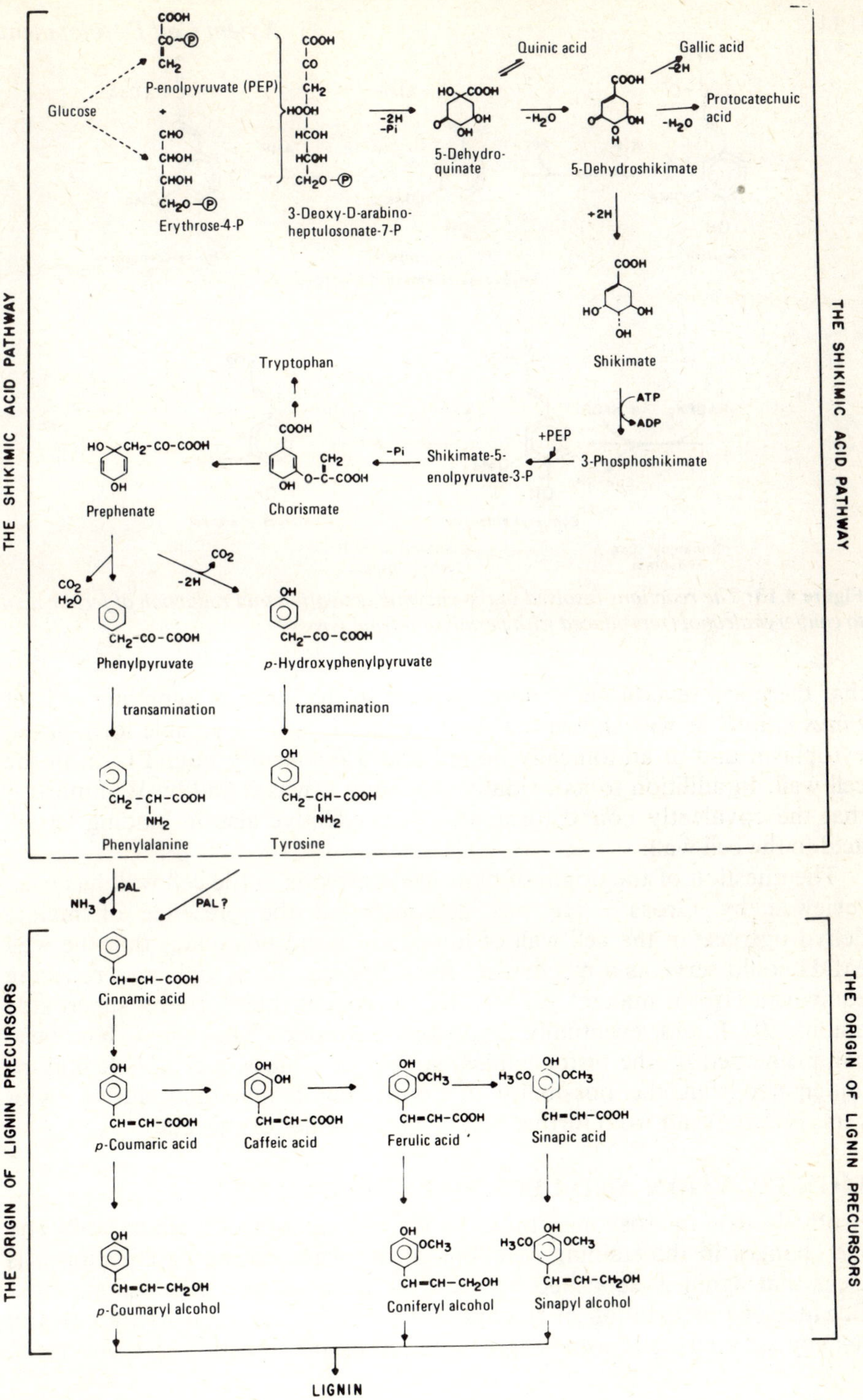
THE SHIKIMIC ACID PATHWAY
Glucose
P-enolpyruvate (PEP)
Erythrose-4-P
3-Deoxy-D-arabino-heptulosonate-7-P
-2H
-Pi
5-Dehydroquinate
Quinic acid
-H2O
5-Dehydroshikimate
Gallic acid
-2H
Protocatechuic acid
+2H
Shikimate
ATP
ADP
3-Phosphoshikimate
+PEP
Shikimate-5-enolpyruvate-3-P
-Pi
Chorismate
Tryptophan
Prephenate
CO2
H2O
-2H
Phenylpyruvate
p-Hydroxyphenylpyruvate
transamination
Phenylalanine
Tyrosine
PAL
NH3
PAL?
THE ORIGIN OF LIGNIN PRECURSORS
Cinnamic acid
p-Coumaric acid
Caffeic acid
Ferulic acid
Sinapic acid
p-Coumaryl alcohol
Coniferyl alcohol
Sinapyl alcohol
LIGNIN

Figure 4.12

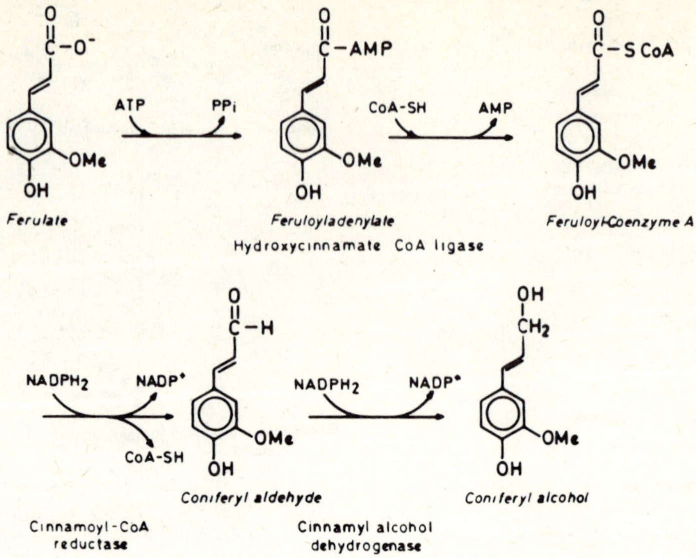

Figure 4.13: *The reactions involved in the enzymic activation and reduction of ferulic acid to coniferyl alcohol (reproduced with permission from Gross*[50]*).*

that there are several peroxidases present in the walls of lignifying cells of *Pinus elliotii*. It was shown that peroxidases exist in a soluble form in the cytoplasm and in an ionically bound and a covalently bound form in the cell wall. In addition to its oxidative capacity it was found by Whitmore[126] that the covalently bound form was most effective also in binding ferulic acid to the cell wall.

The question of the origin of hydrogen peroxide in the cell wall has been reviewed by Gross[49]. He has demonstrated the presence of malate dehydrogenase in the cell wall of horseradish and postulates that the wall MDH could serve as a mechanism for supplying the cytoplasmic reducing equivalents (from malate) via NADH to oxygen thus forming superoxide anions (0_2^-) and eventually hydrogen peroxide. This work has been complemented by the histochemical studies of Catesson *et al.*[13] It may be anticipated that the possibility of controlling lignification during xylogenesis may result from further biochemical studies of this kind.

LIGNIFICATION AND CELL WALL PROPERTIES

From electron microscopic studies of stained sections of lignified walls and the changes in the staining reactions of the walls during lignification it is clear that lignin is deposited between the fibrils of the cellulose skeleton. Because of this location of the lignin it is understandable that in cases of heavily lignified walls some degree of swelling may occur[38, 96]. Conversely, lignified walls tend to shrink less than unlignified ones and it is of interest that when 'artificial lignin' is deposited in unlignified walls, their shrinkage is markedly decreased[110].

Lignification does not appear to affect the tensile properties of the wall but the possible influence of the process on the compression strength of lignified walls has been discussed by Frey-Wyssling[37]. Frey-Wyssling envisaged a model in which the cellulose fibril is considered to be loaded axially in compression. At the initiation of failure the fibril would tend to buckle laterally. However, it was argued that if lignin is packed between the fibrils then the tendency to buckle would be reduced and the load in compression necessary to cause the buckling would be increased. In this way lignin could contribute to the compressive strength of the cell wall. It may be noted that for xylem the moisture content is inversely related to the amount of lignin present[16].

This proposed effect of lignin in increasing the compressive strength of the cell wall is relevant to the proposed role of lignin in resisting the compressive forces acting on the walls of vessels due to tension of the water columns they contain (see below). It is also of interest that when lignification of fibres does not occur the mechanical properties of the stems and branches of trees are markedly changed as in the formation of rubbery wood referred to later in this discussion. The capacity of animals to digest plant cell walls is also influenced by the phenolic acids or lignin they contain[56].

DELIGNIFICATION OF CELL WALLS

In relation to the phenomenon of swelling of walls during lignification it is of interest that in one paper, Alexandrov and Djaparidze[4] observed that the highly lignified sclereids of the fruit of quince apparently lost their lignin during ripening of the fruit and supported this view by the diagram shown in Fig. 4.14. However, in more recent observations in this laboratory the conclusion that delignification occurs is in some doubt because cells

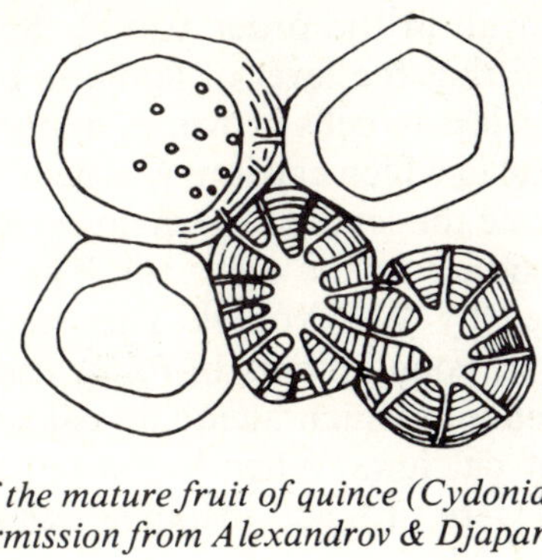

Figure 4.14: *Sclereids of the mature fruit of quince (Cydonia) showing natural delignification (reproduced with permission from Alexandrov & Djaparidze[4]).*

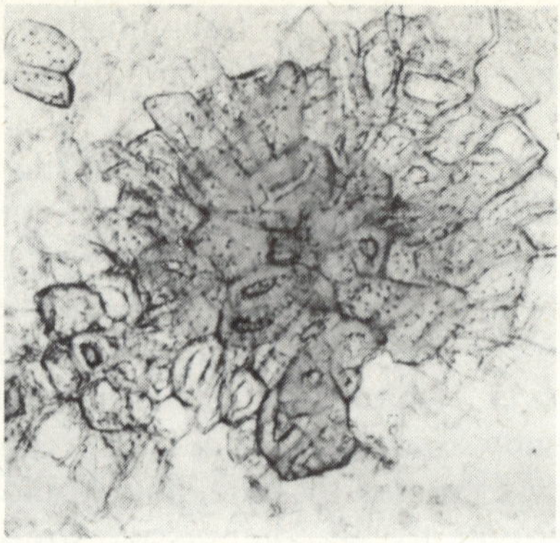

Figure 4.15: *Sclereids of a fruit of quince (Cydonia) 1 cm in diameter showing an appearance similar to that in Figure 4.14.*

showing this structure, (Fig. 4.15), have been observed in the ripening fruit of quince and pear measuring less than 1 cm in diameter and the question arises whether the appearance shown in Fig. 4.14 simply represents the formation of cells of different wall thickness and level of lignification. However, the change in the pattern of pitting of the cell wall illustrated in Fig. 4.14 would appear to require further investigation.

FACTORS GOVERNING LIGNIFICATION

Nutrition

Several elements have been considered to be implicated in the lignification process, in particular calcium and boron.

Calcium In a study of the effects of calcium deficiency on the differentiation of xylem it was shown[21] that the level of calcium available to plants of *Pinus taeda* caused changes in their anatomy and in the dimensions of the lignified cells present. Further observations which suggested that calcium may be implicated in lignification were made by Lipetz[77] and Lipetz and Garrow[78] that various ions were effective in releasing peroxidase from the cell wall in the order $Ca^{++} > Sr^{++} > Ba^{++} > Mg^{++} > NH_4^{++}$. It was also shown that the level of lignin in the walls of lignifying cells of tissue cultures was less in cells grown in a medium with a full complement of calcium than one in which calcium was deficient. It thus appeared that calcium could moderate the level of lignification in the wall through its function of binding peroxidase to the cell wall. The possible influence of calcium was reinvestigated by Wardrop[121]. This study involved two sets of observations. In the first, young plants of *Pinus radiata* and *Helianthus annuus* were grown in media in which adequate calcium was supplied and also in media deprived of calcium. When harvested after 90 days it was observed that the tissues were, in each instance lignified, but the zone of differentiation in xylem, that is the number of partially differentiated cells, was markedly less in a plant deprived of calcium. In addition, in the case of *Pinus radiata*, resin canals appeared more numerous and the calcium deficient plants showed evidence of chlorosis of the leaves. On analysis of the stele the lignin content in *Pinus radiata* was identical in the calcium deficient plants and those with a full complement of calcium. In the case of *Helianthus* there were slight differences and the lignin content appeared somewhat lower in the plants with a full complement of calcium but it was recognised that the results were not as reliable as those obtained with *Pinus radiata* since the cortex was not as readily removed and the cuticle was included in the analysis. In a second approach in this study the calcium level in lignified and unlignified tissues was determined using the technique of x-ray energy dispersive analysis. For this purpose the calcium level in differentiating xylem of *Liriodendron tulipifera* was determined. It was observed that during the progressive lignification of the differentiating xylem the calcium level decreased by as much as 20-fold. Furthermore, the calcium level in lignified phloem fibres varied between 22 and 5-fold that in the walls of adjacent parenchyma cells. Similar observations were made on the lignified sclereids of *Prunus*. An example of the type of result obtained by this procedure is shown in Fig. 4.16. These results would appear consistent, at least superficially, with the observations of Lipetz in that the lignified tissues contained less calcium and that in the tissue culture

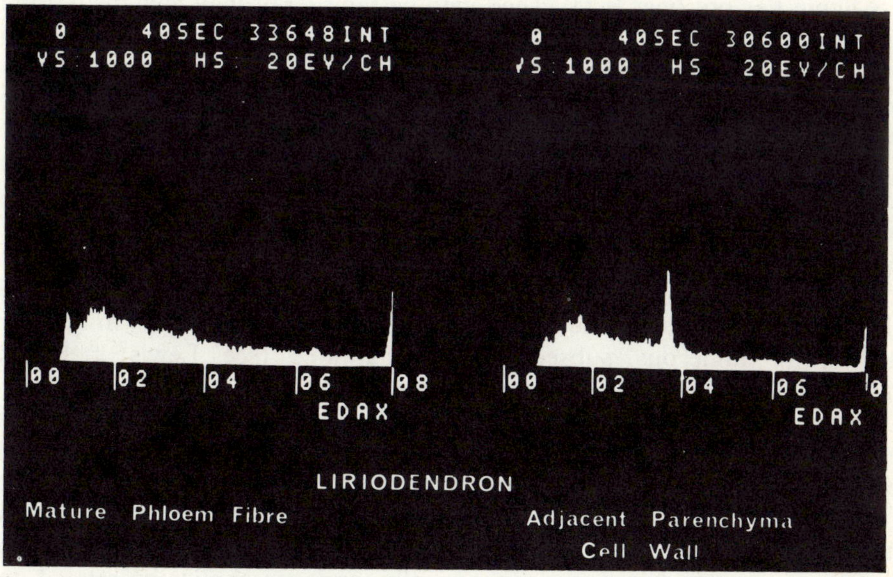

Figure 4.16: *Calcium levels in different cells as indicated by characteristic CaK_{α} emission at 3.69 KeV. These differences in calcium levels were observed in cells such as those illustrated in Figure 4.11 (reproduced with permission from Wardrop[121]).*

experiments, deprivation of calcium resulted in lignification. However, the gross analyses do not support any precise correlation between calcium level and the level of lignification. A possible explanation of the results obtained by x-ray dispersive analysis is that the sites occupied by the lignin in the cell wall are those previously occupied by calcium, so that as the process of lignification proceeds the calcium level declines. This could be related to the possible occurrence of ester linkages between cell wall polysaccharides and lignin (Fig. 4.3).

The role of calcium is further complicated in that although it appears to be involved in binding peroxidase to the cell wall, its removal during lignification could conceivably result in more effective binding of the peroxidase to the wall polysaccharides. As shown by the results of Whitmore[126], covalently bound peroxidase is more effective in binding possible lignin precursors than ionically bound peroxidase. It is of interest to speculate whether the peroxidase(s) binding possible precursors to the wall are the same as the peroxidase(s) affecting oxidation of the precursors and their subsequent polymerisation to lignin. It is also of interest that in a study of the effects of the herbicide dichlobenyl (dichlorobenzyl nitrile) it was observed by Verloop[115] that this substance inhibits cell division in the meristem of roots and causes excessive deposition of phenolic components in cells of parenchyma which would normally not lignify. Further, the differentiation of the xylem is inhibited. Whether the compounds deposited are lignin has not been established with certainty, but they give the normal staining reactions with phloroglucinol-HCl and with chlorine water-sodium sulphite. In the present context it is of interest that it was observed by Verloop that the effect of this herbicide could in part, be inhibited by the addition of calcium to the medium.

Boron The role of boron in lignification has been discussed in considerable detail by Lewis[75], both in relation to its biochemical effects and also in the broader context of the significance of lignification in the evolution of the land flora. That boron is an essential element for plant growth stems from the initial observations of Agulhon[2] who observed that boron stimulated the growth of some vascular plants in culture and from numerous subsequent investigations (see Lewis[75]). Boron has been established as an essential element for plant growth but this is so only in the Tracheophyta, although there are some exceptions in the marine green algae where it may function as a stabilizing agent for some cell wall polysaccharides. The symptoms of boron deficiency in plants include the poor overall lignification and differentiation of the xylem. Differentiation may occur early and is accompanied by the accumulation of lignin precursors or of compounds chemically and metabolically related to them[75]. The manner in which boron mediates in lignification is not fully understood. It may affect growth and differentiation of the whole plant; the physiological activities of a plant such as membrane permeability and translocation; and biochemically may be involved in the metabolism of carbohydrates, polyphenols, auxins and nucleic acids. The manner of the possible participation of boron in these activities has been reviewed by Lewis[75] as well as its role in diseased plants[76]. One investigation illustrating the biochemical role of boron in the formation of phenolic compounds was carried out by Lee and Aranoff[74] who presented evidence that the borate ion is involved in the partitioning of glucose between glycolysis and the pentose phosphate shunt pathway (see Fig. 4.17).

The borate ion forms a complex with 6-phosphogluconate. The 6-phosphogluconate borate complex functions by inhibiting the action of 6-phosphogluconate dehydrogenase, an enzyme which controls the pentose phosphate shunt pathway. The activity of this pathway affects the supply of erythrose-4-phosphate which combines with phosphoenyl pyruvate at the beginning of the shikimic acid pathway (Fig. 4.12). However, some dihydroxy phenolic acids compete for the available borate, thereby decreasing the formation of the inhibitory 6-phosphogluconate borate complex and alleviating the inhibition of 6-phosphogluconate dehydrogenase. Consequently, regulation of the shikimic acid pathway by borate is lost, causing phenolic acids to accumulate. These compounds can also complex with the remaining borate. The accumulation of the phenolic

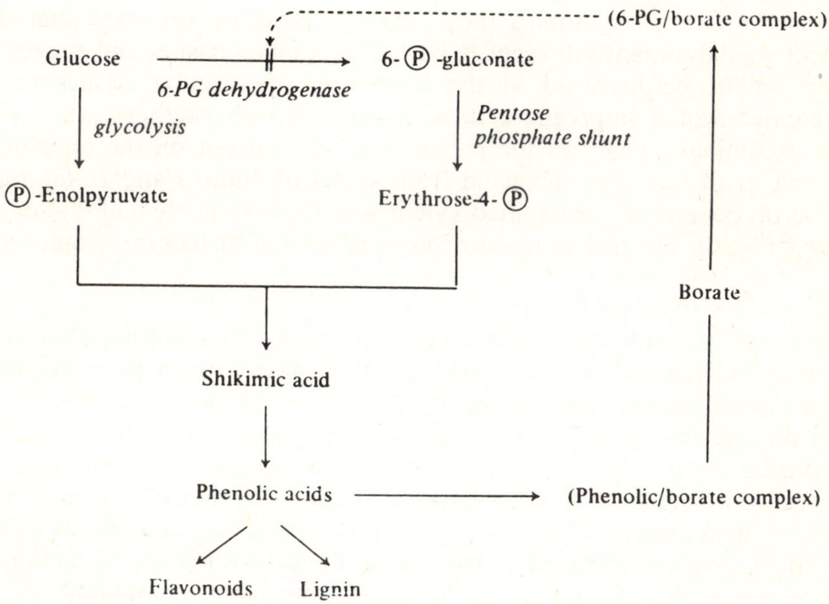

Figure 4.17: *A diagram indicating the biochemical effects of boron deficiency (reproduced with permission from Walker[116]) and based on the observations of Lee and Aronoff[74].*

compounds in the cells of the boron deficient plants can result in their death.

It is important to note, as pointed out by Lewis[75] that the branch points in the pathway of the final lignin precursors are p-coumaric acid and ferulic acid. Both substances can be either reduced or hydroxylated. On reduction p-coumaric acid yields p-coumaryl alcohol. Ferulic acid on hydroxylation yields 5-hydroxyferulic acid before undergoing methylation to form sinapic acid. Both caffeic acid and 5-hydroxyferulic acid are 0-diphenols, and each is capable of complexing with boron so that this element could theoretically influence the lignification process at critical points in the precursor pathway. In the original study, Agulhon[2] demonstrated that plants grown on a boron deficient medium contained less boron than plants with adequate levels of the element and so established that it was essential for plant growth. This observation is similar to that relating to calcium[121] in which it was shown that plants grown on a medium deficient in calcium contained less calcium than plants grown in a medium with an adequate supply of the element.

The question as to whether boron accumulates in lignified tissues appears somewhat confused. Thus, it is stated by Lewis[75] that Agulhon[2] had shown that 'boron was abundant in lignified tissue'. However, reference to the original paper shows only that there is an optimal concentration of boron for plant growth. On the other hand, Dutta and McIlrath[24] quoting data from Agulhon's thesis stated: 'Agulhon observed that this element was abundantly present in bark and lignified tissues and suggested that it might be involved in the formation of vascular tissues or of compounds which impregnate these tissues'. To check these statements some preliminary analyses for boron were carried out on the unlignified pith and lignified xylem dissected from stems of *Datura* and *Helianthus*. The boron content of the lignified xylem was approximately half that of the unlignified pith. The question is obviously in need of further investigation.

Auxin, Gibberellin and Kinetin

While the effects of boron on processes governing lignification appear to be relatively well defined, it is also probable that other substances which may interact with boron may be involved in the lignification process, in particular auxins and cytokinins. Some of these interactions have been discussed in the review by Lewis[75]. An early investigation relating auxin to the level of lignification of differentiating tissues was that of Wershing and Bailey[125] who observed that when lanoline paste containing IAA was applied to stems of *Pinus* at physiological concentration the formation of the reaction xylem of conifers was induced. This xylem is characterised by its high lignin content compared with normal xylem of the same growth ring and is formed when stems or branches are bent out of their normal orientation. Nečesány[85] studied the distribution of auxins in stems bent horizontally both in *Pinus sylvestris* and in *Populus alba*. Ether extracts were made of the differentiating xylem from the upper and lower sides of the bent stems and the effect of components separated by chromatography on the inhibition of growth of coleoptile segments of *Triticum aestivum* was determined. It was observed that substances stimulating the growth of coleoptile sections were present mainly on the lower side in both species and of these IAA appeared to be the most important of the substances affecting elongation. In these observations the concentration of IAA was greatest in the region of most intense lignification. In the observations on *Populus* the characteristic reaction xylem (tension wood) of angiosperms is formed on the upper side of the bent stems, i.e. on the side of lower concentration of IAA. It is consistent with these observations that in the work of Kennedy and Farrar[65] and Morey and Cronshaw[83] the application of the anti-auxin, tri-iodobenzoic acid to vertical stems of angiosperms resulted in the formation of tension wood.

In addition to studies on whole plants the effects of auxins and cytokinins have been studied in tissue cultures. Vennerloo[114] carried out a most useful bridging study on lignification of xylem in *Populus nigra* and in tissue cultures from the same tree. In carrot tissue cultures Koblitz[68] observed that auxin stimulated lignification of differentiating tracheary elements. In *Pisum sativum* Cheng and Marsh[15] observed that gibberellin increased the level of lignification and PAL activity and in an extensive quantitative investigation of the influence of kinetin Bergmann[9] demonstrated enhanced lignification in cultures of *Nicotiana tobacum*. It is of interest that in the cultures studied by Bergmann (Fig. 5, p. 225, Bergmann[9]), following cell division only one of the daughter cells developed into a tracheid. The influence of other compounds on lignification and cytodifferentiation, including ethylene have been reviewed by Roberts[100] but to this stage there is no overall view as to the way in which these different compounds influence lignification.

Light It was shown by Phillips[93] that light may have a direct or indirect effect on the lignification of the xylem. In many deciduous species cambial activity is initiated before the opening of the buds, when the plants presumably utilize nutrient reserves. Phillips shaded roots of *Fraxinus* for a period of 14 days at a stage when the leaves were *ca.* 5 cm in length (May). After this period the shoot was allowed to grow for one month under normal conditions of illumination and then shaded for an additional period of 21 days. When transverse sections of the xylem were examined they showed alternate zones of unlignified and lignified fibres in the current growth ring. These observations show a strong correlation between the level of illumination and the degree of lignification of the xylem and may reflect the level of auxin activity which may vary with the degree of leaf development. Extensive biochemical studies of the effect of light on enzymes such as PAL involved in the lignification process have been reviewed by Gross[50] but it is pointed out that these enzymes are also involved in flavonoid biosynthesis so that it would appear that the effect of light on lignification is indirect.

In an attempt to relate the influence of light to the development of lamellae in the secondary cell wall and also its influence on lignification an extensive investigation was carried out by Nečesāńy[86]. In this work trees of *Populus nigra*, *Pinus strobus* and *Eucalyptus radiata* were used. In each case one tree was chosen with at least twelve branches 3-6 years old and was examined periodically beginning at 08.00 hours. The internal surface of the differentiating fibres or tracheids was examined using replica techniques and the fibril orientation was determined. The difficulty in such a procedure is that the precise stage of development of the wall when the examination was made, was not known. However, in the electron micrographs published it was apparent that the surface examined showed little evidence of cellulose fibrils at 08.00 hours, but these became progressively more apparent in replicas prepared in subsequent time intervals and exhibited a change in orientation until *ca.* 24.00 hours. After this time the fibrils became obscured again by an amorphous material. It is suggested by Nečesāńy that the deposition of lamellae of cellulose fibrils in the secondary cell wall exhibits a circadian rhythm and that the deposition of lignin causes the amorphous appearance of the inner wall surface apparent in replicas prepared during the hours of 02.00 hours to 08.00 hours. This conclusion would appear to be supported by measurements of changes in various characteristics of the cell such as the isoelectric point of the cytoplasm, in osmotic potential, the carbohydrate content, and by changes in the content of extractable lignin components such as aldehydes and aromatic acids. The changes in the IAA content in the cambial zone, and of peroxidase, and catalase were also determined. The maximum concentration for the extractable lignin components, aldehydes and aromatic acids, occurred between 20.00 and 04.00 hours. These observations would be consistent with a rhythmic deposition of lignin in the cell wall related to the illumination of the plant. It is obviously difficult, however, to ensure that in each sample the cells examined were at a similar stage of differentiation.

Genetic Factors

The first evidence of genetic factors governing lignification was obtained by Jorgensen[63] in brown-rib maize; a mutant first observed in 1924. Jorgensen noted that the pigment characteristic of the mutant is found in the stem, root, leaves, the tassel and the cob of the plant and is present only in lignified tissue and is most apparent in the mid-rib. The formation of the pigment appeared to be influenced by light. Subsequently other mutants were recorded by other investigators and this work has been described by Kuc and Nelson[69]. The mutants have been designated brown-rib mutants 1–4. All four mutants are Mendelian recessives. Kuc and Nelson have investigated the chemical nature of the lignin of brown mid-rib 1 mutant (bml) in comparison with that of the non mutant maize (Bml). It was shown that on nitrobenzene oxidation of the lignin of bml there was a lower yield of aromatic aldehydes and a greater fraction resistant to oxidation than in lignin from the normal maize (Bml) and also the lignin from the mutant was lower in methoxyl content. It was shown further that when normal plants and the mutant were fed with uniformly labelled L-phenylalanine or L-tyrosine the incorporation of either compound into the lignin of mutant plants was considerably less than that incorporated into the lignin of normal plants. These results were interpreted by Kuc and Nelson to mean that the gene of the normal Bml plants at least in part, controls the manner in which the lignin monomers are incorporated into the lignin and that in the bml the lignin has fewer sites at which p-coumaric acid can be esterified.

It will be appreciated that these conclusions are at variance with some earlier suggestions of a spontaneous polymerization of free radical precursors, suggested by Freudenberg and co-workers[12]. Subsequent cytological work on the cytoplasmically compartmentalized, and cell wall localized enzymes, such as peroxidase, participating in the lignification process also indicate cytoplasmic participation in the process[20]. The complexity of the problem is illustrated further by the observation made in this laboratory that different tissues in the Bml and bml maize showed both similar and differing reactions to the phloroglucinol-HCl reaction. In the normal maize, vessel elements, surrounding sclerenchymatous fibres and subepidermal sclerenchyma gave a positive reaction with phloroglucinol-HCl reagent. In the mid-rib of mutant maize the protoxylem vessels stained purple, the metaxylem vessels were unstained, the sclerenchyma fibres surrounding the vascular strand stained red, and the subepidermal sclerenchyma was stained brown. Thus it would seem that at least as indicated by this reaction, some of the lignin in the mutant is similar to that in the normal maize whereas the lignin in other cells, particularly the sub-epidermal sclerenchyma, is different. These observations could affect the analytical results reported by Kuc and Nelson, but the ^{14}C precursor incorporation studies indicate fundamental metabolic differences.

LIGNIFICATION AND DISEASE RESISTANCE IN PLANTS

In the concluding section of this discussion attention is directed to the question of what adaptive advantage the development of the lignification process may have conferred on a developing land flora. One suggestion has been that it may have created a barrier to the growth of pathogenic organisms[67].

Many higher plants when infected by a fungus respond by secreting cell wall material around the penetrating hyphae. This material frequently contains substances of a phenolic nature and in many instances gives all the characteristic reactions of lignin. The lignin was detected using colour reactions such as those outlined above[109] and also by ionization difference spectra determined in alkaline and neutral solutions of extracts of leaves[111, 113]. In reed canary grass (*Phalaris arundinacea*) it was shown that leaf spot fungi which were non-pathogenic to reed canary grass but pathogenic to other plants formed appressoria and penetration pegs. On the outer epidermal wall below the penetration peg the wall was modified around the site of attempted penetration by the hyphae. Lateral walls were modified chemically and swollen below the points of attempted penetration by the hyphae. These walls gave reactions for lignin with phloroglucin-HCl and with chlorine water and sodium sulphite. The modified wall was not penetrated by the fungus[109]. In a further study, lignin content and the enzymes involved in lignification were measured in leaf discs inoculated with *Helminthosporium arveni* and were floated on water and on solutions of the protein synthesis inhibitor, cycloheximide[112]. After 18 hours, inoculated discs on water had a high lignin content and higher activity of the enzymes PAL, hydroxycinnamate-CoA ligase and peroxidase than non-inoculated discs floating on water. When inoculated discs were floated on cycloheximide solution the lignin content was depressed, the enzyme activities associated with lignin biosynthesis were inhibited and the tissue was susceptible to fungal penetration. Thus, from these observations it would seem that lignin biosynthesis is important in governing the response of the leaf tissue to infecting fungi. Again, in studies of the infection of discs of susceptible and resistant varieties of potatoes by spores of *Phytophthora infestans*, increases in the level of PAL, chlorogenic acid, and lignin were observed[40]. It was observed that lignification in the resistant strain took place more rapidly than in the non-resistant one. Similar observations were made by Asada *et al.*[5] in studies of infection in radish roots.

LIGNIFICATION IN RELATION TO THE EVOLUTION OF VASCULAR PLANTS

In his book *The Life of Plants*, Corner[17] considered that there are four main 'inventions' required for the evolution of a land flora: the development of a cuticle; an intercellular space system; lignification, and the seed. In the existing flora lignin is known to occur only in the Tracheophyta although as pointed out above, analogous compounds have been found in some mosses. This observation seems to form the basis of most ideas advanced as to what part the development of the lignification process has had in the evolution of the land flora. Neish[87] and Freudenberg and Neish[34] have discussed the formation of lignin against the background of the general biological function of secondary products proposed by Frankael[32]. According to this view lignin is regarded as having been formed as a product of detoxification for excess quantities of phenols produced by the tracheary cells. If initially a mutation or series of mutations occurred leading to the formation of phenolic substances and eventually lignin, and the accumulation of such substances conferred upon the plant some survival advantage, then it could be envisaged that this would lead to the development of selection pressure in favour of those plants with lignified tissues. Another suggestion is that the accumulation of phenolic substances in plants might initially have been related to defence of the organism against pathogens[98]. If as in the hypothesis of Neish[87] the soluble phenolic substances were effective in defence of the organism, it could then be postulated that polymers of these substances (lignin) might well accumulate in sufficient quantities to have a structural role. As pointed out by Raven[97] it would thus be possible to account for the evolution of lignin without having to postulate the unregulated synthesis of phenolic components in the cell wall. That the polymerization of phenolic substances to form lignin is genetically and cytoplasmically regulated is supported by the evidence of Kuc and Nelson[69] and that reviewed by Grisebach[47], and on the basis of evidence previously described that precursors of lignin arise within the lignifying cell. It has been pointed out by Raven that taking into account the structure of what are considered to be the most primitive land plants and the distribution of lignin in the cells of existing land flora that the initial function of lignin was probably not so much that of mechanical support but that it has had a role in those cells involved in water transport. Thus, it is well known that the water of the transpiration stream is in a state of tension and this stress could be translated as a compressive stress to the walls of the conducting xylem. As pointed out above the presence of lignin is thought to increase the resistance of the cell wall to compressive forces. It may be noted further that in plants with vessels, although the walls may contain layers of differing orientation of cellulose fibrils, the predominant orientation is transverse to the axis of the vessel element and this too would enhance resistance of the vessel elements to compression. That lignification would help to resist the compressive forces applied to the cell wall has been discussed in a theoretical discussion on water stress in cotton by Molz and Klepper[82]. It is consistent with this view that in circumstances when unlignified vessel elements are formed they show signs of collapse. This can be seen in extreme instances of reaction xylem formation in angiosperms and in the vessels of the so-called ‘rubbery wood’ of apples which appears to result from infection of the plant by a mycoplasma and is especially common in the xylem of the variety Lord-Lambourne[88]. The branches of infected trees do not display the rigidity of those of normal trees. This observation would seem consistent with the role of lignin contributing to the compressive strength of lignified cells referred to previously[37]. The anatomical characteristics of this tissue are very similar to those of angiosperm reaction xylem, namely the fibres are unlignified, or only slightly so, the vessels are fewer in number and lack lignification. It may be noted further that in 'rubbery wood' it has been shown that the mycoplasma appears to interfere with the methylation pattern of lignin and that monomers with an abnormal number of unmethylated phenolic groups are present. It has been suggested that the lignin is less firmly bound to the polysaccharide framework than in normal wood[107].

It may be noted that in the most primitive land plants such as *Rhynia* and *Asteroxylon* the stele occupied a relatively small proportion of the cross section of the axis. The xylem tracheids had annular thickenings (*Rhynia*) or were predominantly helically thickened (*Asteroxylon*). The pro-gymnospermous plants of the Middle Devonian period showed cambial activity such as that observed in present-day conifers and angiosperms[1]. Extensive secondary xylem was present with tracheids and parenchyma and mechanical fibres were present in the phloem and cortex. In these plants the tracheids of the xylem functioned to give mechanical support to the aerial leaf system as well as providing pathways for the conduction of water. With the development of angiosperms the functions of water conduction and mechanical support were separated in the vessels and xylem fibres. In examining fossil remains the occurrence of tracheary elements is usually taken as an indication that these were lignified but this cannot always be shown to be true.

CONCLUSION

Lignification is part of the cytodifferentiation sequence of many plant cells, especially those involved in the functions of transpiration and mechanical support. The process of lignification appears to confer properties on the cell wall which may be regarded as an adaptation in resistance to compressive stress. Considerations of this type are of interest in studies of the evolution of the Tracheophyta.

The site of lignin deposition in cell walls and between them determines the properties of many plant products but little progress has been made in controlling the amount of lignin formed or the site of its deposition. However, because of its scientific interest and economic importance the process of lignification is one which should continue to attract investigation.

REFERENCES

[1] Adler, E., Bjorkquist, K. J., Haggroth, S. Über die Ursache der Farbreaktionen des Holzes. *Acta Chem. Scanda.* **2,** 93-94 (1948)

[2] Agulhon, H. Emploi du bore comme engrais catalytique. C.R. *Hebdomadairs Séances de l'Acad. Sci.* **150,** 288-291 (1910)

[3] Albersheim, P. Concerning the structure and biosynthesis of the primary walls of plants. *Int. Rev. Biochemistry*. Biochemistry of Carbohydrates **II 16,** 127-150 (1978)

[4] Alexandrov, W. G., Djaparidze, L. I. Uber das Entholzen und Verholzen der Zellhaut. *Z. Wiss. Biol. Abt. E. Planta* **4,** 467-475 (1927)

[5] Asada, Y., Oguchi, T., Matzumoto, I. *Rev. Plant Protect. Res.* **8,** 104 (1975) (—quoted by Grisebach (1977))

[6] Bailey, A. J. Lignin in Douglas Fir—composition of the middle lamella. *Ind. Eng. Chem.* **8,** 52-55 (1936)

[7] Barghoorn, E. S. Evolution of cambium in geologic time. In: *The Formation of Wood in Forest Trees*, pp. 3-17, M. H. Zimmermann (ed.) (New York: Academic Press 1964)

[8] Barskaya, E. I. Ontogenetic changes in the viscosity of the protoplasm of certain herbaceous plants in relation to drought resistance. *Bot. Zhur.* **46,** 802-807 (1962)

[9] Bergmann, L. Der Einfluss von Kinetin auf die Ligninbildung und Differenzierung in Gewebekultusen von *Nicotiana tabaccum*. *Planta* **62,** 221-254 (1964)

[10] Berlyn, G. P., Mark, R. E. Lignin distribution in wood cell walls. *Forest Prod. J.* **15,** 140-141 (1965)

[11] Bland, D. E. Colorimetric and chemical identification of lignins in different parts of *Eucalyptus botryoides* in relation to lignification. *Holzforschung* **20,** 12-16 (1966)

[12] Brown, S. A. Lignin. *Ann. Rev. Plant Physiol.* **17,** 223-244 (1966)

[13] Catesson, A. M., Czaninski, Y., Monties, B. Caractères histochimiques des peroxydases pariétales dans les cellules en cours de lignification. *C.R. Acad. Sci. Paris*, C. 286, Ser. D. 1787-1790 (1978)

[14] Chang, H. M., Allan, G. G. Oxidation. In *Lignins: Occurrence, Formation, Structure and Reactions*, pp. 433-478, K. V. Sarkanen and C. H. Ludwig (eds.) (New York: Wiley Interscience 1971)

[15] Cheng, C. K. C., Marsh, H. V. Gibberellin acid-promoted lignification and phenyl-alanine ammonia-lyase activity in a dwarf pea (*Pisum sativum*). *Plant Physiol.* **43,** 1755-1759 (1968)

[16] Clarke, S. H., Pettifor, C. B. Influence of cell wall composition on the moisture relations of hardwood timbers. *Nature* **145,** 425 (1940)

[17] Corner, E. J. H. *The Life of Plants*. (New York: New American Library, Mentor Books 1968)

[18] Côté, W. A., Timell, T. E., Zabel, R. A. Studies on compression wood. Distribution of lignin in compression wood of red spruce (*Picea rubens*). *Holz. Roh. Werkst.* **24** (10), 432-438 (1966)

[19] Crocker, E. C. An experimental study of the significance of 'lignin' colour reactions. *J. Ind. Chem.* **13,** 625-627 (1921)

[20] Czaninski, Y., Catesson, A. Activités peroxydasiques d'origines diverses dans les cellules d'acer *Pseudoplatanus* (tissus conducteurs et cellules en culture). *J. Microscopie* **9** (8), 1089-1102 (1970)

[21] Davis, D. E. Some effects of calcium deficiency on the anatomy of *Pinus taeda*. *Amer. J. Bot.* **36,** 276-282 (1949)

[22] Dinwoodie, J. M. Timber—a review of the structure-mechanical property relationship. *J. Micr.* **104,** 3-32 (1975)

[23] Duchaigne, A. The diverse types of collenchyma in dicotyledons: their ontogeny and

their lignification. *Ann. Sci. Nat. Bot.* **16,** 455-479 (1955)
[24] Dutta, T. R., McIlrath, W. J. Effect of boron on growth and lignification in sunflower tissue and organ cultures. *Bot. Gaz.* **125,** 89-96 (1964)
[25] Eriksson, O., Goring, D. A. I., Lindgren, B. Structural studies in the chemical bonds between lignins and carbohydrates in spruce wood. *Wood Chem. Technol.* (in press)
[26] Eriksson, O., Lindgren, B. O. About the linkage between lignin and hemicelluloses in wood. *Svensk. Papperstidning* **2,** 59-63 (1977)
[27] Esau, K. *Plant Anatomy.* (New York: John Wiley 1953)
[28] Feder, N., O'Brien, T. P. Plant microtechnique: Some principles and new methods. *Amer. J. Bot.* **55,** 123-142 (1968)
[29] Fergus, B. J., Procter, A. R., Scott, J. A. N., Goring, D. A. I. The distribution of lignin in spruce wood as determined by ultraviolet microscopy. *Wood Sci. Technol.* **3,** 117-138 (1969)
[30] Fergus, B. J., Goring, D. A. I. The location of guaiacyl and syringyl lignins in birch xylem tissue. *Holzforschung* **24,** 113-117 (1970)
[31] Fergus, B. J., Goring, D. A. I. The distribution of lignin in birch wood as determined by ultraviolet microscopy. *Holzforschung* **24,** 118-124 (1970)
[32] Fraenkel, G. S. The Raison d'être of secondary plant substances. *Science* **129,** 1466-1470 (1959)
[33] Freudenberg, K. The formation of lignin in the tissue and in vitro. In: *The formation of Wood in Forest Trees*, M. H. Zimmermann (ed.) (New York: Academic Press 1964)
[34] Freudenberg, K., Neish, A. C. *Constitution and Biosynthesis of Lignin.* (Berlin: Springer-Verlag 1968)
[35] Freudenberg, K., Reznik, H., Fuchs, W., Reichardt, M. Untersuchung über die Entstehung des Lignins und des Holzes. *Naturwiss.* **42,** Part 2, 29-35 (1955)
[36] Freudenberg, K., Harkin, J. M., Reichardt, M., Fukuzumi, T. Enzymes participating in lignification—dehydrogenation of sinapin alcohol. *Chem. Ber.* **91,** 581-590 (1958)
[37] Frey-Wyssling, A. *Die Pflanzliche Zellwand.* (Berlin: Springer-Verlag 1959)
[38] Frey-Wyssling, A. *The Plant Cell Wall.* (Berlin: Gebruder Borntraeger 1976)
[39] Frey-Wyssling, A. Ultraviolet and fluorescence optics of lignified cell walls. In: *The Formation of Wood in Forest Trees*, 153-167, M. Z. Zimermann (ed.) (New York: Academic Press 1964)
[40] Friend, J., Reynolds, S. B., Avegand, M. A. Phenylalanine ammonia lyase, chlorogenic acid and lignin in potato tuber tissue inoculated with *Phytophthora infestans. Physiol. Plant Pathol.* **3,** 495-507 (1973)
[41] Fujita, M., Saiki, H., Harada, H. The secondary wall formation of compression wood tracheids. II. Cell wall thickening and lignification. *Mokuzai Gakkaishi* **24,** 158-163 (1978)
[42] Fujita, M., Saiki, H., Harada, H. The secondary wall formation of compression wood tracheids. III. Cell organelles in relation to cell wall thickening and lignification. *Mokuzai Gakkaishi* **24,** 353-361 (1978)
[43] Fujita, M., Harada, H. Autoradiographic investigations of cell wall development. II. Tritiated phenylalanine and ferulic acid assimilation in relation to lignification. *Mokuzai Gakkaishi* **25,** 89-94 (1979)
[44] Gibbs, R. D. The Mäule reaction, lignins and the relationships between woody plants. In: *The Physiology of Forest Trees* Ch. 13, pp. 269-312, K. V. Thimann (ed.) (New York: Ronald Press 1958)
[45] Gijzel, P. Polychromatic ultraviolet fluorescence microphotometry of fresh and fossil plant substances, with special reference to the location and identification of dispersed organic material in rocks. *Colloque International—Pétrographie de la*

Matière Organique des Sediments. Paris, C.N.R.S. pp. 67-91, 1973

[46] Griffoen, K. On the origin of lignin in the cell wall. *Recueil Trav. Bot. Néerl.* **35,** 323-380 (1938)

[47] Grisebach, H. Biochemistry of lignification. *Naturwiss.* **64,** 619-625 (1977)

[48] Gross, G. G. (1976)—(quoted by G. G. Gross, (1977))

[49] Gross, G. G. Biosynthesis of lignin and related monomers. *Rec. Adv. Phytochem.* **11,** 141-184 (1977)

[50] Gross, G. G. Recent advances in the chemistry and biochemistry of lignin. *Rec. Adv. Phytochem.* **12,** 177-220 (1978)

[51] Hägglund, I. *The Chemistry of Wood.* (New York: Academic Press 1951)

[52] Harada, H. Ultrastructure and organization of gymnosperm cell walls. In: *Cellular Ultrastructure of Woody Plants*, pp. 215-233, W. A. Côté (ed.) (Syracuse New York: Syracuse University Press 1965)

[53] Hardell, H-L., Leary, G. J., Stoll, M., Westermark, U. Variations in lignin structure in defined morphological parts of spruce. *Svensk papperstidning* **83,** 44-49 (1980)

[54] Hardell, H-L., Leary, G. J., Stoll, M., Westermark, U. Variations in lignin structure in defined morphological parts of birch. *Svensk papperstidning* **83,** 71-74 (1980)

[55] Harkin, J. M., Obst, J. R. Lignification in trees: indication of exclusive peroxidase participation. *Science* **180,** 296-297 (1973)

[56] Hartley, R. D. Carbohydrate esters of ferulic acid as components of cell walls of *Lolium multiflorum. Phytochem.* **12,** 661-665 (1973)

[57] Hartley, R. D., Jones, E. C., Wood, T. M. Carbohydrates and carbohydrate esters of ferulic acid released from cell walls of *Lolium multiflorum* by treatment with cellulolytic enzymes. *Phytochem.* **15,** 305-307 (1976)

[58] Hasegawa, M., Higuchi, T., Ishikawa, H. Formation of lignin in tissue culture of *Pinus strobus. Plant Cell Physiol.* **1,** 173-185 (1960)

[59] Hepler, P. K., Fosket, D. E., Newcomb, E. H. Lignification during secondary wall formation in *Coleus:* electron microscopic study. *Amer. J. Bot.* **57** (1), 85-96 (1970)

[60] Higuchi, T. Biochemical studies of lignin formation. *I. Physiol. Plantarum* **10,** 356-371 (1957)

[61] Imagawa, H., Fukazawa, K., Ishida, S. *Res. Bull. College Exp. For. Hokkaido Univ.* **33,** 127 (1976) (quoted by Fijita *et al.* (1978))

[62] Imamura, Y., Harada, H., Hiroshi, S. Electron microscopic study on the formation and organization of the cell wall in coniferous tracheids. *Bull. Kyoto Univ. Forests* **44,** 183-193 (1972)

[63] Jorgensen, L. R. Brown midrib in maize and its linkage relations. *J. Am. Soc. Agron.* **23,** 549-557 (1931)

[64] Kawamura, L., Higuchi, T. *Chim Biochim Lignine Cellulose Hemicelluloses.* Grenoble p. 469 (1964)

[65] Kennedy, R. W., Farrar, J. L. Induction of tension wood with the antiauxin 2-3-5 tri-iodobenzoic acid. *Nature* **208,** 406-407 (1965)

[66] Kerr, A. J., Goring, D. A. I. The ultrastructural arrangement of the wood cell wall. *Cellulose Chem. and Technol.* **9,** 563-573 (1975)

[67] Kirk, T. K. Effects of microorganisms on lignin. *Ann. Rev. Phytopath.* **9,** 185-210 (1971)

[68] Koblitz, H. Some observations on the biochemistry of the formation of the components of the cell wall in plant tissue cultured in vitro. *Int. Congr. Biochem.* **6,** 516 (1964)

[69] Kuc, J., Nelson, O. E. The abnormal lignins produced by the brown-midrib mutants of maize. *Arch. Biochem. Biophys.* **105,** 103-113 (1964)

[70] Kuo, J., O'Brien, T. P. Lignified sieve elements in the whole leaf. *Planta* **117,** 349-353 (1974)

[71] Lai, Y. Z., Sarkanen, K. V. Isolation and structural studies in lignins. In: *Lignins: Occurrence, Formation, Structure and Reactions,* 165-230. K. V. Sarkanen, C. H. Ludwig (eds.) (New York: Wiley-Interscience 1971)
[72] Latif, M. A., Saleh, T. M., Sarkanen, K. V. Synthesis of tritium labelled lignin precursors. *Holzforschung* **21,** 46-50 (1967)
[73] Lange, P. W. In *Fundamentals of paper making fibres*. Technical Section of the British Paper & Board Makers Association (Inc.) pp. 147-185 (1958)
[74] Lee, S., Aronoff, S. Boron in plants: a biochemical role. *Science* **19,** 798-799 (1967)
[75] Lewis, D. H. Boron, lignification and the origin of vascular plants—a unified hypothesis. *New Phytol.* **84,** 209-229 (1980)
[76] Lewis, D. H. Are there inter-relations between the metabolic role of boron, synthesis of phenolic phytoalexins and the germination of pollen? *New Phytol.* **84,** 261-270 (1980)
[77] Lipetz, J. Calcium and the control of lignification in tissue cultures. *Amer. J. Bot.* **49,** 460-464 (1962)
[78] Lipetz, J., Garrow, A. J. Ionic effects on lignification and peroxidase in tissue cultures. *J. Cell Biol.* **25,** 109-116 (1965)
[79] Luft, J. H. Ruthenium red and violet. I. Chemistry, purification, methods of use for electron microscopy and mechanisms of action. *Anat. Rec.* **171,** 347-368 (1971)
[80] Meeuse, A. D. J. Development and growth of the sclerenchyma fibres and some remarks on the development of the tracheids in some monocotyledons. *Rec. Trav. Bot. Néerl.* **35,** 288-321 (1938)
[81] Mische, G. E., Yasada, S. Zur Struktur des Lignins der Büche (*Fagus sylvatica*). *Liebig's Ann. Chem.* 1323-1332 (1976)
[82] Molz, F. J., Klepper, B. On the mechanism of water-stress-induced stem deformation. *Agron. J.* **65,** 304-306 (1973)
[83] Morey, K. P. R., Cronshaw, J. Developmental changes in the secondary xylem of *Acer nubrum* induced by gibberellin acid, various auxins, and 2-3-5 tri-iodobenzoic acid. *Protoplasma* **65,** 315-326 (1968)
[84] Mühlethaler, K. Electron micrographs of plant fibres. *Biochim. Biophys. Acta* **3,** 15-25 (1949)
[85] Nečesāńy, V. Effect of β-indole acetic acid on the formation of reaction wood. *Phyton* **11,** 117-127 (1958)
[86] Nečesāńy, V. Circadian physiological rhythm of secondary cell wall formation. *Drevársky Výskum* **2,** 55-69 (1974)
[87] Neish, A. C. Cinnamic acid derivatives as intermediates in the biosynthesis of lignin and related compounds. In: *The Formation of Wood in Forest Trees*, pp. 219-239, M. H. Zimmermann (ed.) (New York-London: Academic Press 1964)
[88] Nelmes, B. J., Preston, R. D., Ashworth, D. A possible function of microtubules suggested by their abnormal distribution in rubbery wood. *J. Cell Sci.* **13,** 741 (1973)
[89] Nimz, H. Beech lignin—proposal of a constitutional scheme. *Angew. Chem.* **13,** 313-321 (1974)
[90] O'Brien, T. P., Feder, N., McCully, M. E. Polychromatic staining of plant cell walls by toluidine blue O. *Protoplasma* **59,** 367-373 (1964)
[91] Parameswaran, N. Zur Wandstruktur von Sklereiden in einingen Baumrinden. *Protoplasma* **85,** 305-314 (1975)
[92] Parameswaran, N., Liese, W. On the fine structure of bamboo fibres. *Wood Sci. Technol.* **10,** 231-246 (1976)
[93] Phillips, E. W. J. Influence of leaf activity on the composition of the wood cell wall. *Nature* **174,** 85-86 (1954)
[94] Pickett-Heaps, J. D. Xylem wall deposition—radioautographic investigations using

lignin precursors. *Protoplasma* **65,** 181-205 (1968)
[95] Preston, R. D. *The Physical Biology of Plant Cell Walls.* (London: Chapman & Hall 1974)
[96] Preston, R. D., Middlebrook, M. Fine structure of sisal fibres. *J. Textile Inst.* **40,** 1715-1726 (1949)
[97] Raven, J. A. The evolution of vascular plants in relation to supracellular transport processes. In: *Advances in Botanical Research,* Vol. 5, pp. 153-219, H. W. Woolhouse (ed.) (London: Academic Press 1977)
[98] Ride, J. P. Lignification in wounded wheat leaves in response to fungi and its possible role in resistance. *Physiol. Plant Pathol.* **5,** 125-134 (1975)
[99] Ritter, G. J. Distribution of lignin in wood. *Ind. Eng. Chem.* **17,** 1194-1197 (1925)
[100] Roberts, L. W. *Cytodifferentiation in plants: Xylogenesis as a Model System.* (Cambridge: Cambridge University Press 1976)
[101] Rubery, P. H., Northcote, D. H. Site of phenylalanine-ammonia lyase activity and synthesis of lignin during xylem differentiation. *Nature* **29,** 1230-1234 (1968)
[102] Saleh, T. M., Leney, I., Sarkanen, K. V. Radiographic studies of cottonwood, Douglas fir, and wheat plants. *Holzforschung* **21,** 116-120 (1967)
[103] Sarkanen, K. V. Precursors and their polymerisation. In: *Lignins, Occurrence, Formation, Structure and Reactions,* pp. 95-163, K. V. Sarkanen and C. H. Ludwig (eds.) (New York: Wiley-Interscience 1971)
[104] Sarkanen, K. V., Hergert, H. L. Classification and distribution. In: *Lignins, Occurrence, Formation and Reactions,* pp. 43-89, K. V. Sarkanen and C. H. Ludwig (eds.) (New York: Wiley-Interscience 1976)
[105] Scott, J. A. N., Procter, A. R., Fergus, B. J., Goring, D. A. I. The application of ultraviolet microscopy to the distribution of lignin in wood. *Wood Sci. Technol.* **3,** 73-92 (1969)
[106] Scott, J. A. N., Goring, D. A. I. Lignin concentration in the S_3 layer of softwoods. *Cellulose Chem. Tech.* **4,** 83-93 (1970)
[107] Scurfield, G., Bland, D. E. The anatomy and chemistry of rubbery wood in apple variety Lord Lambourne. *J. Hort. Sci.* **38,** 297-306 (1963)
[108] Scurfield, G. A., Wardrop, A. B. The nature of reaction wood. VII. Lignification in reaction wood. *Aust. J. Bot.* **11,** 107-116 (1963)
[109] Sherwood, R. T., Vance, C. P. Histochemistry of papillae formed in reed canary grass leaves in response to non-infecting pathogenic fungi. *Phytopath.* **66,** 503-510 (1976)
[110] Siegel, S. M. *The Plant Cell Wall.* (Oxford: Pergamon Press 1962)
[111] Vance, C. P., Sherwood, R. T. Regulation of lignin formation in reed canary grass in relation to disease resistance. *Plant Physiol.* **57,** 915-919 (1976)
[112] Vance, C. P., Sherwood, R. T. Cycloheximide treatments implicate papillae formation in resistance of reed canary grass to fungi. *Phytopath.* **66,** 498-502 (1976)
[113] Vance, C. P., Anderson, J. D., Sherwood, R. T. Soluble and cell wall peroxidases in reed canary grass in relation to disease resistance and localized lignin formation. *Plant Physiol.* **87,** 920-922 (1976)
[114] Vennerloo, C. J. The lignin of *Populus nigra* L.C.V. (Italica). A comparative study of the lignified structures in tissue cultures and the tissue of the tree. *Acta Bot. Néerl.* **18,** 241-314 (1969)
[115] Verloop, A. Fate of the herbicide dichlobenzil in plants and soil in relation to its biological activity. *Residue Rev.* **43,** 55-103 (1972)
[116] Walker, J. R. L. *The Biology of Plant Phenolics.* (London: Edward Arnold 1975)
[117] Wardrop, A. B. The phase of lignification in the differentiation of wood fibres. *Tappi* **40,** 225-243 (1957)
[118] Wardrop, A. B. Morphological factors involved in the pulping and beating of wood fibres. *Svensk. Papperstidning* **66,** 1-17 (1963)

[119] Wardrop, A. B. Cellular differentiation in xylem. In: *Cellular Ultrastructure of Woody Plants,* pp. 61-97, W. A. Côté (ed.) (Syracuse, New York: Syracuse University Press 1965)

[120] Wardrop, A. B. The structure of the cell wall in lignified collenchyma of *Eryngium* sp. (Umbelliferae). *Aust. J. Bot.* **17,** 229-240 (1969)

[121] Wardrop, A. B. Lignification of the plant cell wall. *Appl. Polymer Symp.* **28,** 1041-1063 (1976)

[122] Wardrop, A. B., Dadswell, H. E. The nature of reaction wood. II. The cell wall organization of compression wood tracheids. *Aust. J. Sci. Res. B.* **3,** 1-13 (1950)

[123] Wardrop, A. B., Bland, D. E. The process of lignification in woody plants. IV. *Int. Congr. Biochem.* Vol. 11, *Biochemistry of Wood*, pp. 92-116 (1959)

[124] Wardrop, A. B., Scurfield, G. The nature of reaction wood. VI. Lignification in reaction wood. *Aust. J. Bot.* **11,** 107-116 (1963)

[125] Wershing, H. F., Bailey, I. W. Seedlings as experimental material in the study of 'redwood' in conifers. *J. Forestry* **40,** 411-414 (1942)

[126] Whitmore, F. W. Binding of ferulic acid to cell walls by peroxidases of *Pinus elliotii. Phytochemistry* **15,** 375-378 (1976)

[127] Wooding, F. B. P. Radioautographic and chemical studies of incorporation into sycamore vascular tissue walls. *J. Cell Sci.* **3,** 71-80 (1968)

5. The Role of the Cell Cycle and Cell Division in Xylem Differentiation

JOHN H DODDS
Department of Plant Biology, The University of Birmingham, Birmingham, England

In the construction of a model for gene control of differentiation many similar questions have been posed by workers studying cell differentiation in both animals[60] and plants[46]. In pursuit of answers to these questions both animal and plant biologists have looked for model systems in which to study the relationship between gene action and differentiation. The differentiation of xylem cells has been proposed for many years now as an excellent system for the study of plant cell differentiation[58, 59, 70, 72]. There are many reasons for this, for example, the differentiated state is terminal, the nucleus and cytoplasm being destroyed in the final stages of xylem cell maturation, thus ruling out the possibility of dedifferentiation once the cytodifferentiation sequence is initiated. Differentiating xylem cells are easily detected microscopically by phase contrast or polarised light optics because of their characteristic secondary wall pattern, and thus may be distinguished from the total cell population at a relatively early stage of differentiation[10]. It is likely that this system has features important to the study of differentiation *per se*.

In the developing root or shoot apex and in the vascular cambium, xylem cells differentiate in a distinct lineage and pattern so that it is possible to find a whole series of developmental stages by following along a cell line[1, 33, 34, 35]. Although the study of intact organs offers the advantage of cell lineage and pattern, control of cellular environment is difficult. A great deal of work has therefore been performed on tissue and cell culture systems which offer a degree of experimental manipulation not available at the whole plant or organ level. The differentiation of xylem cells is associated with regions of active cell division such as root and shoot apices, cambial zones and actively dividing tissue cultures. This association led to the hypothesis that cell division was a prerequisite for differentiation, and that gene reprogramming took place during passage of the nuclear material through the synthetic or mitotic phases of the cell cycle. In recent years new evidence that some cells can differentiate without DNA synthesis or cell division has cast doubt on this hypothesis. In this chapter I will discuss the evidence for and against a need for cell division as a prerequisite for gene reprogramming at the onset of cytodifferentiation, and indicate how a subdivision of the G_1 phase of the cell cycle may help us to rationalise the existing data.

EVIDENCE THAT DNA SYNTHESIS AND/OR CELL DIVISION ARE REQUIRED FOR CYTODIFFERENTIATION

Holtzer worked with a number of model systems of cell differentiation in animal development, including red blood cells[30], muscle[31, 67] and bone[32] and formulated a general model in which cell division was required for cell differentiation. This is now known as the 'quantal mitosis' model[31]. In this model cell cycles are designated as one of two types. Those which yield daughter cells with the same genetic programme as the original parent cell are termed proliferative cell cycles, whereas those producing daughter cells having different genetic programmes have been termed 'quantal' cell cycles. It can be hypothesised that a similar situation exists in xylem cells[62, 63] where differentiation follows a 'quantal' cell division in a suitable inducing environment.

There are many reports of correlation between cell division and differen-

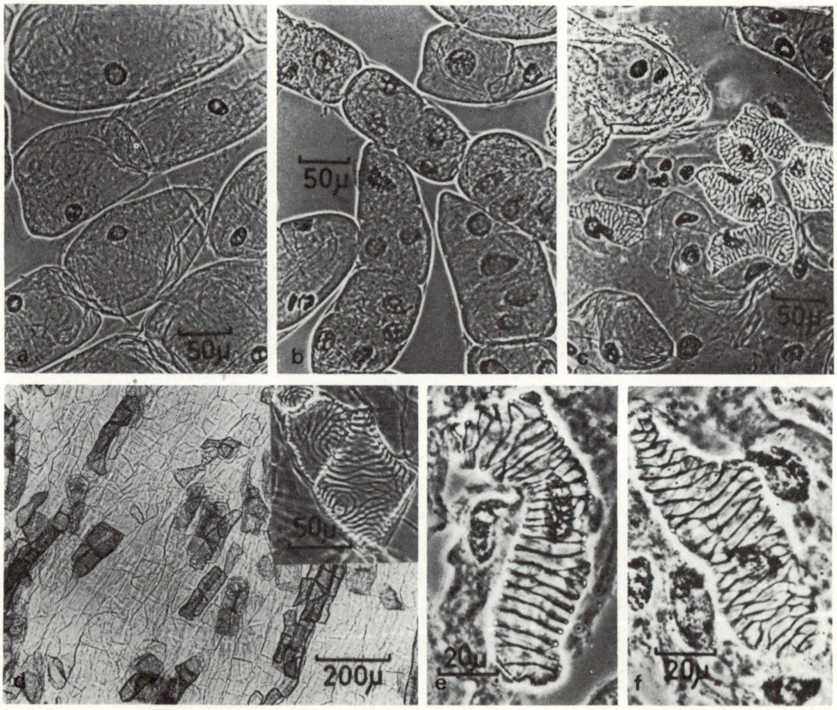

Figure 5.1: *Feulgen stained squash preparations from cultured explants of Jerusalem artichoke tissue.* (a) *Initial cell population prior to culture.* (b) *Cells after 24 hours of culture undergo rapid and partially synchronous cell division.* (c) *After 48 hours of culture young nucleate tracheary elements can be easily identified using phase contrast techniques.* (d) *Using a Safranin staining technique tracheary elements can be seen in a 'cleared' explant.* (e), (f) *Immature nucleate tracheary elements after 48 hours of culture detected by phase contrast optics.*

tiation. In intact plants, xylem differentiation is closely associated with areas of high mitotic activity such as apical meristems and cambial zones[36, 37]. When plants are wounded, xylem cell differentiation often follows the wave of rapid cell division induced at the surface of the wound. In plant tissue cultures most calli contain a significant population of tracheary elements, (word used to describe xylem-like cells formed in culture) and because of the ease with which the cellular environment may be manipulated tissue culture systems have been used widely in xylem cell/cell division studies[4, 5, 11, 12, 38, 39, 42, 44, 47, 66, 76].

If pea root cortical explants are cultured on a nutrient medium supplemented with cytokinins, the cells of the cortex are stimulated to divide and after 3–4 days of culture xylem cells are seen to have developed[3, 43, 56, 57, 71]. If cytokinin is omitted from the culture medium then cell division is prevented and xylem cells are not formed. Meins[46] used a mathematical modelling technique to analyse the data obtained by Phillips and Torrey[56] on xylem differentiation in the pea cortical explant system and concluded that cell division was the 'rate limiting' factor controlling xylem cell formation. Shininger[61], using the same system found that when explants were cultured on a medium containing a cytokinin together with the DNA synthesis inhibitor 5-fluorodeoxyuridine (FUdR) the explants did not undergo any DNA synthesis or cell division. No xylem cells were detected in these cultures even after 14 days. The use of FUdR as a DNA synthesis inhibitor for this type of study is widespread. Fosket[24, 25] showed that *Coleus* stem explants cultured in the presence of 10^{-5}M FUdR for 14 days would not differentiate xylem cells, while similar results were obtained with the Jerusalem artichoke tuber explant system studied by Phillips and Dodds[55]. It should always be kept in mind, however, that no inhibitor is specific, and the possibility exists that FUdR in addition to inhibiting DNA synthesis, is also having some other unknown inhibitory effect on xylem development.

Clear evidence supporting the hypothesis that cell division must precede differentiation comes from the studies by Malawer and Phillips[45] using the cultured Jerusalem artichoke tuber explant system. For a number of reasons this offers an excellent model system for studying relationships of cell division to differentiation. The cells at the beginning of the culture period consist of a homogeneous population of large uninucleate storage parenchyma cells (Fig. 5.1*a*), which are all diploids in G_1 of the cell cycle (see later). When transferred to xylogenic media the cells of the explant divide partially synchronously after about 24 hours[48, 77, 78] (Fig. 5.1*b*) and after 48 hours xylem cells can be detected easily by phase contrast microscopy (Fig. 5.1*c*) or by conventional clearing techniques (Fig. 5.1*d*)[54]. A great advantage of this system is that immature xylem cells (Figs. 5.1*e* and *f*) still containing an intact nucleus can be observed readily in cell squash preparations and can be used for histochemical or autoradiographic studies prior to destruction of the nucleus and cytoplasm during maturation.

Malawer and Phillips cultured the explants on a xylogenic medium for 72 hours in the continuous presence of tritiated (H^3) thymidine (a radioactive DNA precursor). Thus, any cell in the explant synthesising DNA incorporates radioactive thymidine into the nucleus, and this can be detected by autoradiography. It was found that when ^{3}H-thymidine was present for the whole 72 hour culture period *all* the differentiating xylem cells analysed by autoradiography were labelled (Fig. 5.2). It can therefore be concluded that all the cells differentiating as xylem cells in the Jerusalem artichoke explant system have undergone at least one round of DNA synthesis. Since all the differentiating xylem cells in the Jerusalem artichoke tuber system are uninucleate and diploid in G_1 of the cell cycle (see later), it follows that the differentiating xylem cells studied by Malawer and Phillips must have undergone not only synthesis of DNA but also mitosis and cytokinesis prior to differentiation.

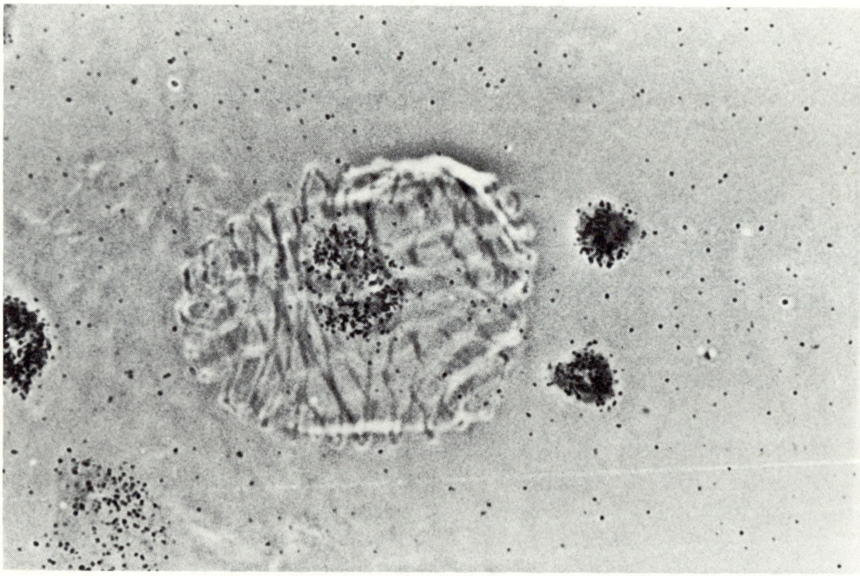

Figure 5.2: *Autoradiograph of immature tracheary element. Jerusalem artichoke explants were incubated from 0 to 48 hours in the presence of 1.0 μCi/ml ^{3}H thymidine. After Feulgen staining, squash preparations were exposed to Ilford K2 emulsion for 5 days. Labelling is seen over the nucleus of the tracheary element and two undifferentiated cells.* (Reproduced with permission from *Malawer and Phillips*, Plant Sci. Letters **15,** 47-55 (1979).)

EVIDENCE THAT CELLS CAN DIFFERENTIATE 'DIRECTLY' WITHOUT PRIOR DNA SYNTHESIS, MITOSIS OR CELL DIVISION

In recent years a number of reports have appeared claiming the occurrence without a prior period of DNA synthesis, mitosis or cell division of the direct transformation of parenchyma or mesophyll cells into xylem cells. I shall discuss each of these reports in detail, in some cases indicating possible alternative interpretations of these results.

The first report of xylem cell differentiation without mitosis and cell division was in the classical irradiation experiments of Haber and Foard[22, 23, 29]. They studied the effects of chronic doses of radiation, (up to 400 Krads) on the growth and development of wheat seedlings and showed that a number of differentiated cell types, including xylem cells, could be detected during subsequent development of the seedlings, without any evidence that either DNA synthesis or cell division had taken place. However, these results are open to the objection that the cells differentiating to give xylem cells were already committed or preprogrammed genetically to do so before the developing seedling was irradiated. Differentiation therefore merely required the completion of a pathway of events already programmed within the cell.

Siebers[64, 65] cultured isolated interfascicular cambium tissue from *Ricinus communis* in the presence of very high concentrations of IAA. (Indol-3yl-acetic acid). He similarly reported that the interfascicular cells differentiated into xylem without any preceding cell division.

As indicated earlier, 5-fluorodeoxyuridine (FUdR), an inhibitor of DNA synthesis, has been used by many workers to provide evidence that DNA synthesis is a requirement for differentiation. Turgeon[73] however, has used FUdR to support the hypothesis that DNA synthesis is not a requirement for reprogramming of the genome prior to xylem cell differentiation. Lettuce pith explants were cultured on a xylem induction medium containing FUdR at 10^{-5}M. After 14 days xylem cells could be detected. Following the use of Feulgen microdensitometry and visual observations, Turgeon concluded that DNA synthesis and cell division were not necessary for differentiation. However, as pointed out recently by Shininger[63], these conclusions may be premature, since it was not shown rigorously that DNA synthesis was prevented. It is not adequate to demonstrate a lack of increase in cell numbers, or to measure DNA microspectrophotometrically in only 0.5% of the cells (when in fact 6% differentiate) to be able to conclude that DNA synthesis is totally prevented. Turgeon's results contradict all previous experiments using 10^{-5}M FUdR. Fosket[23, 24], using Coleus stem explants, Shininger[61], using pea root explants, and Phillips[53] using Jerusalem artichoke tuber explants, were all unable to detect any differentiating xylem cells using this concentration of FUdR in their culture systems. Obviously, it would be desirable to confirm independently Turgeon's results as this study offers a useful example of the separation of the differentiation phase from the DNA synthesis phase.

Another example of so-called 'direct' differentiation has been described by Torrey[69], who isolated cells from a suspension culture of *Centaurea cyanus* L. and plated them out on a xylogenic medium solidified with 1% agar. The development of a large number of marked individual cells was then followed photographically over a 28 day culture period. In an extremely small number of cases, single cells were seen to differentiate into xylem cells without any preceding mitosis or cell division. As Torrey is careful to point out, the previous history of the plated-out cells is uncertain as they were derived from a cell suspension culture which had previously produced many small clumps of cells containing a significant proportion of xylem cells. It is possible therefore, that the small number of 'directly' differentiating cells observed may have been genetically committed to differentiation before being plated out. Torrey did not make any measurements of the DNA content in these embedded cells during the culture period, so the possibility that DNA synthesis has occurred in these cells prior to differentiation cannot be ruled out. Once again, as with the work of Foard and Haber[23, 29] the previous history of the differentiating cell remains uncertain.

Direct transformation of isolated mesophyll cells from mature leaves of *Zinnia elegans* into xylem cells was reported by Kohlenbach and Schmidt[40]. When cultured in a suitable hormonal environment a high proportion of these isolated cells are induced to differentiate into xylem cells without either mitosis or cell division. Further study of the *Zinnia elegans* isolated cell culture system has been made by Fukuda and Komamine[27, 28]. They followed the procedures used previously by Torrey, and traced photographically the development of a number of individual cells during the culture period. Figure 5.3 shows a series of time lapse photographs of a single field of view. The cell which is arrowed differentiates directly from a mesophyll cell to a xylem cell without a preceding mitosis. This type of cell was designated as a 'Type 1' tracheary element and was the most predominant type, forming over 60% of the total. 'Type 2' tracheary elements were those formed after one or more rounds of cell division in culture. As further evidence that Type 1 xylem cells do not undergo mitosis prior to differentiation, Komamine treated cultures of isolated *Zinnia* cells with 10^{-5}M colchicine, an alkaloid which destroys microtubules and inhibits mitosis. After varying times, up to 96 hours, in culture, tracheary elements were counted and identified as either Type 1 or Type 2. The results of an experiment using increasing concentrations of colchicine is shown in Fig. 5.4. The continuous presence of

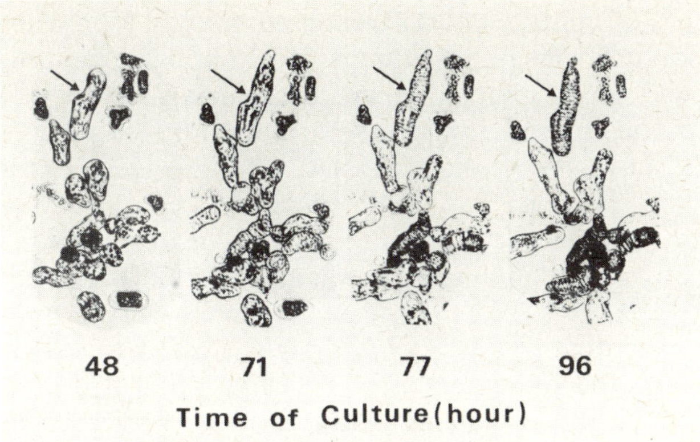

Figure 5.3: *A serial observation of the process of tracheary element differentiation from single cells isolated from the mesophyll of* Zinnia elegans. *Photographs were taken after 48 hr, 71 hr, 77 hr and 96 hr of culture.* (Photograph kindly supplied by Professor A. Komamine.)

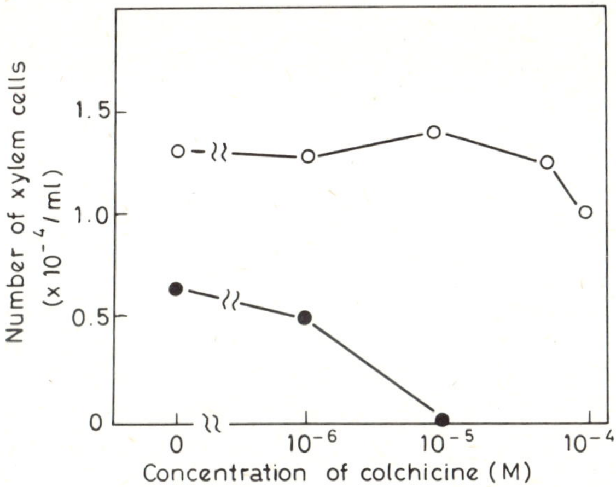

Figure 5.4: *Effect of colchicine on the formation of Type 1* (○) *and Type 2* (●) *tracheary elements. Isolated mesophyll cells of* Zinnia elegans *were cultured in medium containing various concentrations of colchicine for 4 days. Number of tracheary elements formed at concentrations above* 10^{-5}*M colchicine includes both normal and abnormal types of tracheary elements.* (Data kindly supplied by Professor A. Komamine.)

colchicine at concentrations up to 10^{-4}M did not reduce the number of Type 1 tracheary elements found, although the formation of Type 2 tracheary elements was completely inhibited at a colchicine concentration of 10^{-5}M. Unlike the *Centaurea* suspension cells studied by Torrey the *Zinnia* cell system readily forms 'direct' differentiation xylem cells. In the system studied by Komamine, 60% of the total tracheary elements formed after 96 hours of culture had differentiated directly from either palisade or spongy mesophyll cells, representing some 15,000 Type 1 cells/ml.

Fukuda and Komamine[41] are currently using the techniques of autoradiography and microdensitometry to determine whether these differentiating cells have synthesised DNA, though clearly they have not undergone cell division. The *Zinnia* isolated cell system deserves much further study. One question which remains to be answered is whether these Type 1 directly differentiating cells will still form in the presence of 10^{-5}M FUdR.

With the exception of this *Zinnia* system it is fair to say that the bulk of reports support the hypothesis that cell division is a prerequisite for cell differentiation. In those systems where division must precede the onset of differentiation it is of interest to consider whether a specific point in the cell cycle is related to initiation of differentiation as is the case in some animal systems.

RELATIONSHIP OF DIFFERENTIATION TO A SPECIFIC POINT IN THE CELL CYCLE

In dividing cells, nuclei are found to exhibit a cyclic behaviour. The two key events in the cycle are mitosis and the synthesis of DNA in preparation for the following mitosis. These two key events are separated by time gaps. The gap prior to mitosis (M) is termed G_2, and the gap before synthesis of DNA (S) is termed G_1. The term cell cycle represents the sequence between one of these events and the same event in the following generation. Thus, the complete cycle consists of G_1, S, G_2, M[8, 49, 68].

In mammals, compounds which promote arrest in either G_1 or G_2 of the cell cycle are produced by cycling cells. These compounds have been termed 'chalones'[7, 26]. In mammalian systems, differentiation is associated only with the G_1 phase, as, for example, in muscle, bone and red blood cell formation.

In many plant tissues, cells often replicate their DNA without undergoing mitotis, thus increasing the DNA content of the nucleus[13, 14, 15]; a process known as polyploidisation. This makes cell cycle analysis in plant systems much more complex than in animal systems. For example, a 4c nucleus (i.e. one possessing 4 × the haploid complement of DNA) could be either a diploid nucleus in G_2 or a tetraploid nucleus in G_1. It can easily be seen, therefore, that plant tissues which remain strictly diploid during growth and development are the most convenient for studying relationships of the cell cycle to differentiation. It must be kept in mind, however, that these strictly diploid plants (which are in the minority) may not be typical of all plants. A number of plant tissues have been shown to remain diploid during growth and development, and in culture, for example; *Lactuca sativa*[6], *Helianthus annuus*[19, 20] and *Helianthus tuberosus*[52, 16, 17]. The dormant shoot apex of *Fraxinus* is also diploid[9]. In a tissue where polyploidy does not occur, it is possible by the use of DNA specific Feulgen staining and scanning microdensitometry to determine whether xylem cells are differentiating from G_1 or G_2 of the cell cycle. Using the Jerusalem artichoke (*Helianthus tuberosus*) tuber explant system described earlier, Dodds and Phillips[16] examined the DNA levels of cells from the initial tuber explant, and of actively dividing cultured cells and immature differentiating tracheary elements. The results are summarised in Fig. 5.5. The storage parenchyma cells prior to culture were exclusively diploids in G_1 of the cell cycle (Fig. 5.5*b*). When cultured, the cells began to proliferate, and two peaks corresponding to 2c and 4c DNA levels are found.

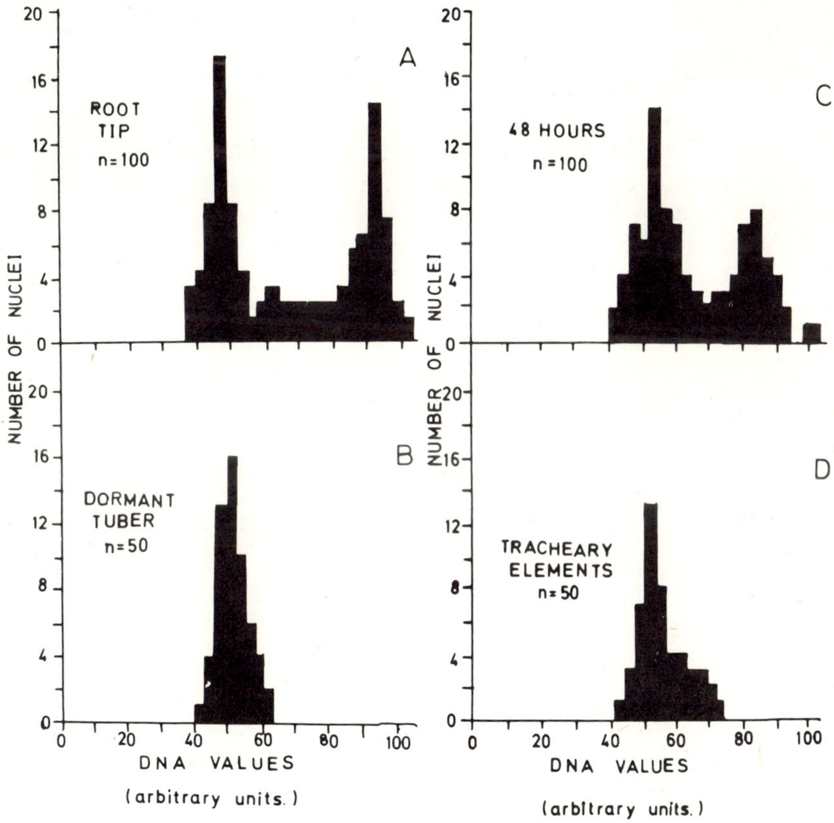

Figure 5.5: *Histograms of the relative amounts of DNA per interphase nucleus in squash preparations of cultured Jerusalem artichoke tuber tissue measured by Feulgen stain microdensitometry.* (A) *Root tips,* (B) *Tissue freshly excised from dormant tubers,* (C) *Undifferentiated cells after 48 hours of culture,* (D) *Immature tracheary elements after 72 hours of culture. n = number of nuclei measured.* (Reproduced with permission from *Dodds and Phillips*, Planta **135,** 206-217 (1977).)

Levels of DNA higher than 4c were not observed, confirming the conclusions of other workers that this system is strictly diploid (Fig. 5.5*c*). The immature xylem cells were exclusively diploids in G_1 (Fig. 5.5*d*). Malawer and Phillips[45] showed that *all* differentiating xylem cells in this system synthesise DNA before differentiating. It follows, therefore, that if all the immature xylem cells are in G_1 then these cells must have undergone at least one round of the cell cycle.

The differentiation of cells from G_1 of the cell cycle, and the significance of this to gene control mechanisms, has been studied in a number of animal cell differentiation systems[30, 32, 51]. Of special interest are the studies of Vonderhaar and Topper[74, 75] on the terminal differentiation of mouse mammary gland *in vitro*. Their model offers a novel interpretation of xylem cell differentiation data obtained to date and may be used to rationalise the conflict which is apparent in existing hypotheses.

Vanderhaar and Topper hypothesise the cell cycle model shown in Fig. 5.6*a*. In their system, G_1 is subdivided into two phases; cell arrest in G_1 can be at either subsection of G_1 depending on developmental age and physiological status of the tissue. Cells arrested in G_1 'primiparous' (G_1 'early') of the cell cycle when given a suitable hormone stimulus can produce milk proteins without going through S, G_2, M because they must merely pass through the 'critical events' period of G_1. If these primiparous cells are treated with FUdR they still produce milk proteins, as the effect of FUdR is to block the cells at the S phase. Cells arrested at the 'mature virgin' stage of G_1 must go through S, G_2 M and the 'critical events' of G_1 before they can produce milk proteins and these cells are inhibited by FUdR.

In Fig. 5.6*b* a similar cell cycle model is proposed for xylem cell differentiation. Cells which have been shown to differentiate directly into xylem cells, for example, *Zinnia* mesophyll cells, may be blocked at G_1 'early arrest' of the cell cycle. These cells are able to go through the 'critical event(s)' section of G_1 and differentiate without having any necessity to go through the S phase. Those cells which must go through the cell cycle to differentiate, such as artichoke tuber cells, pea cortical cells, etc., must be arrested at G_1 'late arrest', and thus need to pass through S, G_2 M and 'critical events' in G_1 before differentiating. It is possible that tissues contain a mixed population of cells arrested in both 'early arrest' and 'late arrest' G_1, thus some cells would differentiate directly while others in the same culture system need to divide. These may be represented by the Type 1 and Type 2 xylem cells described in the *Zinnia* mesophyll cell system. If this hypothesis follows the same basis as the mammary gland model then *Zinnia* cells should still differentiate Type 1 xylem cells in the presence of 10^{-5}M FUdR.

What is now required is to identify a marker for the 'critical event(s)'

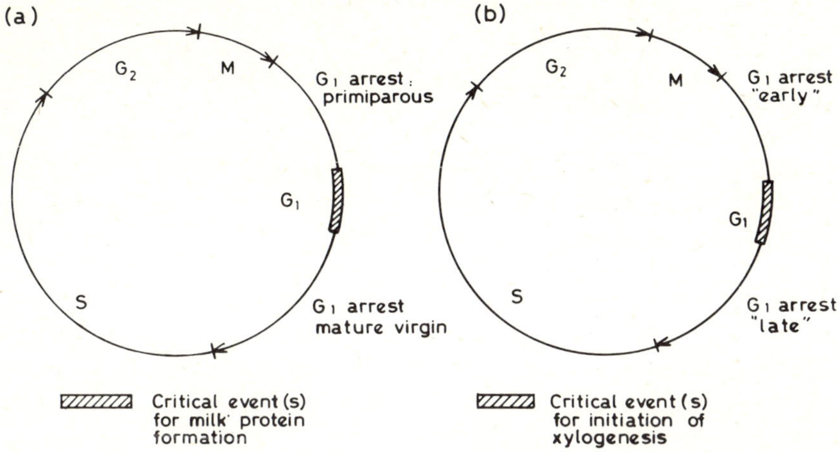

Figure 5.6: *Cell cycle models for* (a) *Terminal differentiation of mammary epithelia cells,* (b) *Xylem cell differentiation. G_1 (Gap 1) G_2 (Gap 2) S (Synthesis) M (Mitosis).*

period in xylogenesis similar to the milk protein marker found in the mammary gland system. The significance of systems which arrest in G_2 is also of interest[18, 21].

GENE AMPLIFICATION AND ITS POSSIBLE ROLE IN CYTODIFFERENTIATION

Amplification of a particular area(s) of the genome, resulting in the production of large numbers of specific genes, has been shown to take place during some developmental processes, and has been suggested as playing a possible role in plant cell differentiation. Amplification of genes results in an increase in the DNA status of the nucleus although this is not in multiples of 2c as in polyploidisation. Innocenti and Avanzi[35] and Avanzi *et al.*[1], have shown that during the differentiation of metaxylem elements in *Allium* roots, gene amplification of the N.O.R. (Nucleolar-organising region) takes place. These genes contain the cistrons for coding rRNA. Nägl and coworkers, experimenting with *Cymbidum protoconns*, have shown that exogenously applied plant growth hormones are capable of inhibiting or enhancing DNA amplification[50]. Many of the differentiation systems studied do not show increases in DNA content when analysed by Feulgen microdensitometry. In view of the limitations of the technique as outline by Bennett and Smith[2], however, it is unlikely that gene amplification can be accurately measured in this way. More detailed biochemical techniques such as DNA hybridisation studies are required. Thus, to date, no clear link exists between xylem cell differentiation and specific gene amplification.

'METABOLIC' DNA

The presence of 'metabolic DNA', that is, DNA subject to rapid turnover, has been reported in the differentiation of *Vicia* roots. The suggestion has been made by a number of animal cell biologists that a special type of DNA may be required for differentiation, and that this DNA is not necessarily produced during the normal 'S' phase of DNA synthesis. At present insufficient is known about the formation and action of metabolic DNA to decide what possible role it may play in xylem cell differentiation.

GENERAL CONCLUSIONS

It can be seen from above that the role of the cell cycle and cell division in relation to xylem cell differentiation is still a very open and controversial problem. A number of useful experimental systems now exist, in particular, the Jerusalem artichoke tuber explant system (where DNA synthesis is required before differentiation) and the *Zinnia* mesophyll cell system (where DNA synthesis is not necessary for differentiation).

We are still ignorant of how hormonal/environmental factors influence the cell cycle and how these stimuli may bring about a reprogramming of the genome to give rise to xylem cells, rays, etc. This is a feature which deserves much more study since the control of cell type formation is critical in the genetic control of 'wood quality'; a feature of xylem cell formation which has enormous economic implications.

REFERENCES

[1] Avanzi, S., Maggini, F., Innocenti, A. M. Amplification of ribosomal cistrons during the maturation of metaxylem in the root of *Allium cepa*. *Protoplasma* **76,** 197-210 (1973)

[2] Bennett, M. D., Smith, J. B. Nuclear DNA amounts in angiosperms. *Phil. Trans. Roy. Soc. Lond.* **(B) 274,** 227-274 (1976)

[3] Bogers, R. J., Hooymans-Klappe, H. T. M., Libbenga, K. R. The first cell cycle in explants from the mature root cortex of 7 day old *Pisum sativum*. *Plant Sci. Letters* **6,** 43-48 (1976)

[4] Bornman, C. H., Ellis, R. P. Differentiation of pseudothallial tracheary elements in *Nicotiana tabacum* tissue cultured *in vitro*. *J. S. Afr. Bot.* **37,** 281-289 (1971)

[5] Bornman, C. H. Cytodifferentiation in tissue culture. In *Tissue Culture and Plant Science* pp. 43-70. H. E. Street (ed.) (London: Academic Press 1974)

[6] Brunori, A. Synthesis of DNA and mitosis in relation to cell differentiation in the roots of *Vicia faba* and *Lactuca sativa*. *Caryologia* **24,** 209-215 (1971)

[7] Bullough, W. S. Epidermal chalone mechanism. In *Proceedings 1st International Chalone Conference*. B. Forscher (ed.) National Cancer Institute, Bethesda (1972)

[8] Cooper, S. A unifying model for the G_1 period in prokaryotes and eukaryotes. *Nature* **280,** 17-19 (1979)

[9] Cottignies, A. The blockage in the G_1 phase of the cell cycle in the dormant shoot apex of ash. *Planta* **147,** 15-19 (1979)

[10] Cronshaw, J. Tracheid differentiation in tobacco pith cultures. *Planta* **72,** 78-90 (1967)

11 Dalessandro, G. Hormonal control of xylogenesis in pith parenchyma explants of *Lactuca. Ann. Bot.* **37,** 375-382 (1973)

12 Dalessandro, G., Roberts, L. W. Induction of xylogenesis in pith parenchyma explants of *Lactuca. Amer. J. Bot.* **58,** 378-385 (1971)

13 D'amato, F. Polyploidy in the differentiation and function of tissues and cells in plants: A critical examination of the literature. *Caryologia* **4,** 311-358 (1952)

14 D'amato, F. Endopolyploidy as a factor in plant tissue development. *Caryologia* **17,** 41-52 (1964)

15 D'amato, F. Nuclear changes and their relationship to histological differentiation. *Caryologia* **17,** 317-325 (1965)

16 Dodds, J. H., Phillips, R. DNA and histone content of immature tracheary elements from cultured artichoke explants. *Planta* **135,** 213-216 (1977)

17 Dodds, J. H. Is cell cycle activity necessary for xylem cell differentiation? *What's New. Plant Physiol.* **10** (4) (1979)

18 Evans, L. S., Romeo, D. P. Promotion of cell arrest in G_2 during cell maturation in roots of *Pisum sativum* by exogenously applied cyclic AMP. *Amer. J. Bot.* **64,** 1170-1179 (1977)

19 Evans, L. S., Van't Hof, J. Dose rate, mitotic cycle duration, and sensitivity of cell transitions from $G_1 \rightarrow S$ and $G_2 \rightarrow M$ to protracted gamma radiation in root meristems. *Rad. Res.* **64,** 331-343 (1975)

20 Evans, L. S., Van't Hof, J. Is polyploidy necessary for tissue differentiation in higher plants? *Amer. J. Bot.* **62,** 1060-1064 (1975)

21 Evans, L. S., Verville, K. M. Is the predominant period of cell arrest and presence of endoreduplicated cells coincident with production of secondary vascular tissue in intact and cultured roots of *Raphanus sativus* L? *Amer. J. Bot.* **65,** 70-74 (1978)

22 Foard, D. E. Differentiation in plant cells. In *Cell differentiation* pp. 575-602. O. A. Schjeide, J. De Villis (eds.) (New York: Van-Nostrand 1970)

23 Foard, D. E., Haber, A. H. Anatomical studies of gamma irradiated wheat growing without cell division. *Amer. J. Bot.* **48,** 438-446 (1961)

24 Fosket, D. E. Cell division and differentiation of wound vessel members in-cultured stem segments of *Coleus. Proc. Nat. Acad. Sci.* (USA) **59,** 1089-1096 (1968)

25 Fosket, D. E. The time course of differentiation and its relation to DNA synthesis in cultured *Coleus* stem segments. *Plant Physiol.* **46,** 64-68 (1970)

26 Frankfurt, O. S. Epidermal chalone, effect on cell cycle of development of hyperplasia. *Exper. Cell Res.* **64,** 140-144 (1971)

27 Fukuda, H., Komamine, A. Establishment of an experimental system for the study of tracheary element differentiation from single cells isolated from the mesophyll of *Zinnia elegans. Plant Physiol.* **65,** 57-60 (1980)

28 Fukuda, H., Komamine, A. Direct evidence for cytodifferentiation to tracheary elements without intervening mitosis in a culture of single cells isolated from the mesophyll of *Zinnia elegans. Plant Physiol.* **65,** 61-64 (1980)

29 Haber, A. H., Foard, D. E. Further studies on gamma irradiated wheat and their relevance to use as mitotic inhibitors for developmental studies. *Amer. J. Bot.* **51,** 151-159 (1964)

30 Holzer, H. Proliferative and quantal mitosis in differentiation. In *Control Mechanisms in the Expression of Cellular Phenotypes.* H. A. Padykula (ed.) (New York: Academic Press 1970)

31 Holzer, H., Abbott, J. Oscillations of chondrogenic phenotypes. In *The Stability of the Differentiated State.* H. Urspring (ed.) (Heidelberg: Springer-Verlag 1968)

32 Holzer, H., Bischoff, R. Mitosis and myogenesis. In *The Physiology and Biochemistry of Muscle as a Food.* E. J. Briskey, R. G. Cassens, B. B. Marsh (eds.) pp. 29-51 (Madison: University of Wisconsin Press 1970)

[33] Innocenti, A. M. Cytophotometric determination of histone content in cell nuclei of proliferating and non-proliferating root meristem cells of *Triticum durum*. *Caryologia* **24,** 457-461 (1971)

[34] Innocenti, A. M. Cyclic changes of histones/DNA ratio in differentiating nuclei of metaxylem cell line in *Allium cepa* root tip. *Caryologia* **28,** 223-228 (1975)

[35] Innocenti, A. M., Avanzi, S. Some cytological aspects of differentiation of metaxylem in the root of *Allium cepa*. *Caryologia* **24,** 283-291 (1971)

[36] Jacobs, W. P. The role of auxin in differentiation of xylem around a wound. *Amer. J. Bot.* **39,** 301-309 (1952)

[37] Jacobs, W. P. Acropetal auxin transport and xylem regeneration—a quantitative study. *Amer. Natr.* **88,** 327-337 (1954)

[38] Jeffs, R. A., Northcote, D. H. Experimental induction of vascular tissues in an undifferentiated plant callus. *Biochem. J.* **101,** 146-152 (1966)

[39] Jeffs, R. A., Northcote, D. H. The influence of indol-3yl-acetic acid and sugar on the pattern of induced differentiation in plant tissue culture. *J. Cell Sci.* **2,** 77-88 (1967)

[40] Kohlenbach, H. W., Schmidt, B. Cytodifferenzierung in Form einer direkten Umwandlung isolierter Mesophyllzellen zu Tracheiden. *Zt. Planzenphysiol.* **75,** 369-374 (1975)

[41] Komamine, A. Personal communication

[42] Landgren, C. T. Patterns of mitosis and differentiation in cells derived from pea root protoplasts. *Amer. J. Bot.* **63,** 473-480 (1976)

[43] Libbenga, K. R., Torrey, J. G. Hormone induced endoreduplication prior to mitosis in cultured pea root cortical cells. *Amer. J. Bot.* **60,** 293-299 (1973)

[44] List, A. Some observations on DNA content and cell nuclear volume growth in developing xylem cells of certain higher plants. *Amer. J. Bot.* **50,** 320-330 (1963)

[45] Malawer, C. L., Phillips, R. The cell cycle in relation to induced xylem differentiation. Tritiated thymidine incorporation in cultured tuber explants of *Helianthus tuberosus L. Plant Sci. Letters* **15,** 47-55 (1979)

[46] Meins, F. Cell division and the determination phase of cytodifferentiation in plants. In *Cell Division and Cell Differentiation*. J. Reinert, H. Holtzer (eds.) (New York: Springer-Verlag 1975)

[47] Minocha, S. C., Halperin, W. Hormones and metabolites which control tracheid differentiation with or without concomitant effects on growth in cultivated tuber tissues of *Helianthus tuberosus L. Planta* **116,** 319-331 (1974)

[48] Mitchell, J. P. DNA synthesis during the early division cycles of Jerusalem artichoke callus cultures. *Ann. Bot.* **31,** 427-435 (1967)

[49] Mitchison, J. M. *The Biology of the Cell Cycle*. (Cambridge: Cambridge University Press 1971)

[50] Nägl, W., Hendon, J., Rucker, W. DNA amplification in *Cymbidium* protocorms *in vitro*. *Cell Diff.* **1,** 229-237 (1972)

[51] Owens, J. N., Molder, M. A study of DNA and mitotic activity in the vegetative apex of Douglas Fir during the annual growth cycle. *Canad. J. Bot.* **51,** 1395-1409 (1973)

[52] Partenen, C. R. On the chromosomal basis for cellular differentiation. *Amer. J. Bot.* **52,** 204-209 (1965)

[53] Phillips, R. Cytodifferentiation. In *Recent Advances in Plant Tissue Culture*. I. K. Vasil, D. E. Murphy (eds.) (New York: Academic Press 1980)

[54] Phillips, R., Dodds, J. H. Rapid differentiation of tracheary elements in cultured explants of Jerusalem artichoke. *Planta* **135,** 207-213 (1977)

[55] Phillips, R., Dodds, J. H. Personal communication

[56] Phillips, R. Torrey, J. G. DNA synthesis, cell division and cytodifferentiation in cultured pea root cortical explants. *Dev. Biol.* **31,** 336-347 (1973)

[57] Phillips, R., Torrey, J. G. DNA levels in differentiating tracheary elements. *Dev. Biol.*

39, 322-325 (1974)
58 Roberts, L. W. The initiation of xylem cell differentiation. *Bot. Rev.* **35,** 201-237 (1969)
59 Roberts, L. W. *Cytodifferentiation in Plants, Xylogenesis as a Model System.* (Cambridge: Cambridge University Press 1976)
60 Rutter, W. J., Piclet, R. L., Morris, P. W. Toward molecular mechanisms of developmental processes. *Ann. Rev. Biochem.* **42,** 601-646 (1973)
61 Shininger, T. L. Is DNA synthesis required for the induction of differentiation in quiescent root cortical parenchyma? *Dev. Biol.* **45,** 137-150 (1975)
62 Shininger, T. L. Hormone regulation of development in plant cells. *In vitro* **14,** 31-50 (1978)
63 Shininger, T. L. The control of vascular development. *Ann. Rev. Plant Physiol.* **30,** 313-337 (1979)
64 Siebers, A. M. Differentiation of isolated interfascicular tissue of *Ricinus communis* L. *Acta Bot. Neerl.* **20,** 343-355 (1971)
65 Siebers, A. M. Vascular bundle differentiation and cambial development in cultured tissue blocks excised from the embryo of *Ricinus communis* L. *Acta. Bot. Neerl.* **21,** 327-342 (1972)
66 Simpson, S. F., Torrey, J. G. Hormonal control of deoxyribonucleic acid and protein syntheses in pea root cortical explants. *Plant Physiol.* **59,** 4-9 (1977)
67 Stockdale, F. E., Topper, J. The role of DNA synthesis and mitosis in hormone-dependent differentiation. *Proc. Nat. Acad. Sci.* **56,** 1283-1289 (1966)
68 Swift, H. The constancy of DNA in plant nuclei. *Proc. Nat. Acad. Sci.* (USA) **36,** 643-654 (1950)
69 Torrey, J. G. Tracheary element formation from single isolated cells in culture. *Physiol. Plant.* **35,** 158-165 (1975)
70 Torrey, J. G. Cytodifferentiation in cultured cells and tissues. *Hort. Sci.* **12,** 138-139 (1977)
71 Torry, J. G., Fosket, D. E. Cell division in relation to cytodifferentiation in cultured pea root segments. *Amer. J. Bot.* **57,** 1072-1080 (1970)
72 Torrey, J. G., Fosket, D. E., Hepler, P. K. Xylem formation: a paradigm of cytodifferentiation in higher plants. *Amer. Sci.* **59,** 338-352 (1971)
73 Turgeon, R. Differentiation of wound vessel members without DNA synthesis, mitosis or cell division. *Nature* **257,** 806-808 (1975)
74 Vonderhaar, B. K., Topper, Y. J. Critical cell proliferation as a prerequisite for differentiation of mammary epithelial cells. *Enzyme* **15,** 340-350 (1973)
75 Vonderhaar, B. K., Topper, Y. J. Role of the cell cycle in hormone-dependent differentiation. *J. Cell Biol.* **63,** 707-712 (1974)
76 Wright, K., Northcote, D. H. Induced root differentiation in sycamore callus. *J. Cell Sci.* **11,** 319-337 (1972)
77 Yeoman, M. M. Early changes in callus cultures. *Inter. Rev. Cytol.* **29,** 383-409 (1970)
78 Yeoman, M. M., Macleod, A. J. Tissue (callus) culture-techniques. In *Plant Tissue and Cell Culture* 2nd ed.

6. Biochemical Changes during Xylem Element Differentiation

P B GAHAN
Department of Biology, Queen Elizabeth College, London, England

INTRODUCTION

It is rare to find discussion of the biochemistry of xylogenesis[155], and when one does, it is related generally to the biochemistry of cell walls and the process of lignification (Chapter 4). Indeed, little mention of the biochemical changes occurring in the differentiating protoplast is given by Roberts[140] in his excellent monograph on xylogenesis. The main reason for this probably lies in the fact that xylem as a tissue has a number of component cell types[47] and is surrounded by other tissues, each with its own complement of cell types. Each cell will have a metabolic state which can differ from that of its neighbours so that any attempt at a gross biochemical measurement will merely yield a mean value for each cell. The concept of the 'average cell' will apply even to such measurements on the meristem, or terminal 0.5 mm, of the root, since this will include the quiescent centre and the root cap. Moreover, the meristem is comprised of a cell population in which individual cells have varying cycling times[4] and many show cytochemical differences depending upon the particular tissue for which they are the precursor[56, 111]. In consequence, this will not give a true indication of the events occurring in a protoplast during the change from a meristematic cell to a fully differentiated xylem element. The rapid development of quantitative cytochemistry has resulted in the possibility of trying to determine such changes in single cells, which may now be examined individually (Gahan, in press).

It is proposed to review the available data concerning the biochemical changes occurring during xylogenesis. In particular, those changes occurring in the protoplast during the formation of xylem elements from meristematic cells will be examined. These include the passage from meristematic to elongating cells, through to senescence and death of the protoplast. Most of the data to be presented relate to primary xylem formation rather than to secondary or wound xylem differentiation, about which there is a paucicity of information. Whilst trying to present a positive viewpoint on the progress being made, it will become apparent that the gaps in our knowledge far outweigh the known events.

Most studies have been performed on the root system since it is a geometrically simple organ containing progressively advanced stages in the

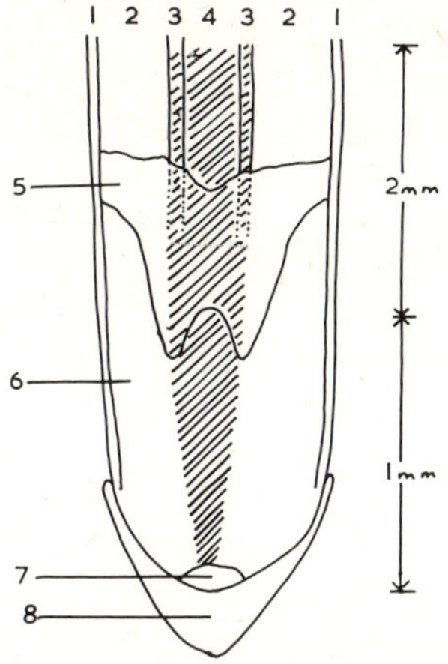

Figure 6.1: *Diagram of a longitudinal section through a 'typical' root tip to demonstrate the zones referred to in the text. (1) rhizodermis, (2) cortical parenchyma, (3) vessels, (4) pith, (5) zone of endoreduplication, (6) zone of mitosis, (7) quiescent centre, (8) root cap. Hatching delineates zone of esterase activity.*

differentiation of xylem elements. These are arranged linearly from the meristem to the mature region of the root. A number of root systems exist in which the diploid state is maintained throughout the root, although it is clear from many studies that endopolyploidisation can occur, raising the number of sets of chromosomes *n* to 4*n*, 8*n*, 16*n* or even higher in some species[11, 42, 113, 122]. The zone in the root in which endopolyploidization occurs appears on average to be between 1 mm and 1.5 mm from the junction of the root tip with the root cap (Tschermak-Woess[169] Fig. 6.1). Any study of the biochemical events during xylogenesis must therefore be interpreted in the light of polyploidization. The various meanings of endopolyploidization have been well discussed by Barlow[5] and Nagl[122], and one important point emerging, concerns the possible impact on cell metabolism of the extra-gene copies so produced. If all of the extra-copies are active in a cell, then this could have implications for the biochemical events within that cell[3, 120]. Thus, increases in protein and enzyme activity during differentiation may, in part, reflect the numbers of active gene copies[18, 148].

NUCLEIC ACIDS

(a) Deoxyribonucleic acid (DNA)

Biochemical studies on serial segments of roots of *Pisum sativum*[87], *Allium cepa* and *Vicia faba*[99] showed an increased DNA content for each 'average cell', occurring especially in the zone of endomitosis. Jensen's measurements showed the DNA content remaining constant for 2 mm in *Allium cepa* and 3 mm in *Vicia faba*, whilst there was a slight increase with distance from the tip in *Pisum sativum* over 12 mm.

More exact measurements have been made using three quantitative cytochemical approaches namely *(i)* an autoradiographic determination of the incorporation of radioactive DNA precursors, e.g. [H^3] thymidine; *(ii)* reaction with specific dyes, e.g. the Feulgen reaction[48]; and *(iii)* reaction with specific fluorochromes, e.g. acridine orange[139]. Application of these methods allows the ready identification of the quiescent zone in which nuclear incorporation of [^3H] thymidine does not normally occur[33], the nuclei remaining diploid and in G_0. The zone of mitosis becomes apparent through the appearance of tetraploid nuclei which have incorporated [^3H] thymidine in preparation for mitosis, when they return to the diploid state. On passage to the zone of endomitosis, nuclei of some species incorporate [^3H] thymidine into DNA, but increase their ploidy level as shown by Feulgen photometry[34, 42, 165] or by fluorescence microscopy[24]. Not all species however show such an increase, e.g. *Bryonia verrucosa*[15]. Where the increase occurs, it may reach a level of 32c or more. Eventually, the nucleus will be lost from the protoplast.

Whilst this approach has been useful in determining nuclear DNA behaviour, little attention has been paid to cytoplasmic DNA. DNA of cytoplasmic organelles labels readily with [^3H] thymidine in the quiescent centre (plates 23-5 in Clowes[33], Hurst[91]). As the cells move into a mitotic phase, so there is an approximate doubling of the level of incorporation of [^3H] thymidine into cytoplasmic DNA in the meristems of young roots when compared with the quiescent centre. However, the level of cytoplasmic DNA labelling does not appear to increase in the differentiating xylem elements in parallel with the increasing values of euploidy (Gahan, unpublished). It is not clear if the incorporation represents synthesis of new DNA for replication of cytoplasmic organelles or merely DNA repair.

Metabolic DNA[129] clearly exists in plants[22, 58, 92, 93, 130, 146]. That it exists in the nuclei of diploid cells which may differentiate to form xylem elements may be inferred from the studies of Pelc and La Cour[130] though this has not been definitely shown (Chapter 5). The difficulty of demonstrating its existence may be due in part to the limitations of the methods employed. Thus, it can be lost readily from nuclei during tissue preparation for Feulgen photometry[43, 106].

(b) Ribonucleic Acid (RNA)

Biochemical studies, again employing the serial segment method show increased levels of RNA with differentiation in root tissue. Jensen[99] found that the RNA content of the 'average cell' in *Vicia faba* was five times as high 2–3 mm from the tip as it was in the meristem while similar studies on *Allium cepa* showed a three-fold increase. Heyes[87] showed a three-fold increase across the first 3 mm of *Pisum sativum* root, the content increasing linearly for 9 mm from the tip up to about five times that of the meristem.

When Jensen re-examined his material autoradiographically through the incorporation of [^{14}C] adenine, he found an average four-fold greater labelling at 2 mm from the tip compared with that of the apical initials. Fluorescence studies of RNA stained with acridine orange in roots of *Pisum sativum* showed only about a doubling of RNA in the stele at 2 mm when compared to the apical initials[24]. The increase reached a peak at about 3 mm. More recently, the autoradiographic studies of Nägl[121] using [^{3}H] uridine in *Allium cepa* roots, indicated that RNA synthesis continued throughout endomitosis. Furthermore, in the developing xylem elements, [^{3}H] uridine incorporation doubled with a doubling of the nuclear DNA content.

Such studies have been made only on species in which xylogenesis is accompanied by endopolyloidization, and it would be of interest to know if this increase in RNA synthesis also accompanies differentiation of the xylem elements in cells which remain diploid. At present, it is difficult to know if the increased RNA synthesis is merely a reflection of the increased number of gene copies, or if it represents an important step for the synthesis of new and additional proteins for the differentiation process. It is likely that both factors are involved, since the studies by Innocenti and Avanzi[96], Avanzi *et al.*[3] and Durante *et al.*[44] indicate that there is an amplification of ribosomal cistrons during the differentiation of metaxylem in roots of *Allium cepa*. Furthermore, the DNA containing the ribosomal cistrons is present in larger amounts during the early stages of metaxylem differentiation (0.5–4.5 mm from the tip) than it is in the first 0.5 mm of root. However, these DNA preparations included DNA from all cells in the root and so it is difficult to reconcile the observed differences to changes concerning only the metaxylem. Also, no amplification of ribosomal genes in the absence of other gene amplification has been demonstrated so that the data may only represent the proportionate increase in the number of r – RNA genes accompanying eupolyploidisation such as occurs in mammalian liver[28, 119]. An increased amount of RNA could also be necessary for the production of new proteins such as those required for the formation of microtubules.

PROTEINS

(a) General studies

The serial segment method has indicated an increased protein content per 'average cell' with increasing distance from the root tip, eventually reaching a constant value. Such overall increases have been reported as eleven-fold[19] or three-fold[89] for *Pisum sativum*, two-fold for *Zea mays*[45] and two-fold for *Allium cepa*[99]. However, it has been shown that there is an increase in the volume of the 'average cell' which is very much greater. A recalculation of the data would indicate a decrease of protein per unit volume or per unit surface area of the cell during differentiation[82]. Hence, on this basis, the increased RNA content discussed above may not be parallelled by an increase in protein.

The application of newer biochemical methods, including disc gel-electrophoresis and chromotography, together with immunocytochemical methods, have failed to advance our understanding of the qualitative changes that occur during observed increases in general protein content during differentiation in general, and xylogenesis in particular[127].

Some quantitative measurements have been made of nuclear histone proteins. The cytochemical method using the Sakaguchi test for arginine in roots from a variety of plants showed a doubling of the histone content of the diploid nucleus at the same time as DNA doubling in preparation for mitosis[112]. Similar studies on DNA and histone content by microdensitometry of the Fast green FC method[1] for histones has indicated an increase with increasing ploidy values[34, 94, 95, 105]. The conclusion was that there was not a constant ratio of DNA to histone with increasing ploidy values during xylogenesis, and that possibly a change in composition occurred. One problem with such studies is that the nuclei measured by microdensitometry are large and yet sections as thin as 8 μm have been employed[105]. In consequence, it is difficult to be sure of the accuracy of the results, since they are based on measurements of sections of nuclei rather than whole ones. This could explain some of the observed variations.

If proteins are studied autoradiographically after the incorporation of radioactive amino acids, then alternative information can be obtained. Thus, Clowes[32] showed that in the stele of roots of *Zea mays*, there was a doubling of the incorporation of [^{14}C] leucine 1–2 mm from the tip as compared to the meristem, rising to 50 times as high at 3 mm, when estimated by grain counts on individual cells. Jensen[98] made similar studies with [^{14}C] phenylalanine on roots of *Allium cepa* and showed results parallelling those of Clowes[32] except that the changes were of a lower order. However, Jensen confined his studies only to the first 2 mm of root.

The autoradiographic data pose problems of interpretation in that as obtained, it is difficult to distinguish between incorporation for genuine protein synthesis, and incorporation due to amino acid exchange. Thus, while it would appear that protein synthesis rates increase with differentiation, it would be of interest to repeat these experiments using alternative systems of xylogenesis, and with specific protein synthesis inhibitors.

Although measurements may indicate an increase in protein content, they do not tell us if this is related to structural proteins or enzyme proteins or both. Nor do they tell us anything about protein turnover rates during xylogenesis. It becomes difficult to know therefore if the extra gene copies present in the differentiating xylem elements of a number of species are expressed in general. Studies such as that by Middleton and Gahan[120], but on developing tracheids, may help to resolve this point.

If studies have not been performed on protein turnover, this may be due in part to lack of suitable methodology[60]. Cytochemical studies on developing xylem elements have yielded frequent reports of increased activities of a number of enzyme proteins (see below) and this has given rise to the question as to whether this activity represents molecules synthesized in advance or newly synthesized molecules. If the latter, then are they synthesized prior to the loss of the nucleus from the xylem element, or is there enough long-lived m-RNA to allow this to occur after nuclear loss? A partial answer comes from a study on acid hydrolase activity in the total absence of protein synthesis[39]. Roots of *Vicia faba* were grown for up to 7 days in the absence of protein synthesis yet the activities of acid β-glycerophosphatases, acid naphthol AS-B1 phosphatases and naphthol esterases were not measurably impaired. Under conditions of normal growth only 2–3 days were required for passage from meristematic cell to mature xylem element. Thus, once formed, the hydrolase molecules appeared to persist for a sufficient length of time for xylem differentiation to occur without requiring renewal and hence synthesis. This, of course, does not rule out the possibility that new enzyme molecules are formed.

(b) Glycolysis, Respiration, Pentose Phosphate Pathway

A number of enzyme activities have been examined by the serial segment method, and the results in general point to increased activities of many of the enzymes with cell elongation and maturation (reviewed in Obroucheva[127] and Heyes[88]; Table 6.1). How relevant this is to the maturation of the xylem elements is difficult to say since xylem elements will lose their protoplasts and so may contribute little to the measured activity. Conversely, there may be a sharp increase in respiratory activity prior to the death of the protoplast, and the contribution to the general measurement could be great[61]. Few biochemical studies have been performed on isolated steles, and then primarily with respect to ion transport[107, 108]. However one study by Hall *et al.*[80] is relevant in that measurements of adenosinetriphosphatase, malate dehydrogenase, glucose-6-phosphate dehydrogenase, 6-phosphogluconate dehydrogenase and alcohol dehydrogenase activities, were made both in the freshly isolated steles from roots of *Zea mays* and also in steles 'aged' for 24 hours. The 'ageing' process resulted in a loss of pentose phosphate pathway, adenosinetriphosphatase and malate dehydrogenase activities, but a five-fold increase in alcohol dehydrogenase activity. This experiment illustrates the problems of using such material for studies on the stele, for it is difficult to know how real are the decreased activities, since the act of wounding when removing the steles most probably will have initially stimulated some enzyme activities. Furthermore, there is no biochemical possibility of comparing activities between isolated steles and those in the intact root.

The earlier biochemical results indicated increased respiratory and pentose phosphate pathway activities but a concomitant decrease in glycolysis as the cells change from the meristem to the elongation zone. These results have been confirmed by Fowler and Ap Rees[52] and by Sutcliffe and Sexton[164]. The latter authors also showed that in pea roots there was an eight-fold increase in respiratory activity during cell enlargement, and that oxygen uptake per unit of protein increased four-fold, as estimated with an oxygen electrode method. An attempt to analyse the nature of the enzyme changes was also made by gel-electrophoresis[164], but few differences could be detected except in the increased activities of the minor bands. An alternative approach, using roots of *Pisum sativum* incubated in a solution of [^{14}C]-amino acids prior to extraction and analysis of the proteins, showed little difference in the protein banding pattern. However, the incorporation of the radioactive precursors revealed a changing labelling pattern from apical to elongating to mature cells, implying both a changing protein composition and synthesis of different or new proteins[170].

Differences in the rates of accumulation and turnover of glyceraldehyde-3-phosphate dehydrogenase, phosphoglycerate dehydrogenase and phosphopyruvate hydratase have been observed during cell maturation[127], and studies on mitochondrial proteins have shown matrix proteins accummulating at more than double the rate of synthesis of the inner membranes. In elongation there is also a differential accumulation of cytochromes with cytochrome *b* increasing relative to cytochrome *a*.

That these biochemically measured changes may be relevant to xylogenesis can be seen from a number of cytochemical studies on marker enzymes for glycolysis, the tricarboxylic acid (TCA) cycle and the pentose phosphate pathway (Gahan and Kalina[65]; Gahan, Dawson, Black and Chapman, in preparation). Pentose phosphate pathway activity increases from the meristem to the zone of elongation, while glycolysis continues at a lower level. The TCA cycle is also elevated.

The elevated pentose phosphate pathway and succinic and malate dehydrogenases can also be seen in the elongated cells of the prostele of the embryo of *Lactuca sativa* although no mature xylem elements are present (Gahan, Dawson, Black and Chapman, in preparation). The activity is very high when compared with that of other tissues of the seed with the exception of the vascular tissue of the cotyledons. On germination, the meristem of the radicle exhibits low pentose phosphate pathway activity, whilst that in the elongating cells is maintained at the elevated level. Thus, the initiation of the biochemically differentiated stele with respect to cell elongation has already occurred in the embryo. It is also noteworthy that this increased enzyme activity is present in cells which are reported to be diploid[23] and hence does not reflect gene amplification. It is hoped that the newer quantitative cytochemical methods for analysing these pathways[25, 63, 65, 100] will permit a more detailed analysis of the changes already observed. This will lead to a better understanding of events during secondary xylem formation when pentose pathway activity increases in the developing xylem elements of roots of *Vicia faba* and *Taraxicum officinale*, and shoots of *Ranunculus bulbosus* and *Salix* sp. (Gahan, unpublished data).

The results so far obtained by both cytochemical and biochemical methods are not easy to explain. For example, it is not clear whether the increases or decreases in activity represent changes in actual amounts of enzyme proteins, or just in the degree of activity induced by activation, inhibition or merely allosteric effects[168] of an existing set of molecules. It is likely, however, that the increased respiratory activity reflects the higher energy requirements of the various synthetic processes involved in cell differentiation. The pentose phosphate pathway poses a slight difficulty in that studies on its role in plant cells have not shown a definite involvement in the production of intermediates for nucleic acid metabolism. It is more likely to be of importance for the production of intermediates for lipid biosynthesis[63], cell wall pentosans and phenolic compounds, all of which would seem to be basic components needed for cell elongation and maturation of xylem elements.

Most of the studies on the tricarboxylic acid cycle have employed malate and succinate dehydrogenases as markers of the pathway's activity. However, neither of these may be good indicators of the flux through the pathway. If an extrapolation may be made from animal studies, the non-equilibrium reactions involving citrate synthase and α-ketoglutarate dehydrogenase may be the flux generating steps in the cycle. Thus, a study of these enzymes is more likely to produce useful information on the behaviour of the pathway as a whole rather than will those of such as the near-equilibrium enzyme, succinic dehydrogenase[123, 124] Studies on glycolysis are probably more reliable since attention is now being paid to measurements of constant-proportion enzymes[103, 171, 174]. Here, the activities of enzymes which are in constant ratio are measured which may give a better index of the behaviour of the pathway rather than the individual enzymes. Thus, if one enzyme of the group of constant-proportion enzymes is elevated the others should be elevated, while maintaining the same ratios of activities to the other enzymes in the group[132].

This may be extendable to other enzyme pathways, and is of importance because it should be valid for both non-equilibrium and near equilibrium enzymes[131]. In the case of the studies of Zeleneva and Khavkin, glyceraldehyde-3-phosphate dehydrogenase, phosphoglycerate kinase and phosphopyruvate hydratase formed such a group of constant-proportion enzymes.

A further complication concerns the seasonal variation in the response of the cambium to the tetrazolium salt reaction for dehydrogenases in *Corylus*, *Populus* and *Prunus*[70]. Maximum activity was seen in spring, with a second, though lower peak, in November. Minimum activity was found in August and September, and in winter. It would be of interest to know if this is typical, or if it represents a seasonal variation pattern peculiar to Italy where the work was performed.

(c) Peroxidase

Peroxidase activity in xylogenesis has been extensively studied both biochemically and cytochemically, a consequence perhaps of the degree of activity apparently associated with lignification[86, 158]. Studies of the enzyme have been in progress since 1863 when its action was first discovered by Schönbein. The review by Van Fleet[171] indicates little evidence for peroxidatic activity in xylem. Subsequent workers found activity in roots of *Vicia faba*[97] and *Pisum sativum*[79], sections of wood[82] and wound vessel members[86]. Other workers either could not find this relationship[41, 73] or found only low activity[49, 50]. It is not always clear if workers take sufficient care to eliminate the possibility of aldehyde groups in the xylem elements reacting with the cytochemical test[138]. In view of its fairly general distribution in most cell types, it is difficult to see peroxidase as a specific marker of lignification or having a role linked only to lignification rather than to other functions in plant cell metabolism.

(d) Invertase

Little information is available on invertase activity, though a review by Sutcliffe and Sexton[163] indicates that invertase activity increases in roots at the stage when cell elongation is maximal. Its presence in cell walls adds the possibility of a role in cell wall changes.

(e) Hydrolases

In general, biochemical studies made for a number of hydrolases (Table 6.1), on serial segments from roots, showed an increased activity with increasing distance from the apical initials, the rate depending upon the enzyme measured[88, 127]. The specific involvement of hydrolases with the developing and maturing stele has been confirmed by cytochemical studies

Table 6.1: *Biochemical studies of enzyme activities possibly associated with xylogenesis*

Enzyme	*Reference*
Invertase	Brown and Robinson[20]
	Hellebust and Forward[85]
	Sutcliffe and Sexton[163]
	Lyne and Ap Rees[109]
Acid Phosphatases	Sexton and Sutcliffe[149]
	Sutcliffe and Sexton[163]
	Popatov *et al.*[137]
	Zeleneva *et al.*[175]
	Fielding and Hall[50]
ATPase	Salamatova and Balajam[145]
	Sexton and Sutcliffe[150]
β-glucosidase	Zeleneva *et al.*[175]
β-galactosidase	Zeleneva *et al.*[175]
Peroxidase	Potapov *et al.*[137]
	Hall *et al.*[79]
	Fielding and Hall[50]
Phenyladenine ammonia lyase	Fielding and Hall[50]
Nucleases	Robinson and Cartwright[144]
Proteolytic enzymes	Robinson and Brown[143]
	Robinson[142]
Dehydrogenases	Gibbs and Earl[71]
	Hadačová[74]
	Khavkin and Varakina[102]
	Fowler and Ap Rees[52]
	Hall *et al.*[81]
	Khavkin[101]
	Sutcliffe and Sexton[164]

(Table 6.2). Much attention has been focussed upon the acid phosphatases; in part, because acid β-glycerophosphatase was originally employed as a marker enzyme of lysosomes in animal cells[40, 90] and was exploited in plant studies to establish the concept of phytolysosomes in a wide range of phanerogams and cryptogams[12, 36, 37, 54, 55, 57, 84, 116, 135]. In particular, the central vacuole in many plant cells appears to behave as a large lysosome containing many hydrolases[14, 117, 125] and materials including cytoplasmic structures apparently passed to it for digestion[35, 36]. Evidence is now available to show the lysosomal system to be based on the plant vacuolar system[12] and to form a Golgi-endoplasmic reticulum-lysosomal (GERL) system[118].

The first cytochemical investigation of acid β-glycerophosphatase activity and xylogenesis was on roots of *Vicia faba* and showed the enzymes to be particle-bound in the meristem and elongating cells, though there were many more active sites present in the cell during elongation. These sites reacted more rapidly than did those of the meristem cells indicating a change in the membranes of the structures with which the enzymes were associated. As the xylem elements differentiated and the protoplast was lost from the cell, so the enzyme activity appeared to be released into the cytoplasm as though contributing to the death and/or removal of the protoplast[66]. These observations have been confirmed for β-glycerophosphatase and have been reported similarly for acid naphthol AS-B1 phosphatases and p-nitrophenylphosphatases in *Vicia faba*[10, 59, 76], *Pisum sativum*[162, 163], *Lepidium sativum*[13], *Phaseolus lunatus*[166], *Phaseolus aureus*[162], *Zea mays*[76, 80, 151], *Hordeum vulgare*[76, 151], *Helianthus annuus*[151], and *Phaseolus vulgaris*[31, 151] plus a variety of bryophytes[83, 84]. Alternative interpretations to that of Gahan and Maple[66] have been proposed in which the increased amount of enzyme end product is regarded as demonstrating an increased production of enzyme sites throughout the differentiating cells[115, 162]. This implies a need for a net synthesis of hydrolytic enzymes during the early stages of elongation at least.

Cytochemical investigations have been made with three broad-spectrum substrates, namely, β-glycerophosphate, p-nitrophenyl phosphate and naphthol AS-B1 (or AS-MX or AS-TR) phosphate. Paranitrophenylphosphate would appear to show those enzymes demonstrable with the two other substrates, including glucose-6-phosphatase, K^+-activated acyl phosphatase[64], nucleotide pyrophosphatase[63, 104, 159], nucleotide phosphotransferase[17] and 5′ (3′)-ribonucleotide phosphohydrolase[136]. Most of these enzymes are membrane-bound, and some are directly related to xylogenesis. Thus, K^+-activated acyl phosphatase is present in the elongating cells, while glucose-6-phosphatase is in the maturing xylem elements[64, 151]. Nucleotide pyrophosphatase activity, which is a major component of the reaction seen with naphthol AS-B1 phosphatase and p-nitrophenylphosphatase activities is very active in the differentiating xylem fibres and tracheids[63, 159]. Unfortunately, it has not been possible to ascribe clear-cut roles to these enzymes in xylogenesis, though a number of authors have attempted to link phosphatase activity with the transport of molecules[16, 51, 67, 72, 83, 160]. ATPase activity has also been thought to be associated with ion transport and other metabolic processes[77], but a careful study by Hall[77, 78] has shown that ATPase is not included in the phosphatase activities shown by acid β-glycerophosphatase. Indeed, no reaction for ATPase activity can be seen in the differentiating xylem from *Zea mays*, *Hordeum vulgare* or *Vicia faba*, or from *Pisum sativum*[163]. Few similar studies have been made as yet on secondary xylem, but prospective workers should beware of the seasonal variation of acid β-glycero-

Table 6.2: *Histochemical studies of enzyme activities possible associated with xylogenesis*

Enzyme	*Reference*
Acid phosphatases	Beneš and Opatrná[10]
	Gahan and Maple[66]
	Catesson and Czaninski[29, 30]
	Hall[76]
	Sexton and Sutcliffe[149]
	McLean and Gahan[111]
	Gahan *et al.*[68]
	Hall and Davie[80]
	Hall[76]
	Sheikh *et al.*[153]
	Berjak and Lawton[13]
	Hébant[83]
	Shaykh and Roberts[151]
	Hébant[84]
	Charvat and Esau[31]
	Sheikh and Gahan[152]
	Thomas *et al.*[166]
	Stewart and Pitt[162]
	Gahan *et al.*[63, 64]
ATPase	Sexton and Sutcliffe[150]
	Hall[76, 77]
	Shaykh and Roberts[151]
Glucose-6-phosphatase	Shaykh and Roberts[151]
	Gahan *et al.*[68]
Esterase	Beneš[7]
	Gahan and McLean[67]
	McLean and Gahan[111]
	Gahan *et al.*[68]
	Hall and Davie[80]
	Sheikh *et al.*[153]
	Beneš[9]
	Sheikh and Gahan[152]
	Thomas *et al.*[166]
Peroxidase	Van Fleet[171]
	Jensen[97]
	Hepler *et al.*[86]
	Harkin and Obst[82]
	Goff[73]
	Hall *et al.*[79]
	Fielding and Hall[49, 50]
β-glucuronidase	Hall and Davie[80]
Leucyl aminopeptidase	Thomas *et al.*[166]
β-glucosidase	Freudenberg[53]
	Ashford and McCully[2]
	Thomas *et al.*[166]
β-galactosidase	Thomas *et al.*[166]
Dehydrogenases	Sato[147]
	Gerola *et al.*[70]
	Sutcliffe and Sexton[164]
	Thomas *et al.*[166]

phosphatase activity such as has been demonstrated for *Robinia pseudoacacia* and *Acer pseudoplatanus*[29, 30].

That most of the phosphatases demonstrated are membrane-bound poses a serious argument against their being released into the cytoplasm to aid autolysis of the protoplast. It is possible that initially, there is an increase in the number of sites of acid phosphatase activity as the meristem cells elongate, and that these represent, at least in part, the provacuolar material which will add to the larger vacuole(s). As the protoplast senesces and autolyses, the general activity may be due to the distribution of membrane fragments together with the enzyme complement throughout the cell. Alternatively, there may be sufficient changes in the membranes at an early stage in senescence to result in activation of all of the membrane-bound phosphatases[59, 61]. The activity could also represent the formation of many autophagosomes associated with autolysis[13, 161]. The loss of the nucleus from the protoplast, the longevity of the acid phosphatase molecules and the increased speed of the cytochemical response when taken together would seem to favour the explanation involving membrane alterations. However with the limited information and methods available at present, it is too early to draw precise conclusions concerning a direct lysosomal involvement in xylogenesis. Indeed, the enzymes, while being good markers of membrane changes, may not indicate a general release of phosphatases.

Other hydrolases have also been studied (Table 6.2) and a similar pattern of increased activity during xylogenesis has been observed. As with the acid phosphatases, no definite roles can be ascribed to the enzymes and hence the specific involvement in xylogenesis is not clear. Thus, strong β-glucosidase activity has been shown cytochemically to be associated with lignifying secondary xylem elements though no activity was present in cells of the mature wood or the cambial zone[53]. Some doubt has been expressed as to the relationship between glycosides and lignification[21] and more recently an involvement with cell wall metabolism, possibly in wall loosening associated with cell elongation, was proposed by Ashford and McCully[2] and supported by Hadačová and Beneš[75]. Thomas *et al.*[166] concur with this idea through studies on β-glucosidases and β-galactosidases, even to the extent of involving cell wall breakdown. This is based on the need for the unlignified primary wall to be hydrolysed so as to establish continuity in the xylem tissue[38, 46]. Although cellulase is known to be involved in this process[126, 154] it has been reported that high galactosidase and glucosidase activities are generally associated with senescing tissues, especially when polysaccharide resorption occurs[172].

Leucyl naphthylamidase activity may be seen as an indication of proteolytic activity (although it is only an amino-peptidase), but cathepsin B activity would seem to be absent as shown by lack of activity against n-α-benzoyl-D.L.-arginine-β-naphthylamide (Gahan, unpublished data).

Whilst most of the hydrolases studied show an increased activity in xylogenesis, a distinction must be made between their involvement in cell metabolism related to elongation and that involved in cell senescence and death. Some activities associated with elongation can also be seen in other tissues such as phloem, and may not be specific to xylem. Some activities will be specifically associated with xylem, e.g. in cell senescence and death, which may well precede the breakdown of membranes. High enzyme activity may be seen to precede loss of cell compartmentalization which may be the final visible, irreversible change leading to cell death[61]. Thus, any subsequent release of hydrolases into the cytoplasm could represent essentially a *post mortem* change and not a primary event in the death of the protoplast[6]. Even when the hydrolases are released, it is not clear if the environment will be at an optimal pH for them to act, though xylem exudate from corn seedlings has been shown to be at pH 5.3[26], an optimal value for many, though not all, hydrolases.

(f) Esterases

Esterase activity has been examined almost entirely by cytochemistry (Table 6.2). Like the other hydrolases, there is increased activity during xylogenesis as shown by a variety of substrates such as α- and β-naphthyl acetate[7], naphthol AS-D acetate[7, 56, 67, 80, 111] and indoxyl acetate[80, 110, 111, 166]. Perhaps the most interesting of these substrates is naphthol AS-D acetate since in the root apex of *Vicia faba* and *Pisum sativum*, the activity can be seen to extend from the stele into the root apex, stopping 5–6 cells from the junction of the root tip with the root cap (Gahan, in preparation). There is little or no activity present in the adjacent meristem or control cells[110, 111] (Fig. 6.1). In the plants studied, the cells between the junction of the root cap and root tip and the start of the enzyme activity constitute the quiescent centre[33, 167]. The enzyme activity starts in the cells immediately outside the quiescent centre and continues up and into the stele (Fig. 6.1). It can be considered, therefore, that the esterase activity constitutes a biochemical continuity between the meristem and the stele so giving an early marker of commitment to differentiation in the meristem cells. As the cambial zone in shoots of *Lycopersicon esculentum*, *Spinacia oleracea*, *Mimosa pudica*, *Salix sp.* and roots of *Pisum sativum*, *Vicia faba*, *Taraxacum officinale*, *Ranunculus bulbosus* and *Hypericon calcynium* also show this activity, it would appear that the cambium is also a committed meristem. The substrate employed for these cytochemical studies is capable of demonstrating three groups of esterases, namely, carboxylesterases (E.C. 3.1.1.1.), arylesterases (E.C. 3.1.1.2) and acetylesterases (E.C. 3.1.1.6.)[128]. Inhibitor studies with eserine, p-chloromercuriphenyl sulphonic acid and E 600 would indicate the enzymes in the cortex to be acetylesterases whilst those of the prostele are 33% acetylesterases and 67% carboxylesterases*, part of which may be an esteroprotease (Gahan, unpublished data).

Light microscopy reveals the activity to be associated with lysosome-like structures[8, 54, 110] and with the cell walls and/or plasmalemma[110]. Electron microscope cytochemical studies showed at least part of the activity to be associated with the GERL and appearing to be exported to the cell walls[67].

That part of the activity is associated with the cell walls can be seen in artichoke tuber explants *in vitro*[134] when esterase activity is increased substantially at an early stage in the transformation of parenchyma cells into tracheary elements (Phillips and Gahan, unpublished data). If auxin is omitted from the culture medium so that cell enlargement occurs in the absence of cell division, then the esterase activity associated with the cell walls is even greater than in the full culture medium.

Preliminary biochemical studies from roots of *Vicia faba* indicate seven esterase bands separating on polyacrylamide gels, though not all of these bands may be involved in xylogenesis. Thus, a similar study on tumours induced in *Lycopersicon esculentum* by *Agrobacterium tumafaciens* showed only three such esterase bands present in old tumours formed mainly of tracheids and sclereids as compared to five bands obtained from the normal tomato tissue[68, 152].

The use of esterase as a marker of commitment to differentiate into stele and tracheids is further confirmed by studies on the regeneration of vascular tissue in wounded pea roots. If the stele of a young main root of *Pisum sativum* is severed, a vascular bridge will form in the remaining cortex. By 24 h after wounding, cell divisions occur on both proximal and distal sides of the wound. Differentiation of the new vascular tissue follows, with first sieve elements and then tracheids appearing by about 3 days[141]. Esterase activity can be seen in those cells which will form the vascular bridge at about the time that mitotic activity is initiated (Rana and Gahan, unpublished data). Thus, activity increases and persists throughout the formation and functioning of the bridge.

The fact that a group of four centrally located meristem cells, immediately outside the quiescent centre is committed to differentiate into stele tissues raises two fundamental questions, namely: "is there really a need for mitosis to occur in order to permit commitment?" and secondly, "how is commitment initiated?". It has been suggested that two or three rounds of cell division are necessary for commitment to xylogenesis to occur with *in vitro* systems (Chapter 5[114, 133]).

In order to leave the quiescent centre, cells *in vivo* would need to undergo mitosis and pass through at least one round of division before entering a

* Gahan, P. B. An early cytochemical marker of commitment to differentiate in meristems from dicotyledonous plants. *Ann. Bot.* (In Press).

zone of commitment. Clearly, they will then undergo a series of divisions prior to differentiating fully to xylem elements. The data may be reconciled in at least two ways. Firstly, if all other events are ready in the *in vivo* system, either no or only a single cell division may be required for commitment to differentiation. Mesophyll cells from leaves of *Zinnia elegans* have been shown to differentiate into tracheary elements without mitosis (Chapter 5) and differentiation without mitosis has also been reported for *Dictyostelium discoideum*[27].

The second explanation would involve a two-step system in which there is a general 'coarse' programming of the genes followed by a 'fine' programming. The 'coarse' programming woud occur immediately in the primary and secondary meristems, the 'fine' programming requiring one or more rounds of division. This implies a first round of programming, possibly independent of mitosis. Although a round of division could be involved in the root apex, i.e. on leaving the quiescent centre, this does not appear to be the case in the secondary meristem where it has not been possible to identify uncommitted cambial initial cells by the esterase reaction. Hence the implication is that all cambial cells are committed (at least 'coarse' programmed) in the absence of cell division. This might be explained if the 'coarse' programming occurs during the division steps in which the cambium is established.

Based upon evidence from *in vitro* systems in which parenchyma cells may be switched to form tracheids in the presence of auxin, cytokinin and sucrose (which systems also give an early stimulation to the esterase system), it is possible to envisage an *in vivo* 'coarse' programming of the committed cells as the result of their being in the pathway of cytokinin, generated from the root cap/quiescent centre, and auxin (and sucrose), generated from the aerial parts and possibly from the root tip. Thus, the secondary meristem should be continuously exposed to these substances as would be the root apical cells which are in a source/sink relationship. The same substances, but in different ratios, might also be responsible for the 'fine' programming for xylogenesis, but if they give a basic commitment to form tissues of the stele, an additional 'fine' programming component may be needed for xylogenesis, e.g. indol-3-yl acetic acid[156, 157].

The commitment situation as presented, implies a continuous triggering mechanism in the form of established gradients. Preliminary investigations on embryos of *Pisum sativum* would suggest that such (a) gradient(s) is(are) initiated by the parent plant during the formation of the seed and embryo (Gahan, unpublished data).

A point of fundamental importance is that the patterns described for dicotyledonous plants would seem to apply to gymnosperms such as *Picea* and *Thuja* but not to monocotyledons such as *Zea mays*, *Hordeum vulgare*, *Avena fatua* and *Allium cepa* (Gahan, unpublished data). This may reflect a difference in the monocotyledons which leads to a different organisation of the vascular tissues. Such organisation might require a cortex which is programmed in a way that would allow a more ready switch to the formation of vascular tissue.

REFERENCES

[1] Alfert, M., Geschwind, I. I. A selective staining method for the basic proteins of cell nuclei. *Proc. Nat. Acad. Sci. Wash.* **39,** 991-999 (1953)

[2] Ashford, A. E., McCully, M. E. Histochemical localization of β-glycosidases in roots of *Zea mays* III β-glycosidase activity in the meristem of lateral roots. *Protoplasma* **77,** 411-425 (1973)

[3] Avanzi, S., Maggini, F., Innocenti, A. M. Amplification of ribosomal cistrons during the maturation of metaxylem in the root of *Allium cepa. Protoplasma* **76,** 197-210 (1973)

[4] Barlow, P. W. Properties of cells in the root apex. *Revista Fac. Agron. La Plata* **XLVII,** 275-301 (1971)

[5] Barlow, P. W. Endopolyploidy: Towards an understanding of its biological significance. *Acta Theoretica* **27,** 1-18 (1978)

[6] Barton, D. Fine structure of mesophyll cells in senescing leaves of *Phaseolus. Planta* **71,** 314-325 (1966)

[7] Beneš, K. Histochemical demonstration of esterase in the root tip of *Vicia faba* L with azo-coupling method. *Biol. Plant* (Praha) **4,** 211-219 (1962)

[8] Beneš, K. Histogenesis and localization of non-specific esterases in root tip. *Biol. Plant* (Praha) **13,** 110-121 (1971)

[9] Beneš, K. The use of derivatives of 2-naphthol as substrates for the demonstration of carboxyl esters *in situ* and the enhancement of the reaction with DMSO. *Histochemistry* **42,** 193-197 (1974)

[10] Beneš, K., Opatrná, J. Localization of acid phosphatase in the differentiating root meristem. *Biol. Plant* (Praha) **6,** 8-16 (1964)

[11] Bennett, M. D., Smith, J. B. Nuclear DNA amounts in angiosperms. *Phil. Trans. Roy. Soc.* London **274,** 227-274 (1976)

[12] Berjak, P. Lysosomal compartmentalization: Ultrastructural aspects of the origin, development and function of vacuoles in roots of *Lepidium sativum. Ann. Bot.* **36,** 73-81 (1972)

[13] Berjak, P., Lawton, J. R. Prostelar autolysis: a further example of a programmed senescence. *New Phytol.* **72,** 625-637 (1973)

[14] Boller, T., Kende, H. Hydrolytic enzymes in the central vacuole of plant cells. *Plant Physiol.* **63,** 1123-1132 (1979)

[15] Brabec, F. Über Polysomatie in der Wurzel von Bryonia. *Chromosoma* **6,** 135-141 (1953)

[16] Braun, H. J., Sauter, H. Phosphatase—Lokalisation in Phloembeckenzellen und Siebröhren der Dioscoraceae und inhre mögliche Bedeutung für den aktiven Assimilattransport. *Planta* **60,** 543-557 (1964)

[17] Brawerman, G., Chargaff, E. Distribution and significance of the nucleoside phosphotransferases. *Biochim. Biophys. Acta* **16,** 524-532 (1955)

[18] Broekaert, D., Van Parijs, R. The relationship between the endomitotic cycle and the enhanced capacity for protein synthesis in leguminosae embryogeny. *Z. Pflanzenphysiol.* **86,** 165-175 (1978)

[19] Brown, R., Broadbent, D. The development of cells in the growing zones of the root. *J. Exper. Bot.* **1,** 249-263 (1950)

[20] Brown, R., Robinson, E. Cellular differentiation and the development of enzyme proteins in plants. In *Biological Specificity and Growth* 93-118. (New Jersey: Princeton University Press 1955)

[21] Brown, S. A. Lignins. *Ann. Rev. Plant. Physiol.* **17,** 223-244 (1966)

[22] Brunori, A. Synthesis of DNA and mitosis in relation to cell differentiation in the roots of *Vicia faba* and *Lactuca sativa. Caryologia* **24,** 209-215 (1971)

[23] Brunori, A., D'Amato, F. The DNA content of nuclei in the embryo of dry seeds of *Pinus pinea* and *Lactuca sativa. Caryologia* **20,** 153-161 (1967)

[24] Bucknall, R. A., Sutcliffe, J. F. The nucleic acids and cellular differentiation in the root apex of *Pisum sativum* II. The nucleic acids and cellular development patterns. *J. Exper. Bot.* **16,** 587-603 (1965)

[25] Butcher, R., Dawson, A. L., Knaab, S., Gahan, P. B. Dehydrogenase activity and loss of formazan from tissue sections. *Histochem. J.* **12,** 591-598 (1980)

[26] Butz, R. G., Long, R. C. L-malate as an essential component of the xylem fluid of corn seedling roots. *Plant Physiol.* **64,** 684-689 (1979)

[27] Cappuccinelli, P., Fighetti, M., Rubins, S. Differentiation without mitosis in *Dictyostelium discoideum. Cell Diff.* **8,** 243-252 (1979)

[28] Carriere, R. The growth of liver parenchymal nuclei and its endocrine regulation. *Inter. Rev. Cytol.* **28,** 201-277 (1969)

[29] Catesson, A. M., Czaninski, Y. Mise en évidence d'une activité phosphatasique acide dans le reticulum endoplasmique des tissus conducteurs de Robinier et de Sycamore. *J. Microscop.* **6,** 509-514 (1967)

[30] Catesson, A. M., Czaninski, Y. Localisation ultrastructurale de la phosphatase acide et cycle saisonnier dans les tissus conducteurs de quelques arbres. *Bull. Soc. Franç. Physiol. Veget.* **14,** 165-173 (1968)

[31] Charvat, I., Esau, K. An ultrastructural study of acid phosphatase localization in *Phaseolus vulgaris* xylem by the use of an azo-dye method. *J. Cell Sci.* **19,** 543-561 (1975)

[32] Clowes, F. A. L. Protein synthesis in root meristems. *J. Exper. Bot.* **9,** 229-238 (1958)

[33] Clowes, F. A. L. *Apical Meristems.* (Oxford: Blackwell Scientific Publications 1961)

[34] Corsi, G., Avanzi, S. Cytochemical analyses on cellular differentiation in the root tip of *Allium cepa. Caryologia* **23,** 381-394 (1970)

[35] Coulomb, C. Mise en évidence de systèmes á fonctions autophagiques dans les méristèmes radiculaires de la courge. *Ann. Sci. Nat. Bot.* **IX,** 541-574 (1968)

[36] Coulomb, C. Processus lytiques dans les vacuoles des cellules radiculaires méristèmatiques de la Scorsinère (*Scorzonera hispanica*). Ph.D. Thesis. University of Marsailles (1978)

[37] Coulomb, P. Mise en évidence de structures analogues aux lysosomes dans le méristème radiculaire de la courge. *J. Microscop.* **8,** 123-138 (1969)

[38] Cronshaw, J., Bouck, G. B. The fine structure of differentiating xylem elements. *J. Cell Biol.* **24,** 415-531 (1965)

[39] Dawson, A. L., Gahan, P. B. Longevity of hydrolase molecules during primary xylem differentiation. *Ann. Bot.* **43,** 251-254 (1979)

[40] De Duve, C. In *'Subcellular Particles.'* 128-159. T. Hayashi (ed.) (New York: Ronald Press 1959)

[41] De Jong, D. W. An investigation of the role of plant peroxidase in cell wall development by the histochemical method. *J. Histochem. Cytochem.* **15,** 335-346 (1967)

[42] Deeley, E. M., Davies, H. G., Chayen, J. The DNA content of cells in the root of *V. faba. Exper. Cell Res.* **12,** 582-591 (1957)

[43] Dini, F., Viola-Magni, M. P. DNA metabolic instability during hydrolysis. *Bioch. Biol. Sper.* **XII,** 63-70 (1976)

[44] Durante, M., Cremonini, R., Brunori, A., Avanzi, S., Innocenti, A. M. Differentiation of metaxylem cell line in the root of *Allium cepa*. I. DNA heterogeneity and ribosomal cistrons of two different stages of differentiation. *Protoplasma* **93,** 289-303 (1977)
[45] Erickson, R. D.,Goddard, D. R. An analysis of root growth in cellular and biochemical terms. *Growth Suppl.* **15,** 89-116 (1951)
[46] Esau, K. *Plant Anatomy* 2nd ed. (New York: Wiley 1965)
[47] Fahn, A. *Plant Anatomy* 2nd ed. (Oxford: Pergamon Press 1974)
[48] Feulgen, R., Rossenbeck, H. Mikroskopisch-chemischer Nachweis einer Nucleinsäure von Typus der Thymonuclein—säure und die darauf beruhende elecktive Färbung von Zeukenen in mikroskopischen präparaten. *Z. phys. Chem.* **135,** 203-248 (1924)
[49] Fielding, J. L., Hall, J. L. A biochemical and cytochemical study of peroxidase activity in roots of *Pisum sativum*. I. A comparison of DAB-peroxidase and guaiacol-peroxidase with particular emphasis on the properties of cell wall activity. *J. Exper. Bot.* **29,** 969-981 (1978)
[50] Fielding, J. L, Hall, J. L. A biochemical and cytochemical study of peroxidase activity in roots of *Pisum sativum*. II. Distribution of enzymes in relation to root development. *J. Exper. Bot.* **29,** 983-991 (1978)
[51] Figier, J. Localisation infrastructurale de la phosphomonoesterase acide dans la stipule de *Vicia faba* an niveau du nectaire. *Planta* **83,** 60-79 (1968)
[52] Fowler, M. W., Ap Rees, T. Carbohydrate oxidation during differentiation in roots of *Pisum sativum. Biochim. Biophys. Acta* **201,** 33-44 (1970)
[53] Freudenberg, K. The formation of lignin in the tissue and *in vitro*. In *The Formation of Wood in Forest Trees*. 203-218. M. H. Zimmerman (ed.) (New York: Academic Press 1964)
[54] Gahan, P. B. The histochemistry of lysosomes. *Int. Rev. Cytol.* **21,** 1-63 (1967)
[55] Gahan, P. B. Lysosomes. In *Plant Cell Organelles*. 228-238. J. W. Pridham (ed.) (London: Academic Press 1968)
[56] Gahan, P. B. The role of acid hydrolases in the differentiation of plant cells. *Cell Differentiation* 223-224. R. Harris, P. Allin and D. Viza (eds.) Munksgaard (1972)
[57] Gahan, P. B. Plant Lysosomes. In *Lysosomes in Biology and Pathology*. **3,** 69-88. J. T. Dingle (ed.) (London: North-Holland 1973)
[58] Gahan, P. B. DNA turnover in nuclei of cells from higher plants. *Riv. Istoch. norm. Patol.* **20,** 108-109 (1976)
[59] Gahan, P. B. A re-interpretation of the cytochemical evidence for acid phosphatase activity during cell death in xylem differentiation. *Ann. Bot.* **42,** 755-757 (1978)
[60] Gahan, P. B. Quantitative cytochemical studies of plant cells. *Proc. Roy. Mic. Soc.* **14,** 229 (1979)
[61] Gahan, P. B. Cell senescence and death in plants. In *Cell Death*. I. D. Bowen and R. A. Lockshin (eds.) (London: Chapman and Hall 1981)
[62] Gahan, P. B. *Plant Histochemistry*. (London: Academic Press) (in press)
[63] Gahan, P. B., Auderset, G., Greppin, H. Pentose phosphate pathway activity during floral induction in spinach. *Ann. Bot.* **44,** 121-124 (1979)
[64] Gahan, P. B., Dawson, A. L., Fielding, J. Para-nitrophenylphosphate as a substrate for some acid phosphatases in roots of *Vicia faba. Ann. Bot.* **42,** 1413-1420 (1978)
[65] Gahan, P. B., Kalina, M. The use of tetrazolium salts in the histochemical demonstration of succinic dehydrogenase activity in plant tissues. *Histochemie* **10,** 81-92 (1968)
[66] Gahan, P. B., Maple, A. J. The behaviour of lysosome-like particles during cell differentiation. *J. Exper. Bot.* **17,** 151-155 (1966)
[67] Gahan, P. B., McLean, J. Subcellular localization and possible functions of acid

β-glycerophosphatases and naphthol esterases in plant cells. *Planta* **89,** 126-135 (1969)
68 Gahan, P. B., Sheikh, K., Maggi, V., Stroun, M., Smith A. R. W. Electrophoretic analysis of hydrolases from crown gall tissues. *F.E.B.S. Letters* **13,** 53-55 (1971)
69 Gahan, P. B., Sierakowska, H., Dawson, A. L. Nucleotide pyrophosphatase activity in dry and germinated seeds of *Triticum* and its relationship to general acid phosphatases. *Planta* **145,** 159-166 (1979)
70 Gerola, F. M., Barbesino, M. L'attivita deidrogenasica del cambio in differenti periodi del anno in *Corylus, Populus,* e *Prunus. Nuovo Giora. Bot. Ital. N.S.* **63,** 37-45 (1956)
71 Gibbs, M., Earl, J. M. Effect of tissue age on hexose metabolism. I. An enzyme study with pea root. *Pl. Physiol.* **34,** 529-532 (1959)
72 Giraud, G. Localisation des activités phosphatasiques chez les végétaux. *Bull. Soc. Franç. Physiol. Végét.* **14,** 149-163 (1968)
73 Goff, C. W. A light and electron microscopic study of peroxidase localization in onion root tip. *Amer. J. Bot.* **62,** 280-291 (1975)
74 Hadačová, V. Der Einfluss einiger Atmungshemmstoffe auf die Atmung der Wurzelzonen von *Vicia faba* L. *Biol. Plant.* **10,** 385-397 (1968)
75 Hadačová, V., Beneš, K. The differentiation of α- and β-glucosidase and α- and β-galactosidase isoenzymes from maize and broad bean root tips using disc electrophoresis in polyacrilamide gels. *Biol. Plant* (Praha) **19,** 436-441 (1977)
76 Hall, J. L. A histochemical study of adenosine triphosphatase and other nucleotide phosphatases in young root tips. *Planta* **89,** 254-265 (1969)
77 Hall, J. L. Cytochemical localization of ATPase activity in plant root cells. *J. Microscop.* **93,** 219-225 (1971)
78 Hall, J. L. Further properties of adenosine triphosphatase and β-glycerophosphatase from barley roots. *J. Exper. Bot.* **23,** 800-808 (1971)
79 Hall, J. L., Al-Azzawi, M. J., Fielding, J. L. Microscopic Cytochemistry in Enzyme Localization and Development. In *Regulation of Enzyme Synthesis and Activity in Higher Plants.* 329-363. H. Smith (ed.) (New York: Academic Press 1977)
80 Hall, J. L., Davie, C. A. M. Localization of acid hydrolase activity in *Zea mays* L. root tips. *Ann. Bot.* **35,** 849-855 (1971)
81 Hall, J. L., Sexton, R., Baker, D. A. Metabolic changes in washed, isolated steles. *Planta* **96,** 54-61 (1971)
82 Harkin, J. M., Obst, J. R. Lignification in trees: Indication of exclusive peroxidase participation. *Science* **180,** 296-298 (1973)
83 Hébant, C. Acid phosphomonoesterase activities (β-glycerophosphatase and naphthol AS-MX phosphatase) in conducting tissues of bryophytes. *Protoplasma* **77,** 231-241 (1973)
84 Hébant, C. On the occurrence of lysosomal acid phosphatase activity in the differentiating water-conducting strand of *Takakia* and its evolutionary significance. *Phytomorphology* **25,** 279-282 (1975)
85 Hellebust, J. A., Forward, D. F. The invertase of the corn radicle and its activity in successive stages of growth. *Canad. J. Bot.* **40,** 113-126
86 Hepler, P. K., Rice, R. M., Terranova, W. A. Cytochemical localization of peroxidase activity in wound vessel members of *Coleus. Canad. J. Bot.* **50,** 977-983 (1972)
87 Heyes, J. K. Nucleic acid changes during cell expansion in the root. *Proc. Roy. Soc. B.* **152,** 218-230 (1960)
88 Heyes, J. K. Metabolic aspects of cell growth and development in the root. In *The Physiology of the Garden Pea.* 151-181. J. F. Sutcliffe and J. S. Pate (eds.) (1977)
89 Heyes, J. K., Brown, R. Growth and cellular differentiation. *Symposium: The Growth*

of Leaves. 31-49. Ed. Milthorpe (1956)
[90] Holt, S. J., Hicks, R. M. The localization of acid phosphatase in rat liver cells as revealed by combined cytochemical staining and electron microscopy. *J. Biophys. Biochem. Cytol.* **11,** 47-66 (1961)
[91] Hurst, P. R. Studies of the metabolic activity and possible functions of cytoplasmic and nuclear deoxyribonucleic acid in dividing, differentiating and fully differentiated cells. Ph.D. Thesis. Council for National Academic Awards (1972)
[92] Hurst, P. R., Gahan, P. B. Turnover of DNA in ageing tissues of *Lycopersicon esculentum. Ann. Bot.* **39,** 71-76 (1975)
[93] Hurst, P. R., Gahan, P. B., Snellen, J. W. Turnover of labelled DNA in differentiated collenchyma. *Differentiation* **1,** 261-266 (1973)
[94] Innocenti, A. M. Cytophotometric determination of histone content in cell nuclei of proliferating and non-proliferating root meristem cells of *Triticum durum. Caryologia* **24,** 457-461 (1971)
[95] Innocenti, A. M. Cyclic changes of histones/DNA ratio in differentiating nuclei of metaxylem cell line in *Allium cepa* root tip. *Caryologia* **28,** 223-228 (1975)
[96] Innocenti, A. M., Avanzi, S. Some cytological aspects of the differentiation of metaxylem in the root of *Allium cepa. Caryologia* **24,** 283-292 (1971)
[97] Jensen, W. A. The histochemical localization of peroxidase in roots and its induction by indole acetic acid. *Plant Physiol.* **30,** 426-432 (1955)
[98] Jensen, W. A. The incorporation of C^{14}-adenine and C^{14}-phenylalanine by developing root tip cells. *Proc. Nat. Acad. Sci. Wash.* **43,** 1038-1046 (1957)
[99] Jensen, W. A. The nucleic acid and protein content of root tip cells of *Vicia faba* and *Allium cepa. Exper. Cell Res.* **14,** 575-583 (1958)
[100] Kalina, M., Gahan, P. B. A gelatin film method for improved histochemical localization of dehydrogenases in plant cells. *Stain Technol.* **43,** 51-57 (1968)
[101] Khavkin, E. E. The metabolism of growing root cells. In *Physiology of Plants.* I. Physiology of the root. 58-106. (Moscow: Viniti An SSSR Publ. 1973)
[102] Khavkin, E. E., Varakina, N. N. Hexose monophosphate pathway of glucose utilization in growing cells of maize root and coleoptile. *Fiziologia rastenii* **16,** 1064-1073 (1969)
[103] Khavkin, E. E., Zeleneva, I. V. Glycolytic enzyme pattern and constant proportion group in plant cells as related to their developmental and functional state. *F.E.B.S. Letters* **21,** 269-272 (1972)
[104] Kole, R., Sierakowska, H., Shugar, D. Novel activity of potato nucleotide pyrophosphatase. *Biochem. Biophys. Acta* **438,** 540-550 (1976)
[105] Lai, V., Srivastava, L. M. Nuclear changes during differentiation of xylem vessel elements. *Cytobiologie* **12,** 220-243 (1976)
[106] Lang, W., Maurer, W. Zür Verwendbarkeit von Feulgen Gefärbten schnitten für quantitative autoradiographie mit markientem thymidin. *Exper. Cell Res.* **39,** 19 (1965)
[107] Laties, A. G., Budd, K. The development of differential permeability in isolated steles of corn roots. *Proc. Nat. Acad. Sci. Wash.* **52,** 462-469 (1964)
[108] Lüttge, U., Laties, G. G. Absorption and long distance transport by isolated stele of maize roots in relation to the dual mechanisms of ion absorption. *Planta* **74,** 172-187 (1967)
[109] Lyne, R. L., Ap Rees, T. Invertase and sugar content during differentiation of roots of *Pisum sativum. Phytochem.* **10,** 2593-2600 (1971)
[110] McLean, J. A histochemical study of acid hydrolases in dividing and differentiating plant tissues. Ph.D. Thesis. University of London (1969)
[111] McLean, J., Gahan, P. B. The distribution of acid phosphatases and esterases in differentiating roots of *Vicia faba. Histochemie.* **24,** 41-49 (1970)

[112] McLeish, J. Comparative microphotometric studies of DNA and arginine in plant nuclei. *Chromosoma* **10,** 686-710 (1959)
[113] McLeish, J., Sunderland, N. Measurements of deoxyribosenucleic acid (DNA) in higher plants by Feulgen photometry and chemical methods. *Exper. Cell Res.* **24,** 527-540 (1961)
[114] Malawer, C. L., Phillips, R. The cell cycle in relation to induced xylem differentiation. Tritiated Thymidine incorporation in cultured tuber explants of *Helianthus tuberosus*. L. *Plant Sci. Letters* **15,** 47-55 (1979)
[115] Matile, Ph. Plant lysosomes. In *Lysosomes in Biology and Pathology* **1,** 406-430. J. T. Dingle and H. B. Fell (eds.) (Amsterdam: North-Holland 1969)
[116] Matile, Ph. The lytic Compartment of Plant Cells. *Cell Biology Monograph* **1.** (New York: Springer-Verlag 1975)
[117] Matile, Ph. Biochemistry and function of vacuoles. *Ann. Rev. Plant Physiol.* **29,** 193-213 (1978)
[118] Marty, F. Cytochemical studies on GERL, provacuoles and vacuoles in root meristematic cells of Euphorbia. *Proc. Nat. Acad. Sci. Wash.* **75,** 852-856 (1978)
[119] Meinders-Groeneveld, J., James, J. Some quantitative data regarding the nucleoli in cell nuclei from rat liver of different ploidy classes. *Z. Zellforsch.* **114,** 165-174 (1971)
[120] Middleton, J., Gahan, P. B. A quantitative cytochemical study of acid phosphatases in rat liver parenchymal cells of different ploidy values. *Histochem. J.* **11,** 649-659 (1979)
[121] Nägl, W. The mitotic and endomitotic nuclear cycle in *Allium carinatum IV* ^{3}H-uridine incorporation. *Chromosoma* **44,** 203-212 (1973)
[122] Nägl, W. *Endopolyploidy and Polyteny in Differentiation and Evolution.* (Amsterdam: North-Holland 1978)
[123] Newsholme, E. A. In *Placenta—A neglected experimental animal.* P. Beaconsfield and C. Viller (eds.) (Oxford: Pergamon Press 1979)
[124] Newsholme, E. A., Crabtree, B. Theoretical principles in the approaches to control of metabolic pathways and their application to glycolysis in muscle. *J. Mol. Cell Cardiol.* **11,** 839-856 (1979)
[125] Nishimura, M., Beevers, H. Hydrolases in vacuoles from castor/bean endosperm. *Plant Physiol.* **62,** 44-48 (1978)
[126] O'Brien, T. P. Further observations on hydrolysis of the cell wall in xylem. *Protoplasma* **69,** 1-14 (1970)
[127] Obroucheva, N. V. Physiology of Growing Roots Cells. In *'The Development and Function of Roots'.* 279-298. J. G. Torrey and D. T. Clarkson (eds.) (New York: Academic Press 1975)
[128] Pearse, A. G. E. *Histochemistry: Theoretical and Applied.* 3rd Edition. Volume 2. (London: Churchill Livingstone 1972)
[129] Pelc, S. R. Metabolic DNA in ciliated protozoa, salivary gland chromosomes, and mammalian cells. *Inter. Rev. Cytol.* **32,** 327-355 (1972)
[130] Pelc, S. R., LaCour, L. F. The incorporation of ^{3}H-thymidine in newly differentiated nuclei of roots of *Vicia faba. Experientia* **15,** 131-133 (1959)
[131] Pette, D., Hofer, H. W. The constant proportion enzyme group concept in the selection of reference enzymes in metabolism. CIBA Found. Symp. 1980 (in press)
[132] Pette, D., Luh, W., Bücher, Th. A constant-proportion group in the enzyme activity pattern of the Embden-Myerhof chain. *Biochem. Biophys. Res. Comm.* **7,** 419-424 (1962)
[133] Phillips, R. Cytodifferentiation. In *Recent Advances in Plant Tissue Culture.* I. K. Vasil and D. E. Murphy (eds.) (New York: Academic Press 1980)
[134] Phillips, R., Dodds, J. H. Rapid differentiation of tracheary elements in cultured

explants of Jerusalem artichokes. *Planta* **135,** 207-212 (1977)

135 Pitt, D. *Lysosomes and Cell Function.* (London: Longman 1975)

136 Polya, G. M., Ashton, A. R. Inhibition of wheat seedling 5′(3′) ribonucleotide phosphohydrolase by adenosine 3′, 5′—cyclic monophosphate. *Plant Sci. Letters* **1,** 349-357 (1973)

137 Popatov, N. G., Obroucheva, N. V., Maroti, M. On physiology of root system growth. In *Plant Growth.* (Lvov: Univ. Press 1959)

138 Reeve, R. M. Histological observations on seed coats of succulent peas. *Food Res.* **14,** 77-89 (1949)

139 Rigler, R. Microfluorometric characterization of intracellular nucleic acids and nucleo-proteins by acridine orange. *Acta Physiologica Scand.* (Supplement) **267,** 1-122 (1966)

140 Roberts, L. W. *Cytodifferentiation in plants. Xylogenesis as a model system.* (Cambridge: Cambridge University Press 1976)

141 Robbertse, P. J., McCully, M. E. Regeneration of vascular tissue in wounded pea roots. *Planta* **145,**.167-173 (1979)

142 Robinson, E. Proteolytic enzymes in growing root cells. *J. Exper. Bot.* **7,** 296-305 (1956)

143 Robinson, E., Brown, R. The development of the enzyme complement in growing root cells. *J. Exper. Bot.* **3,** 356- (1952)

144 Robinson, E., Cartwright, P. M. Nucleolytic enzymes in growing root cells. *J. Exper. Bot.* **9,** 27 (1958)

145 Salamatova, T. S., Balajan, E. N. Adenosine triphosphatase of cell fractions from growth zones of lupine roots. *Fisiologia rastenii* **13,** 76-81 (1966)

146 Samson, M., Davies, D. D. Metobolically labile DNA in mitotic and non-mitotic cells of *Zea mays. Life Sciences* **5,** 1239-1247 (1966)

147 Sato, S. The distribution of succinic dehydrogenase and mitochondria in the embryos of *Phaseolus vulgaris. Bot. Mag. Tokyo* **69,** 137-141 (1956)

148 Scharpé, A., Van Parijs, R. The formation of polyploid cells in ripening cotyledons of *Pisum sativum.* L. in relation to ribosome and protein synthesis. *J. Exper. Bot.* **24,** 216-222 (1973)

149 Sexton, R., Sutcliffe, J. F. The distribution of β-glycerophosphatase in young roots of *Pisum sativum. Ann. Bot.* **33,** 407-419 (1969)

150 Sexton, R., Sutcliffe, J. F. Some observations on the characteristics and distribution of adenosine triphosphatases in young roots of *Pisum sativum*, cultivar Alaska. *Ann. Bot.* **33,** 683-694 (1969)

151 Shaykh, M. M., Roberts, L. W. A histochemical study of phosphatases in root apical meristems. *Ann. Bot.* **38,** 165-174 (1974)

152 Sheikh, K. M., Gahan, P. B. Soluble proteins and hydrolases during crown gall induction in the tomato, *Lycopersicon esculentum. Histochem. J.* **8,** 87-92 (1976)

153 Sheikh, K., Gahan, P. B., Stroun, M. A cytochemical study of acid phosphatases and esterases during the induction of crown gall in the tomato. *Histochem. J.* **3,** 179-183 (1971)

154 Sheldrake, A. R. Cellulase and cell differentiation in *Acer pseudoplatanus. Planta* **95,** 167-178 (1970)

155 Shininger, T. L. The control of vascular development. *Ann. Rev. Plant Physiol.* **30,** 313-337 (1979)

156 Siebers, A. M. Differentiation of isolated interfasicular tissues of *Ricinus communis* L. *Acta Bot. Neerl.* **20,** 343-355 (1971)

157 Siebers, A. M. Vascular bundle differentiation and cambial development in cultured tissue blocks excised from the embryo of *Ricinus communis.* L. *Acta Bot. Neerl.* **21,** 327-342 (1972)

[158] Siegal, S. M. On the biosynthesis of lignins. *Physiol. Plant.* **6,** 134-139 (1953)
[159] Sierakowska, H., Gahan, P. B., Dawson, A. L. The cytochemical localization of nucleotide pyrophosphatase activity in plant tissues, using naphthyl esters of thymidine-5-phosphate. *Histochem. J.* **10,** 679-693 (1978)
[160] Smith, D. L. Localization of phosphatases in young gametophytes of *Polypodium vulgare.* L. *Protoplasma* **74,** 133-148 (1972)
[161] Srivastava, L. M,., Singh, A. P. Certain aspects of xylem differentiation in corn. *Canad. J. Bot.* **50,** 1795-1804 (1972)
[162] Stewart, P., Pitt, D. Ultrastructural localization of acid phosphatase activity in root tips by the p- (acetoxymercuric) aniline diazotate reagent and a comparison with a Gomori procedure. *J. Cell Sci.* **26,** 19-29 (1977)
[163] Sutcliffe, J. F., Sexton, R. Cell differentiation in the root in relation to physiological function. In *Root Growth.* W. J. Whittington (ed.) 80-102 (London: Butterworths 1969)
[164] Sutcliffe, J. F., Sexton, R. In *Structure and Function of Primary Root Tissues.* 203-219. J. Kolek (ed.) (Bratislava: Veda 1974)
[165] Swift, H. The constancy of desoxyribose nucleic acid in plant nuclei. *Proc. Nat. Acad. Sci. Wash.* **36,** 643-654 (1950)
[166] Thomas, C. V., Esterly, J. R., Ruddat, M. Light microscopical histochemistry of hydrolases in the root tip of Lima beans, *Phaseolus lunatus.* L., in relation to cell differentiation and wound regeneration. *Protoplasma* **91,** 229-242 (1977)
[167] Torrey, J. G., Zobel, R. Root growth and morphogenesis. In *The Physiology of the Garden Pea.* 119-152. J. F. Sutcliffe and J. S. Pate (eds.) (New York: Academic Press 1977)
[168] Truman, D. E. S. *The biochemistry of cytodifferentiation.* (Oxford: Blackwell Scientific 1974)
[169] Tschermak-Woess, E. Karyologische Pflanzenanatomie. *Protoplasma* **46,** 798-834 (1956)
[170] Tulett, M. H. Ph.D. Thesis. University of Edinburgh (1967)
[171] Van Fleet, D. S. Histochemical localization of enzymes in vascular plants. *Bot. Rev.* **18,** 354-398 (1952)
[172] Wiemken-Gehrig, V., Wiemken, A., Matile, Ph. Mobilisation von Zellenandstoffen in der welkenden Blüte von *Ipomoea tricolor* Cav. *Planta* **115,** 297-307 (1976)
[173] Zeleneva, I. V., Khavkin, E. E. Glycolytic enzymes in growing cells of maize roots. *Dokl. AN SSSR* **199,** 481-484 (1971)
[174] Zeleneva, I. V., Khavkin, E. E. The formation of enzymic pattern of respiration in growing cells. I. The glycolytic enzymes in maize seedlings. *Ontogenez* **2,** 311-320 (1971)
[175] Zeleneva, I. V., Reimers, F. E., Khavkin, E. E. The organisation of enzyme systems in the growing root cells. *Fisiologia rastenii* **19,** 1298-1305 (1972)

7. Plant-Growth Regulators and the Differentiation of Vascular Elements

R A SAVIDGE and P F WAREING
Department of Botany and Microbiology, University College of Wales, Aberystwyth, Wales

INTRODUCTION

A vast literature has relevance to the topic of this chapter. Regrettably, the authors cannot hope to cite, much less discuss, all of the research reports which have been published since the 1850's. Instead, the reader's attention will be directed to a large number of monographs and reviews which have treated one or more aspects of the subject; otherwise, only those research reports which bear upon the points at hand will be cited.

Various aspects of plant-growth regulators were reviewed recently in a 2-volume treatise[115], and a current review of analytical techniques used to purify and estimate endogenous levels of plant-growth regulators is in preparation[85]. Several recent reviews on the role of plant-growth substances in the regulation of xylem differentiation precede this one[20, 68, 147, 156, 166], while there are many older, but still useful reviews[9, 14, 20, 24, 38, 65, 74, 105, 112, 149, 150, 154, 157, 176, 185, 194, 211, 226].

XYLEM DIFFERENTIATION IN UNORGANIZED TISSUES, AND RELATED ASEPTIC-CULTURE SYSTEMS

Extensive studies have been made in recent years on the role of plant-growth substances in vascular differentiation; various types of aseptic-culture systems being used to study a variety of problems. The several objectives of sterile-culture investigations might be categorised as follows:

1. Identification of factors necessary for growth of the explanted vascular cambium[47, 62, 63, 90, 91, 136, 153, 177].
2. Identification of factors necessary for vascular-cambium induction, with the subsequent differentiation of xylem and phloem elements, in:
   (*a*) Disorganized parenchymatous calli[12, 26, 33, 62, 93, 155, 209, 210].
   (*b*) Explants of organized tissues[168, 186, 187].
3. Identification of the factors directly affecting tracheary-element differentiation, or sieve-cell differentiation, in:
   (*a*) Disorganized parenchymatous calli[19, 39, 42, 55, 59, 92, 128-130, 148, 183].
   (*b*) Cell-suspension cultures[21, 60, 61, 100, 182].
4. Investigations into the hypothesis that the cell cycle has a role in regulating tracheary-element differentiation.

5. Investigations into the hypothesis that a cell's ploidy state influences its ability to differentiate into a tracheary element, particularly in response to plant-growth regulators.
6. Investigations into the hypothesis that cell division must immediately precede tracheary-element differentiation if differentiation is to occur at all.

Categories 4, 5 and 6 above are dealt with in Chapter 5; accordingly, only the first three categories will be considered here.

The vascular cambium was one of the first plant tissues to be explanted and grown in sterile culture[63], and it has long been known that the cambium of both angiosperms[26, 63, 91, 136, 177] and conifers[26, 47, 63, 90, 91, 153, 164] becomes disorganized when explanted to media which will support growth. Consequently, explants of the cambium have seldom been investigated in aseptic culture from the standpoint of xylem or phloem differentiation[90]; instead, disorganized, parenchymatous tissues have been used. Camus[33] explanted a growing shoot apex onto callus derived from herbaceous roots, and subsequently observed clusters of polarized vessel elements which had differentiated beneath the apex within the callus tissue. Ball[12], Wetmore and Sorokin[210], and Wetmore and Rier[209] confirmed Camus' observation. Moreover, the application of indol-3yl acetic acid (IAA) to the callus, resulted in similar differentiation of xylem and phloem strands beneath the auxin application point as had occurred beneath an inserted apex. Some evidence suggests that sucrose may be interacting with auxin, and possibly other plant-growth substances, in influencing xylem- or phloem-cell differentiation in the latter type of sterile-culture system. Wetmore and Rier[209], Jeffs and Northcote[92], Rier and Breslow[155] and Fadia and Mehta[55] have all reported sucrose concentration to be a factor in determining whether cells differentiated into xylem or phloem elements. Relatively high sucrose concentrations were necessary for observable phloem differentiation, whereas low concentrations appeared to permit only xylem differentiation. J. and P. M. Warren Wilson[203, 204] have proposed that auxin and sucrose gradients control the position of regenerating cambia and subsequent xylem and phloem differentiation, but more experimental evidence will be necessary before the latter proposal will be on firm ground.

Some earlier evidence[33, 12, 210, 209] suggests that auxin alone, in the absence of other plant-growth regulators, is indeed the factor regulating vascular development. The proliferation of explanted vascular cambium on an auxin-containing medium, and failure to induce tracheary-element differentiation in callus cultures, at will, with auxin alone, has, however, led subsequent investigators to postulate that the ploidy state of each cell's nucleus is a determining factor in differentiation (see Dodds, Chapter 5; also Shininger[166]) and that gibberellins, cytokinins, ethylene, abscisic acid

(ABA), and several other substances, as well as auxins, are involved in the process of xylem differentiation[156].

Bergmann[19] found that kinetin increases the number of tracheary elements and the extent of lignin production in tobacco callus cultures. Torrey[183] demonstrated that cytokinins increased the proportion of tracheary elements in several different callus cultures, and Fosket and Torrey[59] found a cytokinin necessary for both cell division and subsequent tracheary-element differentiation in soybean callus cultures, the extent of tracheary-element differentiation increasing with increasing levels of cytokinin. Minocha and Halperin[128–130] reported that a cytokinin, in the presence of an auxin, specifically induced as much as 35% tracheary-element differentiation in the cell population of tuber slices of Jerusalem artichoke *(Helianthus tuberosa)*. Both ABA and gibberellic acid (GA_3) were found to inhibit the cytokinin-stimulated xylogenesis. Phillips and Torrey[148] found as much as 20% tracheary-element differentiation in the presence of an auxin and a cytokinin in pea-root cortical explants, whereas negligible tracheary-element differentiation was observed in the absence of cytokinins. Cytokinins have been found to stimulate the extent of tracheary-element differentiation in cultures of lettuce pith[42], carrot-root slices[132], and *Zinnia* mesophyll cells[100, 60, 61].

The effectiveness of cytokinins in inducing tracheary-element differentiation may reside in their ability to stimulate phenylalanine ammonia lyase (PAL) activity, leading to the availability of phenylpropanoid lignin precursors. Rubery and Northcote[159] reported that differentiating xylem elements may synthesize lignin directly from phenylalanine in the presence of PAL without the requirement for a supply of lignin precursors from the cambium (*cf.*[126]), while Bevan and Northcote[21] have shown that kinetin caused a rapid stimulation of the rate of PAL induction in suspension cultures of *Phaseolus*, and overcame an inhibitory effect of naphthalene acetic acid (NAA) on PAL induction.

Cytokinins have been implicated as necessary for cambial development, which is thought to be a process essentially unrelated to vascular-bundle differentiation[168]. Torrey and Loomis[187] reported a cytokinin necessary for normal cambial development in excised roots, and Siebers[168] made a similar observation on embryonic hypocotyl blocks. Omission of a cytokinin from the medium prevented cambial development in both studies.

The sterile-culture literature suggests that several plant-growth substances may be interacting in the regulation of differentiation of individual cells into tracheary elements. In no instance, however, has it been demonstrated unequivocally that direct regulation of xylogenesis can be attributed to one or more of these substances.

Both Torrey[182] and Fukuda and Komamine[60, 61] have presented evidence for tracheary-element differentiation in single, isolated cells, in the absence of immediately preceding cell division. Torrey's results, however, were obtained on an incompletely-defined medium. Fukuda and Komamine studied differentiation in leaf mesophyll cells, the biosynthetic capacity of which complicates interpretation of whether the tracheary elements differentiated in direct response to growth regulators supplied in the culture medium, or in response to cellular metabolites. Claims that plant-growth regulators are directly involved in tracheary-element differentiation would be more convincing if isolated single cells were to differentiate directly into tracheary elements following explanting onto a defined medium.

Sterile callus- and suspension-culture systems being used to investigate xylogenesis lie somewhere between fully organized tissues and single, isolated cells. Callus-culture investigations such as those first carried out by Camus[33], and later extensively used by Wetmore and Rier[209], clearly suggest a role for auxins in initiating and polarizing vessel elements and cambia which thereafter produce vascular tissues within an otherwise non-polar, non-organised mass of cells. It is reasonable to assume that results of such studies are also applicable to the organized polarisation of the vascular system within the whole plant. However, the contribution made by the numerous more recent callus- and suspension-culture investigations toward understanding the factors regulating xylem-cell differentiation remains uncertain. That both of the latter types of aseptic-culture systems have been multi-cellular, yet only a relatively small proportion of the cell populations have differentiated into xylem or phloem elements might be because polyploid cells respond to plant-growth regulators in a way different from diploid cells. Alternatively, this might mean that the direct roles of plant-growth regulators in xylogenesis have not been clearly separated from their general effects on cell growth, cell division, and associated metabolic processes because of the complexity of the experimental systems used.

Nodulation in callus cultures and clumping in suspension cultures are features frequently associated with xylogenesis, and it has been demonstrated on numerous occasions that manipulation of the auxin: cytokinin ratio in culture systems can be used to induce either root or shoot formation. These considerations again raise the question of whether auxins and cytokinins induce the formation of tracheary elements by acting directly on the differentiation processes of those cells (whether diploid or polyploid) which do differentiate, or whether their primary effect is to establish an environment conducive to the production of organized regions, followed by tracheary-element differentiation in response to substances, or gradients, produced by those regions.

Although it is possible to postulate that varying levels of auxins, gibberellins and cytokinins interact to initiate xylogenesis directly, it remains equally plausible that other as-yet-unidentified regulatory factors interact with plant-growth regulators or act singly as specific differentiation factors. Certainly, auxins, cytokinins, and gibberellins have each been implicated in both phloem- and xylem-cell differentiation (see below) suggesting that none of those substances acts directly to initiate differentiation.

XYLEM AND PHLOEM DIFFERENTIATION IN HERBACEOUS SPECIES

Beginning in 1935, Snow[169], Gouwentak and Hellinga[71], and Fishnich[58] each reported independently that exogenous IAA applied in lanolin paste to decapitated plants effectively substituted for elongating shoots and developing leaves in stimulating cambial activity (cell division) and a limited amount of new xylem differentiation. Since then, much evidence has accumulated to show that application of exogenous IAA will stimulate (1) the initiation and regeneration of vascular tissue, (2) cambial division, and (3) the differentiation of new xylem and phloem. Some of the effects of IAA on initiation of vascular tissue have been described in the preceding section.

Jacobs[87, 88] found that in *Coleus* stems, parenchyma cells centripetal to a severed vascular bundle differentiated into wound-vessel members when the leaves above were left intact. If the leaves above the wound were removed, however, wound-vessel members did not differentiate. When exogenous IAA was substituted for the leaves, wound-vessel member differentiation proceeded normally. Wangermann[191] reported that exogenous auxin could replace the developing leaf in its ability to maintain continuing xylem differentiation within stem tissue below the leaf petiole. However, Young[227], working with *Lupinus*, found that vascular-bundle differentiation did not proceed when the very young leaf primordium was removed and replaced with exogenous IAA. Sachs (personal communication) has confirmed Young's[227] observations in *Pisum*. Thus, although IAA appears capable of replacing the developing leaf in stimulating xylem differentiation, whether it is the factor derived from very young leaves which stimulates vascular-bundle differentiation below them remains unresolved.

Wound-phloem regeneration also occurs around severed vascular bundles, and several investigators[86, 108, 180] have shown that IAA effectively replaces leaves in stimulating its formation. Aloni[6, 7] found mature leaves to be necessary for the differentiation of phloem fibres below them, and that IAA replaced the leaf in that function.

Using auxin bioassay to estimate endogenous auxin levels, Jacobs and Morrow[89] found diffusible auxin levels from developing leaves to show a direct relationship to the rate of production of new xylem in petioles below leaves of similar development.

A number of studies have been carried out on the effects of gibberellins and cytokinins, both alone and in combination with auxins, on cambial division and vascular differentiation. Following the work of Wareing[195] and Digby and Wareing[49] with seedlings and stem cuttings of woody plants (p. 201), in which it had been found that GA_3 acted synergistically with IAA in stimulating cambial-cell division and xylem differentiation, Cutter[40] made applications of IAA and GA_3 to *Phaseolus* internodes which had only primary vascular tissues present at the beginning of the experiments. The following results are evident from her photomicrographs (Cutter[40], p. 93): GA_3 alone induced a cambium and stimulated cambial-cell division, but negligible differentiation of the cambial derivatives. IAA alone induced cambial organisation, cambial-cell division and xylem-fibre and vessel-element differentiation. IAA and GA_3 together induced cambial organisation, cambial-cell division, phloem-fibre differentiation, xylem-fibre and vessel-element differentiation, and also enhanced the number of cambial derivatives produced.

Harrison and Klein[77] found that NAA replaced the apex for continuing xylem production in decapitated plants of *Phaseolus* and *Coleus* but did not affect secondary phloem-cell production. GA_3 or the cytokinin, benzyl-amino purine (BAP), each could partly replace the stem apex for phloem-cell production, and a mixture of auxin and GA_3 synergistically stimulated phloem-cell production.

Using decapitated *Phaseolus* plants, each having only two mature leaves, Hess and Sachs[83] observed that in the absence of applied growth substances, the xylem which formed below a mature leaf contained few vessels, consisting rather of other lignified elements, primarily fibres. In contrast, when young leaves were left intact, together with the pair of mature leaves, a relatively small amount of xylem was produced and it contained more vessels. When young leaves were replaced with a lanolin paste containing auxin, the vessel-fibre composition was similar to that which occurred in plants containing intact young leaves. Auxin applications also increased the area of xylem occupied by vessels below a mature leaf.

Hess and Sachs[83] observed that only vessel formation occurred immediately below the site of auxin application, and Sachs[161] has presented additional evidence that auxin is directly involved in vessel-element differentiation. The question then arising concerns the factor controlling xylem-fibre differentiation. Hess and Sachs[83] found that although gibberellin applications by themselves had no noticeable effect, GA_3 in combination with auxin enhanced cambial activity, and the xylem which differentiated contained fewer vessels, thus resembling that produced in the presence of mature leaves.

In pea *(Pisum)* epicotyls, Sachs[161] found exogenous auxin to be effective in stimulating xylem differentiation when apices and young leaves were both removed. In addition, cytokinins and gibberellins both enhanced the number of tracheids and vessels produced. Neither cytokinins nor gibberellins individually stimulated xylem production, an observation which has been confirmed by numerous investigators.

Shininger[167] observed vascular development in decapitated *Xanthium* stems at a point below two mature leaves and found exogenous NAA, but not IAA, to be effective in stimulating secondary xylem-fibre differentiation. In the absence of exogenous auxin, cambial-cell division and production of new vessel elements amongst parenchyma cells continued. GA_3 enhanced the production of cambial derivatives but did not affect the differentiation of xylem fibres. Similarly, Aloni[6] found GA_3 alone did not stimulate phloem-fibre differentiation, although it did when combined with IAA.

From the foregoing evidence, it would appear that in several plant systems GA_3 acts synergistically with IAA in stimulating cambial division and, in some cases phloem- and xylem-fibre differentiation. Auxin is necessary for differentiation of both phloem fibres and xylem cells, especially vessel elements.

There is much evidence to suggest that cytokinins, in addition to auxins and gibberellins, may also participate in the regulation of xylem and phloem differentiation within the whole plant. Torrey and Fosket[184] presented evidence that both auxins and cytokinins are necessary for tracheary-element differentiation in isolated root segments. Within the intact parasitic vascular plant *Cuscuta campestris*, Tsivion[188] found that tracheid differentiation, necessary for host-parasite union, was limited by cytokinin alone. Sorokin and colleagues[174, 175] reported enhanced xylem differentiation following application of exogenous cytokinins on pea apical meristems. It remains unclear, however, whether auxins and cytokinins act directly to determine xylem-cell differentiation. Houck and LaMotte[86] observed, for example, that most wound-phloem regeneration occurred around severed phloem bundles of *Coleus* when auxins and cytokinins were supplied together, while wound-vessel differentiation did not occur in the cells adjoining the regenerated wound phloem.

The evidence based upon experiments with intact parts of herbaceous plants clearly implicates auxins, gibberellins and cytokinins in both xylem and phloem differentiation. However, when exogenous growth substances are applied to organised vascular systems, (whether to intact, disbudded, or defoliated plants, or to isolated stem or root sections), and new vascular tissue is produced, several considerations complicate determination of positive or negative effects on xylem or phloem differentiation. Although there is good evidence that young leaves are centres of auxin and gibberellin production, there is much evidence that these growth substances are also formed in mature leaves. Consequently, even if a plant is disbudded, auxins and/or gibberellins may still be produced in significant amounts by the mature leaves, and a negative response to applied exogenous GA_3 may not necessarily indicate that gibberellins are not involved in cambial division, or differentiation of xylem or phloem elements. Similarly, there is good evidence that endogenous cytokinins are produced by roots, and consequently, a negative response when exogenous cytokinin is applied to a rooted plant may, again, not indicate that this type of hormone is not involved.

In addition, application of growth substances to decapitated stems frequently leads to changes in cortical and pith cells, to phloem proliferation, and to callus formation at the site of application. Such abnormal cells and tissues may themselves be sources of endogenous growth substances which stimulate cambial division and vascular differentiation. Consequently, effects obtained following application of exogenous growth substances to intact or mutilated plants may be indirect or secondary, and it becomes difficult to decide what is the precise role of a given type of growth substance from such evidence. Moreover, the nature of cambial response to applications of exogenous growth substances no doubt is concentration dependent; however, the concentrations tested have been variable amongst the numerous investigators, and this fact may at least partially account for the existing controversy about the role of plant-growth substances in vascular-element differentiation.

ASPECTS OF SECONDARY VASCULAR DEVELOPMENT IN HARDWOODS

Although cell division in the vascular cambium is conceptually a separate problem from that of xylem-cell, or phloem-cell, differentiation, the possibility that meristematic activity influences differentiation, or *vice versa*, cannot be discounted. Thus both problems will be considered in the pages that follow.

Hartig[78] reported in 1853 that 'cambial reactivation' began in the spring at the bases of extending buds of deciduous hardwoods in Germany. From the bud bases, cambial reactivation spread basipetally. In 1883, Russow[160] contradicted Hartig: "Der Beginn der Cambiumthätigkeit lässt keine Beziehung zum Schwellen oder Aufbruch der Knospen, auf Blatt- oder Blütenentfaltung erkennen." In support of this statement, Russow cited oaks, ashes, elms, spruces and pines, amongst others, as examples. Later Hastings[79] confirmed Hartig's observation in several diffuse-porous hardwood species. Priestley and Scott[151] were apparently first to propose that a fundamental difference existed between ring- and diffuse-porous deciduous hardwoods in their patterns of cambial reactivation. Today, it is well known that both cambial reactivation and new xylem formation first become apparent at the bases of extending shoots in diffuse-porous, deciduous hardwoods. From there, initiation of both processes proceeds basipetally at a rate which agrees with the known rate of basipetal transport of IAA[8, 10, 23, 45, 51, 69-71, 84, 97, 133, 150, 154, 178].

In deciduous hardwoods which have ring-porous xylem, opinions continue to differ as to the site of first cambial reactivation (cell division) in the tree. Some investigators have stated that it begins simultaneously throughout the tree; others, that it begins in the trunk prior to its occurrence in the 1-year-old twigs; others again, that it begins at bud bases and proceeds basipetally[150, 154, 198]. Wareing[198] and Balatinecz and Farrar[10] demonstrated clearly that new xylem formation in ring-porous hardwoods begins in the spring at bud bases and then proceeds basipetally. Consequently, it may well be that the continuing controversy over the pattern of cambial reactivation in ring-porous species will be resolvable only when new phloem formation, cambial-cell division, and new xylem differentiation are carefully distinguished.

The view that cambial reactivation and earlywood differentiation in hardwood species is controlled by auxin is supported by a vast literature. Jost[97, 98] observed a relationship between the development of new shoots and leaves, and the commencement of cambial activity, which led him to propose that a stimulus from the new shoots was responsible for cambial reactivation. Coster[38] revived Jost's[97] 1891 hypothesis, following investigations into cambial activity in a large number of tropical species. Gouwentak and Hellinga[71] demonstrated that exogenous IAA reactivated the cambium of disbudded, non-foliated shoots in hardwoods, and stimulated some new xylem differentiation. Using Went's[207] *Avena* coleoptile bioassay for auxin, Zimmermann[231], Avery *et al.*[8] and Söding[171-173] reported obtaining high concentrations of diffusible auxin from swelling buds and expanding leaves, whereas during the winter months none was found. Gouwentak[70] reported growth hormones in *Fraxinus* buds prior to any evidence for bud swelling.

Further confirmation that extending shoots of hardwood species produce quantities of auxin in the spring was provided by Hatcher[80, 81], Dörffling[51], and Digby and Wareing[50]. Wareing *et al.*[201], Wareing[195], Digby and Wareing[49], and Balatinecz and Farrar[10] have confirmed Gouwentak's[69-71] finding that exogenous auxin reactivates hardwood cambia and results in some new xylem differentiation.

In reviewing the literature about the role of auxin in reactivating the cambia of hardwoods, Reinders-Gouwentak[154] made two important points: (1) hardwood cambia exhibit dormancy which, like buds, must be broken (e.g. by exposure to ethylene or by satisfaction of a chilling requirement) before cambial reactivation will occur, and (2) a large body of evidence clearly suggests that cambial reactivation following dormancy break in spring is limited in deciduous hardwoods by auxin availability.

Wareing[195] and Digby and Wareing[49] found applications of exogenous IAA alone caused some cambial-cell division and vessel differentiation, but no new phloem formation. GA_3 alone caused cambial-cell division with differentiation of new phloem cells, but cambial derivatives on the xylem side remained incompletely differentiated. Application of IAA and GA_3 together gave a large synergistic effect on cambial-cell division and the differentiation of apparently normal xylem and phloem elements, suggesting that IAA and GA_3 act synergistically in seedlings and younger cambia of hardwood cuttings to promote cambial-cell division, and that IAA is necessary for xylem differentiation whereas phloem differentiation is promoted by GA_3 in the absence of IAA.

There appears to be a difference between diffuse-porous and ring-porous temperate deciduous hardwood species not only in their pattern of initiation of cambial activity in the spring, but also in the duration and cessation of cambial activity later in the growing season. In those diffuse-porous species in which the problem has been studied, there is a close correlation between the duration of extension growth and of cambial activity in the annual shoots, with xylem development ceasing soon after the cessation of extension growth[8, 50, 151]. On the other hand, in ring-porous trees, such as *Robinia*, *Ulmus*, *Fraxinus* and *Castanea*, extension growth frequently stops early in the growing season, but cambial activity continues for a longer period. The cessation of extension growth is, however, correlated with the transition from earlywood to latewood production[50, 198]. Anatomical studies on *Robinia pseudacacia*[50, 202] indicated that the latewood vessels in the main stem are continuous with the leaf traces of mature leaves, suggesting that a cambial stimulus is produced in the latter. Latewood continues to be produced in seedlings of *R. pseudacacia* and *Ailanthus altissima* in which extension growth has ceased, so long as they are maintained under long photoperiods; but if the seedlings are transferred to short days, xylem production ceases, although active phloem formation continues for a further two weeks[50, 202]. Waisel and Fahn[189] reported that at high temperatures and short days (conditions which will not normally occur in its natural geographical distribution), the cambium remains active in seedlings of *R. pseudacacia*.

Bioassay of diffusates and extracts for auxins has shown that there is a marked decrease in the levels of endogenous auxins in the shoots of diffuse-porous species when extension growth stops. This is consistent with the observed cessation of cambial activity[8, 50]. Similarly, in the ring-porous species *A. altissima* and *R. pseudacacia*, a decrease in the levels of endogenous auxins, in association with the cessation of xylem development, was found when seedlings were transferred to short days. On the other hand, the levels of endogenous gibberellins in seedlings of *A. altissima* did not decline significantly when transferred from long to short photoperiods, again consistent with the observation that phloem production continues under short days[50].

THE PATTERN OF CAMBIAL REACTIVATION IN CONIFERS

Precise knowledge of the position in the tree where cambial reactivation (cell division), phloem differentiation, and xylem differentiation first occur in the spring may be used as circumstantial evidence that commencement of those activities is regulated by chemical factors derived from nearby organs, such as buds or leaves. Such observations have been used in the formulation of a hypothesis to explain how cambial reactivation and annual-ring formation are regulated in the whole tree[38, 97, 98, 112, 150, 198].

Although direct microscopical observation of the cambial zone (terminology of Wilson *et al.*[212]) is the only technique which can establish definitively the physiological state of the vascular cambium, less definitive techniques have been employed frequently to the same purpose. These include (1) observations on the presence or absence of newly differentiated xylem elements without concomitant observations on the meristematic zone (because of the extra work required to obtain clear sections of the cambial zone), (2) bark peelability, and (3) outside-bark circumference or stem diameter measurements. Phloem-cell differentiation in the spring is, however, associated with both diameter increase and bark peelability, and commonly precedes cambial-cell division by two or more weeks[1, 4, 29, 30, 73, 99, 107, 164]. Furthermore, cambial-cell division precedes xylem differentiation by another 3-6 weeks[4, 14, 24, 107, 135, 164, 211, 226]. The value of indirect techniques to physiological studies on the pattern of cambial reactivation can, therefore, only be considered as negative.

Russow[160] was aware that cambial reactivation in conifers showed no obvious relationship to bud burst. The first systematic microscopical observations which attempted to establish the pattern of this reactivation appear to have been made by Hastings[79] at Cornell University (Ithaca, New York). In *Pinus strobus*, it was found that the first reactivation in 2- to 3-year-old material occurred before that in 1-year-old twigs and before the buds had opened.

Brown[29, 30], also near Ithaca, N.Y., used direct microscopical observation of the cambium to investigate systematically the renewal of cambial activity in two *Pinus* spp. Radial expansion in both phloem and cambial cells occurred from March to early April, preceding cell division in the cambial zone in both species by several weeks. Cambial reactivation occurred in the mid-portion of the main stem earlier than it did at its base or in the 1-year-old shoots. Radial growth in the main stem and branches commenced some distance behind the apical shoots and spread gradually both upwards and downwards. Brown concluded that no correlation existed between shoot extension and the time of commencement of ‘radial thickening' in pine trees.

Knudson[99], also at Ithaca, N.Y., used microscopy to study seasonal cambial reactivation in *Larix;* like Brown, he found cambial reactivation to occur in the mid-portions of the trunk, independently of the branches and developing shoots. Knudson also noted that new phloem differentiation preceded xylem differentiation in the spring.

Priestley[150], working in England, presented arguments in favour of basipetal cambial reactivation beginning at elongating shoot bases, in all tree species, whether temperate or tropical, evergreen or deciduous. Priestley, however, avoided direct confrontation with the contradictory evidence reported by Russow[160], Hastings[79], Brown[29, 30], and Knudson[99] on conifers. Instead, he implied that the observed pattern of reactivation in conifers was probably due to early cessation and over-wintering of otherwise active cambia, as a consequence of freezing temperatures which he presumed may have occurred in North America prior to the onset of dormancy.

Münch[134] (p. 38, Engl. transl.), working in Germany, disagreed with Priestley's[150] generalization that cambial growth was dependent upon shoot growth in all species. He pointed out that in disbudded pine trees, mature needles were sufficient for cambial reactivation and new xylem differentiation. Later, the extensive investigations of Onaka[143, 144] on *Pinus thunbergii* confirmed Münch's observations. Wareing[194] reviewed the Priestley-Münch contradiction and emphasized the need for further careful and detailed observations on the pattern of cambial reactivation in conifers, without which the question must remain open.

Recent observations by Kutscha *et al.*[107] have produced another report of cell division in both normal and compression wood cambial zones of *Abies balsamea* 4-5 weeks prior to bud burst. In contrast, Denne[44, 45] has made detailed studies on the pattern of new xylem formation and presented evidence for strictly basipetal progression of the first earlywood formation in main stems and branches of *Picea* beginning in association with bud burst.

To determine where cambial reactivation first commences, one of us[164] has made detailed microscopical observations of cambia harvested throughout the length of the main stem, and over the full length of one living branch from each whorl, in 20-year-old lodgepole pine *(Pinus contorta* var. *latifolia)* trees. All trees were growing on the same site, at an altitude of 500 m ASL, near Aberystwyth. Two healthy and uninjured trees were destructively harvested on each of three dates: March 13, April 7, and April 29, 1980. All samples for microscopy were taken from the same aspect in both main stem and branches. The diagrammatic representation in Fig. 7.1 illustrates the pattern of cambial reactivation and new xylem formation observed (for purposes of clarity, half of the observations made on April 29 have not been included, but their presence would not alter the interpretation). On April 7, both trees showed new phloem formation occurring throughout the full length of the main stem and branches (phloem mother cells were radially expanded but showed incomplete sieve-area callose development, while the bark readily separated from the wood, leaving the cambial zone adhering to the mature xylem). Cambial reactivation (discerned by the presence of mitotic figures, or the presence of new periclinal dividing walls) was apparent in the lower region of the main stem and at the bases of the older branches, but was not evident in the younger cambia of either main stem or branches.

The first secondary-wall formation and lignification was observed in the 3-year-old main stem cambial region of one of the two trees harvested on April 7. At no other location in that same tree or in the other tree was a similar advanced stage of development observed. Cambial reactivation had occurred in the main stems of both trees by April 7, and radial expansion of differentiating xylem elements was most evident in main-stem cambia between 3 and 9 years of age. In contrast, there was little evidence of radial enlargement of the first earlywood tracheids at any position in any of the branches by April 7. Cambial reactivation at branch-bud bases was not apparent, although 2-year-old and older branch cambia showed reactivation. Bud burst began near the end of April, and by that time several earlywood tracheids had differentiated in 2- to 4-year-old main-stem cambia, as well as in the bases of adjoining branches (Fig. 7.1). In numerous instances, no secondary-wall formation was evident by April 29 in derivatives of 1-year-old cambia, although it had clearly commenced in 2-year-old and older cambia within the same branches.

Cross sections of the entire circumference of cambia were observed in 1-year-old material only in this study. If observations of the cambial region around the full circumference of the main stem and branches had been made, it is possible that a slightly different pattern of reactivation may have been observed in the tree. Nevertheless, the observations suggest that neither cambial-cell division or xylem-cell differentiation began at the bud base, but rather at a considerable distance below the 1-year-old shoot.

Further evidence that the conifer cambium may reactivate, and produce new xylem, independently of new shoot and leaf growth, was obtained by applying electric heat tape to the main stems of potted lodgepole-pine trees of either 5 or 10 years of age. Trees were transferred from out-door winter conditions, in early January, to a refrigerated (6-8°C) growth room with a 9-hr (10,000 lux) photoperiod. Pliable electric heat tape, enclosed in lampwick 2.5 cm in width, was wrapped twice around the circumference of either foliated or non-foliated main-stem regions, at a single location on each tree tested. The temperature between the outer bark and the heat tape was recorded by a thermometer inserted between bark and heat

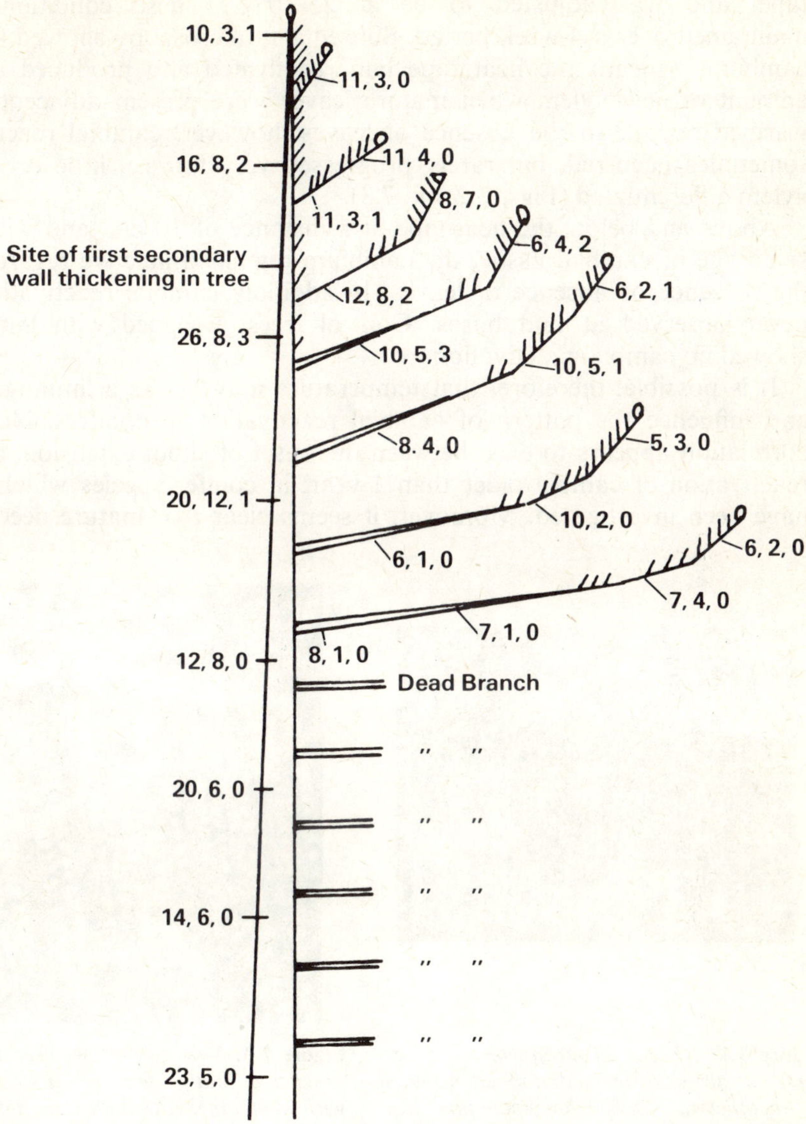

Figure 7.1: *Cambial reactivation in* P. contorta *var.* latifolia, *showing the state of the cambium in a 4.5 m tall tree on April 29, 1980; all buds had extended approximately 1 cm on that date. The numbers refer to the* width of the meristematic zone, *the* width of radially-enlarged, primary-walled, differentiating xylem, *and the* width of the zone of secondary-wall formation and lignification, *respectively; all three parameters are* no. cells/radial file. *The site where first secondary-wall thickening in the tree was observed on* April 7, 1980, *is also indicated. Similar trees harvested March 13, 1980, showed cambial reactivation in both main-stem and branch bases, whereas more apical regions had cambia which were inactive.*

tape, and was adjusted to be at 25-30°C. These conditions were maintained over a 4-week period. Subsequent microscopy showed that the cambium beneath the heat tape had reactivated and produced a small amount of new xylem when mature leaves were present adjacent to the warmed region. In the absence of leaves, however, cambial reactivation sometimes occurred, but rarely progressed very far; i.e. little or no new xylem differentiated (Figs. 7.2 and 7.3).

Above and below the heat-tape at a distance of 10 cm, and within the same age of cambial tissue, the cambium remained inactive regardless of the presence or absence of leaves. In addition, cambial reactivation was never observed at bud bases. Control trees, wrapped with lampwick, showed no cambial reactivation.

It is possible, therefore, that temperature may act as a limiting factor, and influence the pattern of cambial reactivation in conifers. No direct correlation appears to exist between the onset of shoot extension, and the reactivation of cambia older than 1-year, in conifer species which so far have been investigated. Moreover, it seems clear that mature needles are

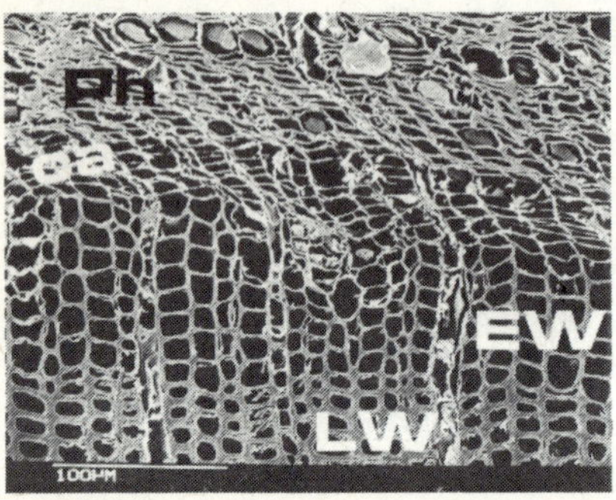

Figure 7.2: *SEM micrograph of the transverse surface of a warmed main-stem region of an otherwise chilled lodgepole-pine tree. Mature needles were present at the site of heat-tape application. LW = late-wood produced the previous year; EW = new early-wood tracheids produced beneath the heat tape; ca = cambial zone; Ph = mature phloem. Some radially-enlarged cells between the cambial zone and the new early-wood tracheids are incompletely differentiated.*

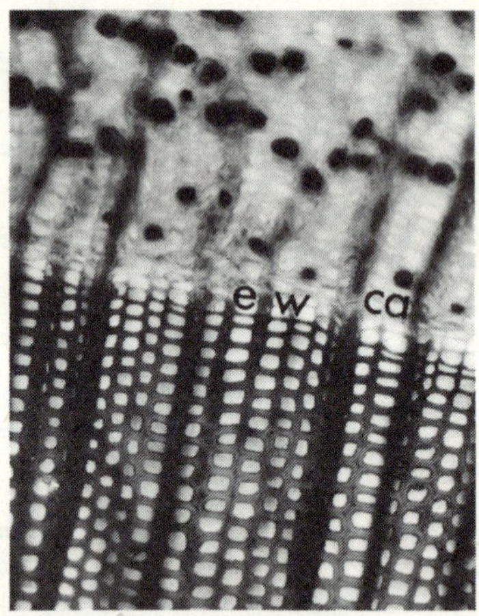

Figure 7.3: *New xylem (ew) produced in the absence of mature leaves in a stem wrapped with heat tape. Although the cambium (ca) did reactivate, only a minimal amount of differentiation occurred.* P. contorta. *(hand section).*

sufficient for initial cambial reactivation and new xylem formation, especially in cambia older than 1 year. Continued cambial activity and xylem differentiation does, however, depend on the presence of extending shoots and developing leaves[56, 104, 134, 143, 144, 164].

PLANT-GROWTH SUBSTANCES AND ANNUAL-RING FORMATION IN CONIFERS

In addition to the differences between diffuse-porous hardwoods and conifers in their patterns of cambial reactivation, correlative seasonal studies on endogenous levels of growth regulators have suggested fundamental differences between angiosperms and gymnosperms of woody habit. The pioneering investigations of Czaja[41], Söding[171-173], Zimmermann[231], Mirov[131], Gunckel and Thimann[76] and Onaka[143, 144] established by means of auxin bioassay, an apparent correlation between shoot extension and increase in amount of endogenous auxin issuing from the bases of new shoots, or extending leaves, in several conifer species and *Gingko biloba*[76].

Brown[25], Brown and Wetmore[27], Allen[5], Larson[110, 112, 114], Giertych and Forward[66], Ogasawara[142], Kopcewicz[101] *et al.*[103], Balatinecz and Kennedy[11], Alden and Eliasson[3], Alden[2] and Dunberg[52] have presented further evidence that auxin is derived from extending conifer shoots in the spring, and that cambial reactivation occurs as a consequence of a wavefront of auxin proceeding basipetally through the cambial tissue.

Larson[109, 110, 112, 114] presented the hypothesis that cambial reactivation and earlywood and latewood differentiation were regulated by seasonal variation in auxin gradients. He proposed that earlywood formation occurs in response to high auxin levels associated with shoot growth while latewood formation occurs following cessation of extension growth in new shoots and leaves as a consequence of the resultant decline in auxin export to the cambium. The relatively thin secondary walls characteristic of earlywood were explained on the basis of competition for assimilate between shoot and diameter-growth meristems, with the shoot being the stronger sink. During latewood formation less radial-wall expansion occurs as a result of decreased auxin levels, and increased secondary-wall thickening results from increased availability of assimilates formerly used for shoot and leaf growth.

Wodzicki[214] analyzed mathematically seasonal variations in auxin levels in the cambial region of conifers and concluded that no clear correlation exists between changes in auxin levels and any of the several aspects of vascular development included within Larson's[112] hypothesis. Moreover, the independent work of Onaka[143], Wodzicki[216, 214], Shepherd and Rowan[165], Wodzicki and Wodzicki[222], Jenkins and Shepherd[95] and Odani[140] has challenged the hypothesis that auxin is derived in the main from extending shoots and developing leaves. Onaka[143] provided evidence that the mature pine leaf produces auxin, and all of the other investigators have reported readily detectable auxin levels in the cambial region of conifers at all times of the year. This suggests that there is no clear correlation between cambial reactivation and an increase in auxin levels, or the onset of cambial dormancy and a decrease in auxin levels. More recently, Wodzicki[213] reported increased auxin levels in the cambial region of mature Scots pine (*P. sylvestris* L.) trees during the season of active radial growth while levels declined as the autumn approached in agreement with Larson's findings[110, 112].

Few of the investigators cited above have attempted to account for the quantity of endogenous IAA lost during extraction and purification procedures (cf.[120]). Moreover, since the auxin was usually detected by bioassay, and was not identified positively in any of their studies, it remains uncertain whether results obtained in separate laboratories are in fact comparable (e.g. see West[208]). Conceivably, the apparent controversy concerning the existence or non-existence of a correlation between endogenous auxin levels and the activity of the vascular cambium could be equally attributed to the semi-quantitative and non-definitive estimation procedures used, as to any other factor. Definitive gas chromatography—mass spectrometry (GCMS) analyses of seasonal variations in levels of endogenous IAA in conifers are few. DeYoe and Zaerr[48] presented evidence of a substantial spring increase in free IAA levels within 1-year-old shoots harvested from 15-year-old Douglas fir *(Pseudotsuga menziesii)*. The presence or absence of associated cambial activity was not, however, investigated. Little *et al.*[120] and Caruso *et al.*[36] reported GCMS measurements of IAA levels in *Picea sitchensis* and *P. menziesii*, respectively. Those reports did not, however, concern themselves with seasonal variations of IAA levels and their correlative significance to cambial activity and new xylem differentiation.

Recently, Little and Wareing[122] have estimated IAA and ABA levels in the cambium of *P. sitchensis* by GCMS; comparisons being made between free, bound, and diffusible levels of both growth substances. Clear maxima of all three forms of both substances were associated with the summer season and new xylem production. Both free and diffusible IAA levels, however, remained high in winter. On the basis of growth-regulator mass per tissue weight, free and diffusible IAA levels at all seasons were generally considerably higher than the corresponding levels of ABA. In contrast, bound ABA levels were always higher than bound IAA levels in the cambial region. Bound ABA levels were highest from mid-June to mid-September, but free ABA levels did not increase as bound ABA levels fell, and no clear relationship was found between the onset of cambial inactivity and the level of ABA in the cambial region.

Little and Wareing[122] found that in ringed trees, IAA accumulated above and decreased below the rings, regardless of whether the trees were ringed in late winter, prior to cambial reactivation, or in mid-October, following the onset of cambial inactivity. Short-day photoperiod induced bud dormancy and cambial inactivity and reduced the levels of free and diffusible IAA, and free, bound, and diffusible ABA. The decline in growth-substance content occurred after the cambium had become inactive, however, and Little and Wareing concluded that ABA probably is not involved in cambial dormancy.

Figures 7.4 and 7.5 illustrate further GCMS results[164] on seasonal

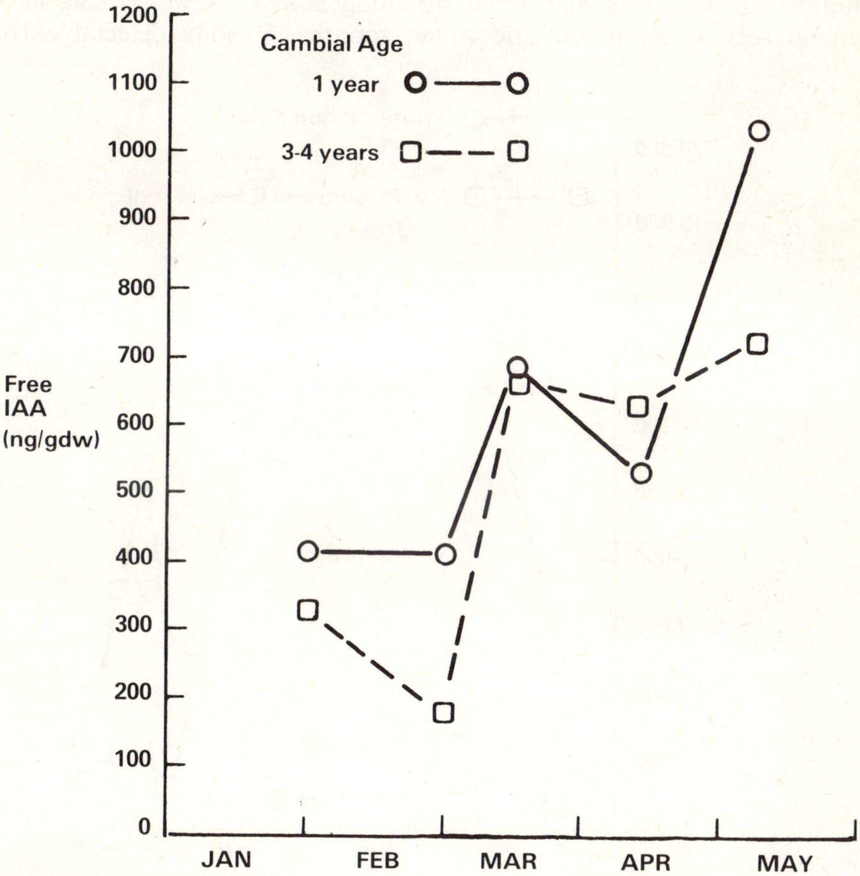

Figure 7.4: *Seasonal variation in free IAA levels in 4-year-old* P. contorta *var.* latifolia. *The main-stem cambial region (phloem + cambial zone + differentiating xylem) from the tops and bases of 5 trees was pooled for extraction at each harvest date. Cambial reactivation occurred between early and mid-April; new earlywood xylem was evident in both top and base of the main stems of the May sample. Values are adjusted for losses which occurred prior to GC-MS estimation; values are accurate ±20%.*

variation in free IAA levels in 4-year-old and 20-year-old lodgepole pine trees. The state of the cambium was assessed in association with the tissue extracts, but will not be reported in detail here. In every case examined (7 in all), a substantial increase in free IAA levels began in association with the onset of extension growth and new xylem differentiation, while IAA levels clearly decreased with the approach of autumn and in association with latewood production. The results obtained from lodgepole pine agree with those of Little and Wareing[122] from Sitka spruce: IAA levels in the cambial region in association with cambial reactivation were not greatly different from those associated with the onset of cambial inactivity in the autumn, and relatively high levels of endogenous IAA were found in the cambial region throughout the winter months. Readily detectable IAA

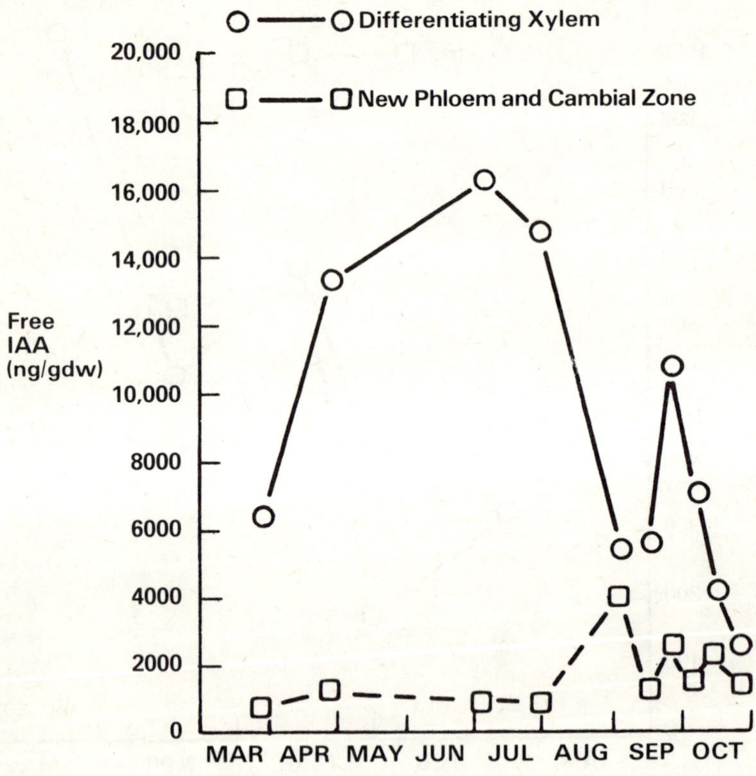

Figure 7.5: *Seasonal variation in free IAA levels in the 10-year-old main-stem region of 20-year-old trees of* P. contorta *var.* latifolia. *The two tissues were separated by bark peeling. IAA levels in mature xylem adjacent to differentiating xylem were also determined, but at no harvest date would those levels be above the origin at the scale shown. Values are expressed in terms of non-extracted tissue dry weight and are accurate ±20%; values are adjusted for losses which occurred prior to GC-MS estimation.*

levels were also evident in dormant buds and mature needles at all times of the year (Table 7.1).

In an effort to determine the source, if any, of the free IAA present in the cambial region during the winter, various experimental treatments were administered from early to late winter to lodgepole-pine trees positioned out-of-doors. The IAA levels obtained in response to a typical set of winter treatments are shown in Table 7.2. Winter disbudding and winter defoliation resulted in substantial declines of free IAA in the cambial region, relative to controls. Winter ringing resulted in the accumulation of free IAA above and a marked decline below the rings, while controls retained relatively high levels of free IAA. Microscopy showed that the cambium remained inactive in all cases. These results suggest that metabolic activity

Table 7.1: *Representative levels of "free" IAA, as determined by GC-MS, in tissues of hardwoods and conifers at Aberystwyth in mid-winter*

Species and Tissue Extracted	*Date Harvested*	*Tissue Age (yrs)*	*IAA* (ng/gdw)*
Acer pseudoplatanus	7 Nov 1978	1	100
15-year-old tree		12	35
Main-stem cambial region			
Quercus robur	7 Nov 1978	1	95
15-year-old tree		12	270
Main-stem cambial region			
Populus robusta	24 Feb 1978	1	230
2-year-old tree			
Main-stem cambial region			
Larix decidua	24 Feb 1978	1-3	1600
17-year-old tree		14	800
Main-stem cambial region			
Pinus contorta	9 Nov 1977	12	2130
20-year-old tree			
Main-stem cambial region			
17-year-old tree	28 Feb 1978	1	250
Main-stem cambial region		12	2100
20-year-old tree	4 Nov 1977	1	325
Needles			
4-year-old tree	19 Dec 1977	3	220
Needles			
20-year-old tree	3 Mar 1978	1	365
Terminal buds			

* Values are expressed in terms of non-extracted tissue dry weight and are accurate ±20%; values are adjusted for losses which occurred prior to GC-MS estimation.

Table 7.2: *Representative levels of 'free' IAA, as determined by GC-MS, found in the innermost phloem and cambial tissues of different ages in lodgepole pine* (P. contorta) *in mid-winter following various treatments*

		*'Free' IAA (ng/gdw)**							
Treatment Period	*Tree Age (yrs)*	*Controls Cambial Age (yrs)*		*Entirely Disbudded Cambial Age (yrs)*		*Entirely Defoliated Cambial Age (yrs)*		*Ringed Cambial Age (yrs)*	
		1	*7*	*1*	*7*	*1*	*7*	*Above* *4*	*Below* *4*
12 Feb to 27 Feb 1978	8	250	2095	70	360	110	340	210	65

* Values are expressed in terms of non-extracted tissue dry weight and are accurate ±20%; values are adjusted for losses which occurred prior to GC-MS estimation.

in dormant buds and mature leaves was responsible for the production and export of free IAA during the winter. The ringing experiments also suggest that IAA was basipetally transported throughout the winter months (cf. Thimann and Kaufman[179], Hollis and Tepper[84], Morris[133]). In addition, analyses of dormant cambia from 1- and 20-year-old regions of the main stem of naturally defoliated *Larix decidua* in mid-winter similarly showed high levels of free IAA, whose nature was confirmed by total-ion-current mass-spectroscopy analysis[164]. In contrast, analyses of dormant cambial regions of untreated mature specimens of *Quercus robur* and *Acer pseudoplatanus* during mid-winter, showed that the older cambial regions of those hardwoods contained relatively low levels of free IAA (insufficient in fact to allow confirmation as IAA), whereas the younger cambial regions of these hardwoods contained levels of free IAA comparable to those present in younger cambia of conifers (Table 7.1)*.

THE ROLE OF THE MATURE EVERGREEN CONIFER LEAF IN CAMBIAL REACTIVATION AND XYLEM DIFFERENTIATION

Münch[134] and Onaka[143] demonstrated that mature leaves of *Pinus* spp. were alone sufficient for both cambial reactivation and new xylem differentiation. Cambial reactivation occurred in disbudded trees, but the distribution of xylem throughout the tree was different; viz. more xylem formed at the bases of the trees and less in the tops than in intact control trees. There was only slight, if any, new xylem formation in defoliated trees, although new shoots grew out. In addition, Onaka[143] presented auxin bioassay evidence that in the 7-year-old intact pine tree, the absolute levels

* Results of this study have been reported in: Savidge, R. A., Wareing, P. F. Apparent auxin production and transport during winter in the non-growing pine tree. Submitted to *Canad. J. Bot.*

of auxin in the cambial region are highest at the main stem base throughout the growth season. Onaka found disbudding in early summer to reduce substantially auxin levels in the cambial region, while defoliation at this time also caused a drop in cambial auxin levels, but to a lesser extent than did disbudding. Ringing resulted in the disappearance of auxin below the girdle, whereas it accumulated above the girdle. Double girdling resulted in the accumulation of auxin above the lower ring when mature leaves were left intact between the two rings, whereas auxin levels in the stem region between two rings with leaves removed were negligible.

The experiments of Münch[134] and Onaka[143] were repeated at Aberystwyth with 9–10-year-old lodgepole-pine trees, growing on a Forestry Commission site. Microscopical observations were made on the treated trees, and levels of free, endogenous IAA were estimated in their cambial tissue[164]. Quantitative results are shown in Table 7.3. New xylem formation occurred in disbudded trees (interfoliar buds were regularly removed during the course of the treatment), although IAA levels dropped substantially by the harvest date. Only a slight amount of new xylem was produced in defoliated trees, although new shoots and new leaves were permitted to develop on them. When living, mature leaves were left intact between two rings, new xylem differentiation occurred. When the mature leaves between the two rings were removed, however, no new xylem differentiated, and IAA levels were relatively high at the harvest date. In addition, IAA levels were found to be highest in trees which were both disbudded and defoliated. A minimal amount of new xylem formation was associated with these high IAA levels.

Applications of exogenous auxins, such as IAA, to young, disbudded shoots, has frequently confirmed the effectiveness of auxins in reactivating the vascular cambium and/or stimulating cambial activity in conifers[10, 28, 46, 82, 94, 112, 118, 138, 201, 223, 228-230]. However, the effectiveness of exogenous auxins in reactivating and stimulating cambial activity in conifer material older than 1 year has received minimal attention[82, 229]. Larson[109] observed that the developing shoot and extending leaves of conifers made little contribution to cambial activity and xylem differentiation in the older parts of the tree when mature needles were shielded from light. Moreover, the 1-year-old or younger stem is organized differently from 2-year-old and older stems; viz. the 1-year-old stem contains apically-derived cortical tissue with reserve materials, whereas the older stem contains phellogens and dead rhytidome[23, 105] which appear to function more in protection than in storage. These considerations suggest that cambial activity should be distinguished on the basis of cambial age.

To investigate possible differences in the responsiveness of conifer cambia of different ages to exogenous IAA, a series of IAA concentrations

Table 7.3: *Cambial response and free IAA levels in the cambial region of* P. contorta *var.* latifolia *following experimental treatments administered April 28*; trees were harvested for IAA extraction and microscopy on August 2. 9-year-old trees*

	Tops (1-yr-old cambium)				*Bases (7-yr-old cambium)*			
TREATMENT	*IAA (ng/gdw)*	*CZ*	*TRC*	*XT*	*IAA (ng/gdw)*	*CZ*	*TRC*	*XT*
Control	520	10	–	74	700	11	–	80
Defoliated**	675	5	3	16	580	5	–	22
Disbudded***	115	8	5	13	275	8	6	77
Disbudded & Defoliated	840	4	4	5	1765	14	–	26
	1-10 cm Above Ring				*1-10 cm Below Ring*			
1 ring in 4-yr-old main stem	785	14	5	115	40	5	5	5
	Mature Leaves Present				*Mature Leaves Removed*			
Between 2 rings 3-yr-old main stem	150	12	4	36	625	14	–	6

CZ = no. cells/radial file in cambial zone
TRC = no. traumatic resin-canal cells/radial file
XT = no. xylem tracheids formed since cambial reactivation/radial file

Endogenous, free IAA levels are expressed in terms of non-extracted tissue dry weight and are accurate ±20%; values are adjusted for losses which occurred prior to GC-MS estimation.

* On April 28, the cambium had reactivated, and 3-5 earlywood tracheids had matured.
** New shoots extended and new leaves developed.
*** Interfoliar buds were removed as they became evident.

(10, 100, 1000, 2000, 5,000 and 10,000 ppm w/w) in anhydrous lanolin, were tested on 15-cm long cuttings of different ages taken from orthotropically growing 12–16-year-old branches of 20-year-old, open-growing lodgepole-pine trees. One-, two-, and three-year-old cuttings were tested with and without their complements of mature leaves while older cuttings naturally lacked living foliage. Applications of IAA were made to cuttings harvested first in January, and the experiment was repeated twice more as the winter progressed. Three replicates were used in each trial. The cuttings were placed upright in 4–6 cm of tap water in a warm (25°C), long-day (18-hr photoperiod) environment. IAA applications to the apical ends of the cuttings were renewed weekly, and each experiment ran for 6 weeks[164].

In control shoots without leaves, supplied with anhydrous lanolin, cambial reactivation was never observed. In 1-, 2-, and 3-year-old foliated control cuttings, cambial reactivation and new xylem formation occurred at their bases, but not at their tops. In contrast, one-year-old cuttings with buds and leaves intact (no lanolin or IAA applied) showed cambial reactivation and new xylem differentiation throughout their length. Similar cuttings, but with leaves removed also showed cambial reactivation, but new xylem differentiation was rare. When exogenous IAA was substituted for the buds of defoliated one-year-old cuttings, cambial reactivation occurred in response to 100 ppm IAA and all higher concentrations. Regardless of the IAA concentration however, new xylem differentiation was uncommon in mid-winter samples. As spring approached, the formation of completely differentiated new xylem in defoliated 1-year-old cuttings increased in the presence of either buds without added IAA, or with exogenous IAA after bud removal.

In 2- and 3-year-old cuttings cambial reactivation and new xylem formation occurred over the length of the cuttings if mature leaves were present and exogenous IAA supplied. Defoliated 2-year-old cuttings supplied with exogenous IAA only rarely exhibited a cambial zone which had doubled in width, and still more rarely showed the onset of a slight amount of new xylem differentiation. Three-year-old leafless cuttings did not show cambial reactivation or new xylem differentiation, regardless of the IAA concentration, or the time of year.

Similarly, older, naturally non-foliated cuttings did not show cambial reactivation or new xylem differentiation. Moreover, callus formation was not observed in response to exogenous IAA in non-foliated cuttings older than 2 years. To confirm these results, a variety of different techniques for supplying exogenous IAA were tested. These included application to laterally-exposed phloem tissue, insertion into the cambial region following bark peeling, and supply in aqueous solution to rubber-tubing wells fitted over the apical ends of 3-year-old, defoliated cuttings. In no instance was cambial reactivation observed. It therefore seems evident that the 1-year-old shoot cannot be considered as representative experimental material for investigating physiological factors regulating the activity of older cambia. When it is considered that 1-year-old shoots contain only a small proportion of the total cambial tissue within mature trees, it becomes apparent that the regulation of cambial activity is more complex than has been thought until now.

When potted 5-year-old lodgepole-pine trees were decapitated 10 cm above soil level in late winter, and the apical surfaces of the non-foliated stumps were supplied with 1000 ppm exogenous IAA in lanolin, cambial reactivation occurred in the region 0–1 cm below the application site in the 3- to 4-year-old cambium[164]. Cambial reactivation was not found, however, further down the same stems. In most instances, new xylem differentiation also occurred 0–1 cm below the IAA application site, but new xylem was not observed at the base (soil level). Application of exogenous IAA also resulted in the production of a substantial amount of green callus tissue at the application site, viz. the transversely-exposed surface of the stump. Metabolic products from this callus tissue may therefore have contributed to new xylem differentiation. It is also possible that the application of exogenous IAA to older stumps with living roots was effective in stimulating cambial reactivation and new xylem formation as a consequence of the mobilization of root-produced or root-stored substances to the application site, e.g. by hormone-directed transport[146, 192]. Decapitated control trees, the stumps of which were supplied with anhydrous lanolin, showed neither cambial reactivation nor new xylem differentiation.

Since leafy pine cuttings harvested in late winter show cambial reactivation and new xylem differentiation regardless of cambial age and whether or not treated with exogenous IAA, it seems clear that the mature leaf, given a suitable environment, is fulfilling an important function in regulating xylem development. Accordingly, the ability of a single pair of mature pine leaves to stimulate xylem differentiation in winter-harvested 1-, 2-, and 3-year-old cuttings of lodgepole pine was investigated. All leaf pairs but one, located 4 to 5 cm below the apical end of the cutting, were removed and the cuttings were placed upright in 4 cm of tap water in a warm, long-day environment, and supplied weekly with exogenous IAA (1000 ppm). Controls were supplied with anhydrous lanolin alone. Other cuttings were entirely defoliated but otherwise treated similarly. In all ages, and both with and without exogenous IAA, the presence of a single leaf pair was found to stimulate new xylem differentiation over a distance 1–10+ mm basipetal to the junction of the needle trace with the vascular cambium[164] (Fig. 7.6). Entirely defoliated cuttings did not exhibit new xylem differentiation at the junction of their needle traces with their vascular cambia. In some instances, new xylem was apparent as much as 1-2 mm acropetal to the trace-cambium junction, but in general, the influence of the mature leaf was more evident below than above the junction. The earlywood xylem elements which differentiated in 2- and 3-year-old single-leaf cuttings, without added IAA, were radially compressed and appeared to have differentiated directly into mature xylem without prior cell division (Fig. 7.7), i.e. without cambial reactivation. In transverse section, these new elements were clearly distinct from the previous year's latewood; they had the shape and dimensions of meristematic fusiform cells in the cambial zone but nevertheless had thickened secondary walls which stained much darker with toluidine blue-0 than did normal latewood xylem. In radial section, many of the newly-differentiated tracheids had the appearance of primary xylem elements; i.e.

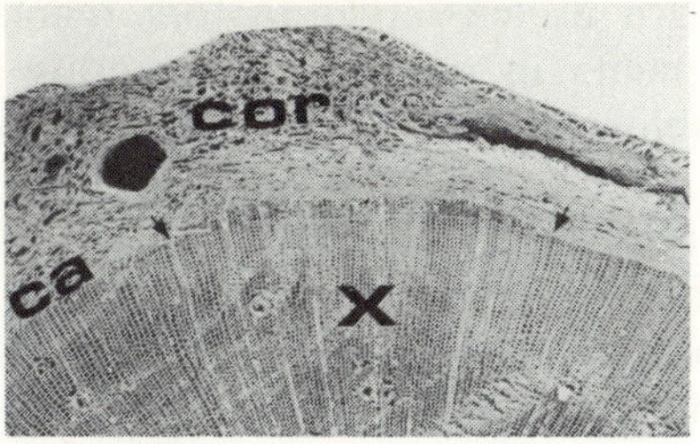

Figure 7.6: *SEM micrograph of the transverse surface of a lodgepole-pine cutting, showing an arc of new tracheids (between the two arrows) which differentiated in response to a single pair of mature needles. IAA was applied to the apical end of the cutting, 5 cm above the position shown. X = xylem which matured prior to the treatment, cor = cortex, ca = cambial zone and adjoining phloem.*

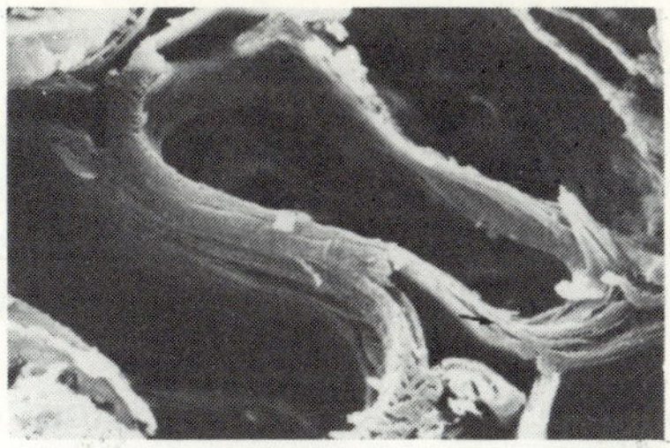

Figure 7.7: *SEM micrograph of the transverse surface of fusiform cambial cells which differentiated into tracheids in response to a mature needle pair on a lodgepole-pine cutting. No exogenous IAA was supplied. Secondary-wall formation is clearly evident (arrow), but neither cell division nor radial expansion were associated with the differentiation of the cambial-zone cells into new tracheids.*

the fusiform cells contained annular or helical secondary thickenings. The trace-cambium junction of entirely defoliated cuttings did not show similar elements associated with the vascular cambium.

Addition of exogenous IAA to single-leaf cuttings resulted in primary-xylem elements below the trace-cambium junction which were noticeably greater in radial dimension than were the primary-xylem elements which differentiated in control single-leaf cuttings. In addition, more primary-xylem elements differentiated per radial file in the presence of exogenous IAA. Although cambial reactivation occurred over the full length of 1-year-old single-leaf cuttings supplied with exogenous IAA, xylem differentiation was rarely observed above the trace-cambium junction.

Clear inter- and intra-cellular differences were observed in the extent of tracheid differentiation which occurred in the vascular cambium of lodgepole pine in response to a single pair of mature needles. At the trace-cambium junction, nearly normal tracheids differentiated, but with increasing distance from the trace-cambium junction, the newly differentiated tracheids became progressively more proto-xylem like, and bordered-pit differentiation became progressively less evident. The latter observations, together with the finding that IAA is necessary for radial enlargement of cambial derivatives, have suggested the hypothesis that whether a cell differentiates into a proto-, meta-, or secondary-xylem late-wood or early-wood tracheid depends on interacting concentrations of IAA and a specific tracheid-differentiation factor, the identity of which remains to be determined (Savidge and Wareing, 1981*).

* Savidge, R. A. and Wareing, P. F. 1981. A tracheid-differentiation factor from pine needles. Submitted to *Planta*.

These results suggest that there are at least two factors regulating differentiation of new tracheary elements in conifers. Exogenous IAA reactivates the 1- and 2-year-old cambium, but has little effect in inducing secondary-wall thickening; whereas a factor from the mature pine leaf, which we have yet to identify, has the ability to induce secondary-wall thickening directly, despite the absence of cambial reactivation. Both IAA and the leaf factor appear to be transported basipetally.

Although a factor from mature pine leaves appears to be necessary for the completion of xylem differentiation, some evidence suggests that IAA may be directly involved in initiating xylem differentiation in a manner such as was previously proposed by Wareing[194, 201] and Larson[110, 112]. During the period of cambial activity and xylem differentiation beginning in early summer, the bark peels in a manner such that differentiating xylem adheres to mature xylem while the meristematic zone adheres to mature phloem. Using this technique to separate the two tissue types at successive dates throughout a summer, it was repeatedly found that IAA levels in enlarging and differentiating xylem were many times greater than those in the meristematic region (Fig. 7.5)[164]. Clearly, a substantial gradient of IAA concentration exists from mature phloem to mature xylem, a distance of less than 1 mm. It appears, therefore, that IAA becomes partitioned toward the centripetal periphery of the cambial zone, rather than being uniformly distributed throughout the tissue or showing passive, physical affinity for the alkaline phloem side of the cambial zone. Nix and Wodzicki[137] have reported similar observations, following the feeding of radioactive IAA to cambial tissues. Most radioactivity was found associated with the differentiating xylem. The mechanism behind this non-uniform distribution of IAA is unknown but it may be attributable to an auxin protecter or an inhibitor of IAA-oxidase[139].

Observations by Rehm[152], Söding[171], Young[227], Sachs[162], and Hess and Sachs[83] also suggested the existence of a factor, different from auxin, which is necessary for secondary-wall thickening and/or lignification. However, the experimental systems used in past investigations have not permitted a clear distinction to be made between auxin and that other factor; i.e. investigations into xylem differentiation in herbaceous species commonly have been made with plants containing both intact mature leaves and green stems, whereas investigations into xylem differentiation in deciduous hardwoods have been unable to take account of the possible contribution of the bark reserves. In contrast, the bark of the older cambium of evergreen conifers has a relatively small reserve component, perhaps because evergreen needles are present[90, 105, 109, 145], thus the conifer cambium has provided a useful experimental system to separate the effects of auxin on cambial-cell division and radial-wall growth from the contribution of the mature leaves to secondary-thickening and lignification.

It is well established that leaves contain abundant amounts of quinic and shikimic acid[32, 67, 126, 141, 163, 206], but it remains unknown whether mature leaves export lignin precursors of the shikimic pathway, or alternatively, whether leaves supply a differentiation factor which triggers the secondary-wall and lignification biochemistry necessary for xylem differentiation.

The importance of the evergreen leaves of conifers to xylem formation extends considerably beyond supplying the mature cambial region with IAA and a factor necessary for secondary thickening. A considerable body of evidence, which will only be considered briefly here, suggests that the regulation of latewood formation and the onset of cambial inactivity in autumn also have their origins in mature leaves. The role of daylength in regulating cambial activity and xylem differentiation has been extensively investigated by Wareing[196, 197, 199, 200], Larson[109, 110, 113, 114], and Wodzicki[216, 217-221]. Wareing demonstrated that following bud burst in the spring in Scots pine, short days produced a cessation of shoot and needle extension and cambial activity, with concomitant latewood production. Exposure of seedling trees to long days resulted in continuing shoot and needle extension and earlywood formation. Larson, working with red pine, and Wodzicki, working with larch, have each confirmed Wareing's observations. In addition, Larson[113] has shown that manipulation of long and short-day photoperiods can be used as an experimental tool with which to induce bands of narrow-diameter, thickwalled latewood-like tracheids within bands of wide-diameter, thin-walled earlywood-like tracheids. Moreover, Larson has presented evidence to show that endogenous auxin levels decline when pine seedlings are transferred to short days. Consequently, he postulates that the onset of latewood formation occurs as a consequence of decreasing IAA and increasing substrate levels within the cambial region[109, 110, 113, 114]. In contrast, Wodzicki's investigations have not confirmed decreasing IAA levels in association with the onset of latewood formation[214] and he postulates an inhibitor, such as ABA[221] as responsible for latewood production[215, 216]. The authors' observations[164] have shown a clear decline in endogenous IAA levels within the cambial region following the completion of shoot and needle extension in mid-summer, but IAA levels have not been found to decline in autumn substantially below values associated with the first earlywood formation (Fig. 7.5)[122, 164]. No correlation has been observed, between increasing free ABA levels and the onset of cambial inactivity.

Evidence[164] that a cambial inhibitor exists became evident from autumn ringing experiments on 20-year-old pine trees. The cambium's position was marked by pin-pricking[224] at time of ringing, and although IAA levels increased above the ring over a period of two months, the cambium above the ring was observed to have become inactive. In addition, the differentiation of latewood went to full completion above the rings, whereas below, cambial activity, secondary-wall formation, and lignification were all interrupted, such that the cambial region gave the impression of continuing to be active in mid-winter.

REQUIREMENTS FOR CELL DIVISION IN THE CONIFER CAMBIUM

Inability to reactivate 3-year-old or older conifer cambium with growth substances alone emphasized the need for suitable sterile-culture techniques for investigating the nutritional, as well as growth-substance, physiology of cambial activity[164]. Brown and Wodzicki[28] developed a sterile-culture technique for supplying liquid media to shoot segments which greatly reduced pressure changes and tissue disruption, features which characterize cambial-strip cultures and which apparently may also influence the process of differentiation[26]. Zajaczkowski[228-230] used the Brown and Wodzicki technique extensively in attempts to identify the factors necessary for normal and continuing xylem production in 1-year-old pine shoots. Although his reports do not make clear whether mature leaves were removed or remained intact on the cultured stem segments, it was nevertheless concluded after testing many defined media, that factors necessary for continuing cambial activity and xylem production remain to be identified. The Brown and Wodzicki technique was used in the authors' laboratory to investigate requirements for cambial reactivation in three-year-old, completely defoliated branch segments, harvested from a single 20-year-old lodgepole pine tree in early March. The apical end of each segment was supplied every second day with autoclaved aqueous media. After several weeks, it was found that neither sucrose alone, nor IAA and sucrose together, were sufficient for cambial reactivation. With the addition of L-glutamine to the medium, however, a small amount of cambial division was observed in all segments. Nevertheless, only a minimal amount of new xylem commenced differentiation, suggesting again that some factors necessary for xylem differentiation were wanting[164].

To the authors' knowledge, the possibility that cambial reactivation in conifers may be regulated by seasonal fluctuations in levels of reduced nitrogen, such as free amino acids transported upward from the roots in the xylem sap, has not hitherto been investigated. However, a considerable body of evidence suggests that substantial seasonal variation of amino-acid levels occurs in the tree, with L-glutamine being the most abundant in conifers and associated with spring shoot growth[15, 16, 17, 22, 37, 53, 75, 145]. It is interesting to note that several observers, including the authors[164], have found that when the cambium reactivates, the first cell divisions occur in those fusiform cells which are adjacent to the xylem and hence the transpiration stream[14, 107, 135, 149, 181]; since IAA has been established to be present throughout the year in the cambial tissue, and sucrose is always present[145], cambial reactivation may be dependent upon the arrival of amino acids from roots or leaves in association with temperatures which are suitable for growth. Preliminary studies[164], using an amino-acid autoanalyser, have shown that although amino acids are present in the mature conifer cambium during the winter months, an increase in the level of free amino acids occurs at the time of cambial reactivation. Investigations are continuing in this area at the authors' laboratory.

In cambia older than three years, the diameter of the segments makes sterile-culture using the Brown and Wodzicki technique impractical, and the technique used by Jacquiot[90] of culturing small blocks containing the vascular cambium sandwiched between mature xylem and phloem was therefore adopted. The blocks have been explanted onto defined media such that the apical end is down. Although new phloem differentiation has been obtained repeatedly using the cambial-block technique, no complete xylem differentiation has yet been observed on any of a variety of defined media containing one or more of all of the known plant-growth regulators. Addition of cytokinins to the medium appeared to enhance the amount of new phloem produced, but had no apparent effect upon xylem differentiation[164]. The growth obtained in sterile culture is reminiscent of the observation of Barnett[18] on *Pinus radiata:* during the winter months, cambial activity and new phloem production continued without concomitant xylem production. Such results suggest that the direction in which cambial derivatives enlarge radially and thereafter differentiate must be regulated by factors other than pressure[26], auxin-sucrose gradients[203, 204], and the factors required for periclinal division within the cambial zone.

Using the cambial-block technique, the previous observations made with 3-year-old stem segments were confirmed; i.e. a medium containing IAA, sucrose, and reduced nitrogen was necessary for cambial reactivation; in addition, IAA was found to be essential for maintaining the organization of the cambial zone itself[164]. In the absence of IAA, regardless of the other components included within the culture media, the fusiform meristematic cells expanded radially and then septated by repetitive transverse divisions; all fusiform cells in each radial file of the cambial zone were reduced to axial parenchyma cells. On the same media with added IAA, the radial dimensions of the fusiform cells remained normal, and transverse septations in those cells were rarely observed (Figs. 7.8-7.11). Thus, one role of IAA in the cambial region of conifers throughout all seasons of the year may be to maintain the cambial fusiform cells polarly elongated and potentially meristematic.

Possible roles of growth substances of other than the auxin class in cambial activity and xylem and phloem differentiation in conifers remain to be clearly demonstrated; however, from the existing literature, it seems

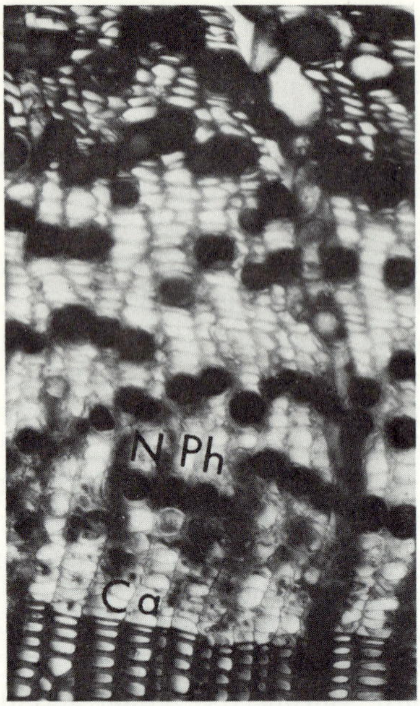

Figure 7.8: *Transverse section showing the 7-year-old cambium of* P. contorta *in sterile culture with added IAA. New phloem (NPh), but not new xylem, has been produced. The cambial zone (Ca) appears normal. (hand section).*

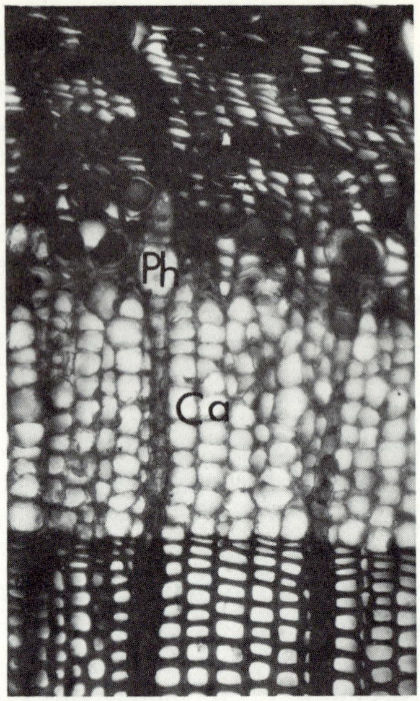

Figure 7.9: *Transverse section showing the 7-year-old cambium of* P. contorta *in sterile culture without added IAA. The cambial zone (Ca) has disappeared, and only mature phloem (Ph) is present. (hand section).*

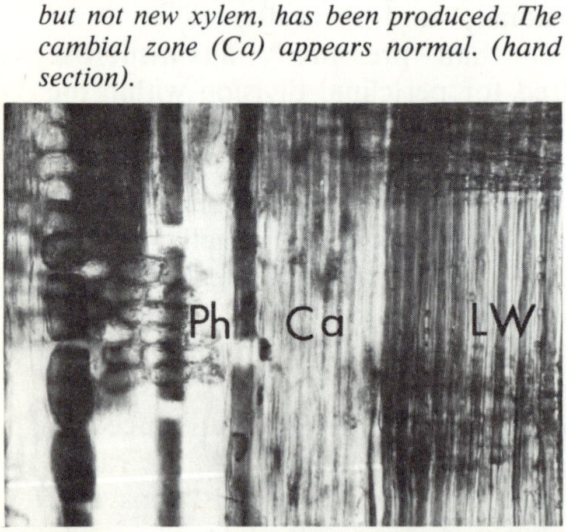

Figure 7.10: *Radial section showing the 7-year-old vascular cambium of* P. contorta *in sterile culture with added IAA. Phloem (Ph), but not xylem, has been produced by the cambium (Ca). The cambial zone is normal, and borders mature latewood (LW). (hand section).*

Figure 7.11: *Radial section showing the 7-year-old vascular cambium of* P. contorta *in sterile culture without added IAA. Mature phloem (Ph) and xylem (LW) border the region which was the cambium (Ca). All fusiform cells of the cambial zone have septated to become axial parenchyma. (hand section).*

very likely that some, if not all, of the known classes of growth substances will be found to have some role in vascular development. Gibberellins may be involved in phloem-cell differentiation, but apparently not cambial activity, in pines[43]. Seasonal variation, as detected by bioassay, of gibberellin-like[52, 102, 104, 124], cytokinin-like[106, 125, 158], and inhibitory[101-104, 127, 142, 215, 221] substances has been reported; in addition, applications of exogenous gibberellins and cytokinins to stem regions of conifers has yielded enhanced amounts of xylem formation[82]. Applications of ABA have been observed to decrease radial expansion and to increase secondary-wall dimensions, but not area, of differentiating conifer tracheids[94]. Ethylene, applied as a chemical precursor to the outer bark of intact trees, has been found to increase the amount of xylem differentiation which occurs beneath the application site[17, 31], a finding which might be associated with butt swell, which occurs in response to wind sway. Applications of exogenous growth regulators also are known to affect the differentiation of compression- and tension-wood cells; however that is the topic of another chapter herein (Wilson, Ch. 10). Some evidence also suggests that ABA may have a role in inhibiting cambial activity during periods of water stress[116, 119].

In addition to IAA and ABA, zeatin[117] and zeatin riboside[117] and indol-3yl butyric acid[164] have all been positively identified by GCMS as growth substances present in the cambial region of conifers; gibberellins and cytokinins[123, 125] have been positively identified in mature leaves of conifers.

CONCLUSIONS

It is clear from the numerous and diverse observations presented above that endogenous growth substances play a critical role in cambial activity and xylem differentiation in both herbaceous and woody plants. Nevertheless, it must be admitted that the information gathered so far seems to raise more problems than it solves.

Thus, auxin appears to be necessary for cambial division and xylem differentiation, while in several woody species GA_3 interacts synergistically with IAA in cambial division and promotes phloem differentiation. But we have no idea as to the role of growth substances in cell differentiation; indeed, we still do not understand the mode of action of any plant growth substance in any system. Each type of endogenous growth substance has a wide spectrum of effects and it is generally held that the specificity of the response in any given cell or tissue resides in the 'competence' of that tissue. Are we to conclude, therefore, that the centripetal cambial derivatives are 'programmed' to differentiate into xylem cells in response to auxin, while the centrifugal derivatives are determined as potential phloem cells, the differentiation of which requires gibberellins? The work of Siebers*, suggesting that the cambial region shows radial polarity, is consistent with this interpretation. However, an alternative interpretation can be postulated, namely, that neither the centripetal nor the centrifugal cambial derivatives are 'determined' with respect to their future pattern of differentiation (i.e. as xylem or phloem cells), but that the growth-substance or other conditions affecting differentiation are different on the 'xylem' and 'phloem' sides of the cambial region.

The nature of the processes controlling cell differentiation present one of the central outstanding problems in biology as a whole, and our lack of understanding of the process of xylem-cell differentiation is therefore part of a vastly broader problem. However, we are in a relatively favourable position in relation to xylem differentiation to the extent that we have known growth substances which will induce specific differentiation patterns in suitable systems. Roberts[156] is therefore justified in asserting that xylogenesis offers an excellent model system for the study of plant-cell differentiation. However, although there seems no doubt that growth substances play a vital role in seasonal control of cambial activity, and in vascular differentiation, the xylem is a complex tissue system and there are many facets to its differentiation, involving the formation of vessels, tracheids, fibres, wood parenchyma and ray cells. In the individual cells which comprise the cambial tissue there are problems relating to the site and plane of cell division, the maintenance of, or change in, the polarity of individual fusiform or ray cells, polar cell extension, cell-division frequency and synchrony, the initiation of ray-cell differentiation following segmentation or lateral division in cambial fusiform cells, and so on. Although growth regulators probably play a role in the control of some of these processes, it would be naive to assume that the only factors involved are hormonal.

Again, although growth substances may well play a central controlling role in the process of secondary-wall formation and lignification in tracheary-element differentiation, it is necessary to understand what determines whether a specific cell differentiates as a vessel, tracheid or fibre. We need to know what controls the diameter of such cells and other specific characteristics of each cell type, including the formation of vessel perforations and specific patterns of wall thickening, and the number and structure of pits. Although growth regulators, such as IAA, may play a central role in the overall process of differentiation, there can be little doubt that other factors determine the specific response of an individual cell to a given hormonal stimulus.

In the final analysis, differentiation of specific cell types would involve

* *Acta Bot. Neerl.* 20: 211-220 (1971)

selective expression of specific parts of the genome; indeed, development can be regarded as essentially a problem of selective gene expression. However, how this selective gene expression is controlled remains one of the central unsolved problems of biology, and until we have a better general understanding of the biophysical and biochemical processes involved in cell differentiation in eukaryotic organisms, we shall not fully understand the specific problems associated with xylem differentiation.

ACKNOWLEDGMENTS

We thank Professor T. Sachs, Professor C. LaMotte, Dr. J. Dodds, and Dr. J. R. Barnett for critical reading of all or parts of the manuscript, and Professor W. P. Jacobs for helpful discussion. The research cited as reference[164] was supported by U.K. Natural Environment Research Council Grant GR3/3703, and we are grateful to the following people for their skilful and knowledgeable assistance in that research: Mr. D. Williams [Zoology SEM unit], Miss L. Rawding [experimentation, sterile culture, and SEM operation], Mr. A. Pugh [photography], Mr. J. K. Heald [GC-MS], Mr. D. Edwards [heat tape and refrigeration experiment], Miss Jane Miners [IAA extraction and purification] and Dr. R. Horgan [organic chemistry procedures].

Unbeknown to the authors at the time of writing this chapter, Professor W. P. Jacobs in Chapter 4 of this Monograph entitled *Plant Hormones and Plant Development*[339] (Cambridge; Cambridge University Press 1979) has presented a useful account of the evidence behind the hypothesis that IAA has a regulatory role in xylem differentiation.

REFERENCES

[1] Abbe, L. B., Crafts, A. S. Phloem of white pine and other coniferous species. *Bot. Gaz.* **100,** 695-722 (1939)

[2] Alden, T. Seasonal variation in the occurrence of indole-3-acetic acid in buds of *Pinus silvestris. Physiol. Plant.* **25,** 54-57 (1971)

[3] Alden, T., Eliasson, L. Occurrence of indole-3-acetic acid in buds of *Pinus silvestris. Physiol. Plant.* **23,** 145-153 (1970)

[4] Alfieri, F. J., Evert, R. F. Seasonal development of the secondary phloem in *Pinus. Amer. J. Bot.* **55,** 518-528 (1968)

[5] Allen, R. M. Changes in acid growth substances in terminal buds of longleaf pine saplings during the breaking of winter dormancy. *Physiol. Plant.* **13,** 555-558 (1960)

[6] Aloni, R. Role of auxin and gibberellin in differentiation of primary phloem fibers. *Plant Physiol.* **63,** 609-614 (1979)

[7] Aloni, R. Polarity of induction and pattern of primary phloem fiber differentiation in *Coleus. Amer. J. Bot.* **63,** 877-889 (1976)

[8] Avery, G. S. Jr., Burkholder, P. R., Creighton, H. B. Production and distribution of growth hormone in shoots of *Aesculus* and *Malus* and its probable role in stimulating cambial activity. *Amer. J. Bot.* **24,** 51-58 (1937)

[9] Bailey, I. W. *Contributions to Plant Anatomy*. pp. 259. (Waltham, Mass: Chronica Botanica 1954)

[10] Balatinecz, J. J., Farrar, J. L. Pattern of renewed cambial activity in relation to exogenous auxin in detached woody shoots. *Canad. J. Bot.* **44,** 1108-1110 (1966)

[11] Balatinecz, J. J., Kennedy, R. W. Mechanism of earlywood-latewood differentiation in *Larix decidua. TAPPI* **51,** 414-422 (1968)

[12] Ball, E. Differentiation in a callus culture of *Sequoia sempervirens. Growth* **14,** 295-325 (1950)

[13] Bandurski, R. S., Schulze, A. Concentration of indole-3-acetic acid and its derivatives in plants. *Plant Physiol.* **60,** 211-213 (1977)

[14] Bannan, M. W. The vascular cambium and tree-ring development. In *Tree Growth*, T. T. Kozlowski (ed.) (New York: Ronald Press, N.Y. 1962)

[15] Barnes, R. L. Glutamine synthesis and translocation in pine. *Plant Physiol.* **37,** 323-326 (1962)

[16] Barnes, R. L. Organic nitrogen compounds in tree xylem sap. *Forest Sci.* **9,** 98-104 (1962)

[17] Barker, J. E. Growth and wood properties of *Pinus radiata* in relation to applied ethylene. *N.Z.J. Forest Sci.* **9,** 15-19 (1979)

[18] Barnett, J. R. Winter activity in the cambium of *Pinus radiata. N.Z.J. Forest Sci.* **1,** 208-222 (1971)

[19] Bergmann, L. Der Einfluss von Kinetin auf die Lignin-bildung und Differenzierung in Gewebekulturen von *Nicotiana tabacum. Planta* (Ber.) **62,** 221-254 (1964)

[20] Berlyn, G. P. Physiological control of differentiation of xylem elements. *Wood and Fiber* **11,** 109-126 (1979)

[21] Bevan, M., Northcote, D. H. The interaction of auxin and cytokinin in the induction of phenylalanine ammonia-lyase in suspension cultures of *Phaseolus vulgaris. Planta* (Ber.) **197,** 77-81 (1979)

[22] Bollard, E. G. Translocation of organic nitrogen in the xylem. *Aust. J. Biol. Sci.* **10,** 292-301 (1957)

[23] Brown, C. L. Secondary Growth. In *Trees, Structure and Function.* M. H. Zimmermann, C. L. Brown and M. T. Tyree. pp. 67-123. (New York: Springer-Verlag 1971)

[24] Brown, C. L. Physiology of wood formation in conifers. *Wood Sci.* **3,** 8-22 (1970)

[25] Brown, C. L. Studies in the auxin physiology of longleaf pine seedlings. In *The Physiology of Forest Trees*. pp. 511-528. K. V. Thimann (ed.) (Harvard: Harvard Univ. Press 1958)

[26] Brown, C. L., Sax, K. The influence of pressure on the differentiation of secondary tissues. *Amer. J. Bot.* **49,** 683-691 (1962)

[27] Brown, C. L., Wetmore, R. H. Auxin transport in the long shoots of pine. *Amer. J. Bot.* **46,** 586-590 (1959)

[28] Brown, C. L., Wodzicki, T. T. A simple technique for investigating cambial activity and the differentiation of cambial derivatives. *Forest Sci.* **15,** 26-29 (1969)

[29] Brown, H. P. Growth studies in forest trees. II. *Pinus strobus. Bot. Gaz.* **59,** 197-241 (1915)

[30] Brown, H. P. Growth studies in forest trees. I. *Pinus rigida. Bot. Gaz.* **54,** 386-404 (1912)

[31] Brown, K. M., Leopold, A. C. Ethylene and the regulation of growth in pine. *Canad. J. Forestry Res.* **3,** 143-145 (1973)

[32] Brown, S. A., Neish, A. C. Shikimic acid as a precursor in lignin biosynthesis. *Nature* **175,** 688-689 (1955)

[33] Camus, G. Recherches sur le rôle des bourgeons dans les phénomènes de morphogenèse. *Rev. Cytol. Biol. Vég.* **9,** 1-199 (1949)

[34] Carlquist, S. *Ecological strategies of xylem evolution*. p. 259. (London: University of California Press, London 1975)
[35] Caruso, J. L., Cutter, E. G. Morphogenetic aspects of a leafless mutant in tomato. II. Induction of a vascular cambium. *Amer. J. Bot.* **57,** 420-429 (1970)
[36] Caruso, J. L., Smith, R. G., Smith, L. M., Cheng, T.-Y., Daves, G. D. Jr. Determination of indole-3-acetic acid in Douglas fir using a deuterated analog and selected ion monitoring. *Plant Physiol.* **62,** 841-845 (1978)
[37] Clark, E. W., Mills, C. L. L. Qualitative seasonal changes in the free amino acids and carbohydrates of loblolly pine inner bark. *Wood Sci.* **3,** 90-93 (1970)
[38] Coster, C. Zur Anatomie und Physiologie der Zuwachszonen- und Jahresringbildung in den Tropen. *Ann. Jard. Bot. Buitzenzorg* **37,** 49-160; **38,** 1-114 (1927)
[39] Cronshaw, J. Tracheid differentiation in tobacco pith cultures. *Planta* **72,** 78-90 (1967)
[40] Cutter, E. G. *Plant Anatomy. Experiment and Interpretation.* II. *Organs.* (London: Edward Arnold 1971)
[41] Czaja, A. T. Der Nachweis der Wuchsstoffes bei Holzflanzen. *Ber. Deut. Bot. Gesell.* **52,** 267-271 (1934)
[42] Dalessandro, G. Hormonal control of xylogenesis in pith parenchyma explants of *Lactuca. Ann. Bot.* **37,** 375-382 (1973)
[43] DeMaggio, A. E. Phloem differentiation: induced stimulation by gibberellic acid. *Science*, N.Y. **152,** 370-372 (1966)
[44] Denne, M. P. Wood structure and production within the trunk and branches of *Picea sitchensis* in relation to canopy formation. *Canad. J. Forest. Res.* **9,** 406-427 (1979)
[45] Denne, M. P. Wood production and structure in relation to bud activity in some softwood and hardwood species. *Leiden Botanical Series* No. 3, pp. 204-211 (1976)
[46] Denne, M. P., Wilson, J. E. Some quantitative effects of indoleacetic acid on the wood production and tracheid dimensions of *Picea. Planta* **134,** 223-228 (1977)
[47] DeTorok, D., Thimann, K. V. The auxin requirement and the effect of an auxin antagonist on tumorous and normal tissues of *Picea glauca. Physiol. Plant.* **14,** 543-547 (1961)
[48] DeYoe, D. R., Zaerr, J. B. Indole-3-acetic acid in Douglas-fir: analysis by gas-liquid chromatography and mass spectrometry. *Plant Physiol.* **58,** 299-303 (1976)
[49] Digby, J., Wareing, P. F. The effect of applied growth hormone on cambial division and the differentiation of the cambial derivatives. *Ann. Bot.* **30,** 539-548 (1966)
[50] Digby, J., Wareing, P. F. The relationship between endogenous hormone levels in the plant and seasonal aspects of cambial activity. *Ann. Bot.* **30,** 607-622 (1966)
[51] Dörffling, K. Über das Wuchsstoff-hemmstoffsystem von *Acer pseudoplatanus* L. I. Der Jahresgang der Wuchs- und Hemmstoffe in Knospen, Blättern, und im Kambium. *Planta* (Ber.) **60,** 390-412 (1963)
[52] Dunberg, A. Changes in gibberellin-like substances and indole-3-acetic acid in *Picea abies* during the period of shoot elongation. *Physiol. Plant.* **38,** 186-190 (1976)
[53] Durzan, D. J. Nitrogen metabolism of *Picea glauca*. I. Seasonal changes of free amino acids in buds, shoots, apices, and leaves, and the metabolism of uniformly labelled ^{14}C-L-arginine by buds during the onset of dormancy. *Canad. J. Bot.* **46,** 909-919 (1968)
[54] Eliasson, L. Growth regulators in *Populus tremula*. I. Distribution of auxin and growth inhibitors. *Physiol. Plant.* **22,** 1288-1301 (1969)
[55] Fadia, V. P., Mehta, A. R. Tissue culture studies on cucurbits: the effect of NAA, sucrose and kinetin on tracheal differentiation in *Cucumis* tissues cultured *in vitro*. *Phytomorph.* **23,** 212-215 (1973)

[56] Farrar, J. L., Rauthan, B. S. The role of buds and foliage in the reactivation of cambium in red pine. *Ann. Rep.* 1971-72. *Res. Forest Bot.* pp. 7-8. (Toronto: Glendon Hall Lab., Fac. For., Univ. Toronto 1972)

[57] Farrar, J. L., Stemberger, A. Resumption of growth in red pine seedlings. *Ann. Rep.* 1968-69. *Forest Res.* p. 6. (Toronto: Glendon Hall Lab., Fac. For., Univ. Toronto 1969)

[58] Fischnich, O. Über den Einfluss von β-Indolylessigsäure auf die Blattbewegungen und die Adventivwurzelbildung von *Coleus. Planta* (Ber.) **24,** 552-583 (1935)

[59] Fosket, D. E., Torrey, J. G. Hormonal control of cell proliferation and xylem differentiation in cultured tissue of *Glycine max* var. *Biloxi. Plant Physiol.* **44,** 871-880 (1969)

[60] Fukuda, H., Komamine, A. Direct evidence for cytodifferentiation to tracheary elements without intervening mitosis in a culture of single cells isolated from the mesophyll of *Zinnia elegans. Plant Physiol.* **65,** 61-64 (1980)

[61] Fukuda, H., Komamine, A. Establishment of an experimental system for the study of tracheary-element differentiation from single cells isolated from the mesophyll of *Zinnia elegans. Plant Physiol.* **65,** 57-60 (1980)

[62] Gautheret, R. J. Factors affecting differentiation of plant tissues grown in vitro. In *Cell Differentiation and Morphogenesis.* W. Beermann (ed.) pp. 55-95. (Amsterdam: North-Holland 1966)

[63] Gautheret, R. J. Culture du tissu cambial. *C.R. Acad. Sci. Paris* **198,** 2195-2196 (1934)

[64] Gee, H. Localization and uptake of ^{14}C-IAA in relation to xylem regeneration in *Coleus* internodes. *Planta (Ber.)* **108,** 1-9 (1972)

[65] Giertych, M. M. Endogenous growth regulators in trees. *Bot. Rev.* **30,** 292-311 (1964)

[66] Giertych, M. M., Forward, D. F. Growth regulator changes in relation to growth and development of *Pinus resinosa Ait. Canad. J. Bot.* **44** (1966)

[67] Goldschmid, O., Quimby, G. R. Lignin precursors: the role of quinic acid. *TAPPI* **47,** 528-533 (1964)

[68] Goodwin, P. B. Phytohormones and growth and development of organs of the vegetative plant. In *Phytohormones and Related Compounds—A Comprehensive Treatise.* Vol. II. D. S. Letham, P. B. Goodwin, T. J. V. Higgins (eds.) (Amsterdam: Elsevier/North-Holland Biomedical Press 1978)

[69] Gouwentak, C. A. Cambial activity as dependent on the presence of growth hormone and the non-resting condition of stem. *Proc. Acad. Sci. Amsterdam* **44,** 654-663 (1941)

[70] Gouwentak, C. A. Kambiumtätigkeit und Wuchsstoff. *I. Meded. Landbouwhoogesch.* (Wageningen) **39,** 3-23 (1936)

[71] Gouwentak, C. A., Hellinga, G. Beobachtungen über Wurzelbildung. *Meded. Landbouwhoogesch.* (Wageningen) **39,** no. 6 (1935)

[72] Gregory, F. G., Hancock, C. R. The rate of transport of natural auxin in woody shoots. *Ann. Bot.* **19,** 451-465 (1955)

[73] Grillos, S. J., Smith, F. H. The secondary phloem of Douglas-fir. *Forest Sci.* **5,** 377-388 (1959)

[74] Grossenbacher, J. G. The periodicity and distribution of radial growth in trees and their relation to the development of "annual rings." *Trans. Wisc. Acad. Sci.* **18,** 1-77 (1915)

[75] Grozdits, G. A., Ifju, G. Nitrogen distribution in eastern hemlock and its relation to wood formation. *Wood Sci.* **6,** 1-8 (1973)

[76] Gunckel, J. E., Thimann, K. V. Studies of development in long shoots and short shoots of *Ginkgo biloba* L. III. Auxin production in shoot growth. *Amer. J. Bot.* **36,** 145-151 (1949)

[77] Harrison, M. A., Klein, R. M. Role of growth regulators in initiation of secondary xylem and phloem cells. *Bot. Gaz.* **140,** 20-24 (1979)

[78] Hartig, T. Über die Entwicklung des Jahresringes der Holzpflanzen. *Bot. Z.* **11,** 553-560, 569-579 (1853)

[79] Hastings, G. T. When increase in thickness begins in our trees. *Plant World* **3,** 113-116 (1900)

[80] Hatcher, E. S. J. Auxin relations of the woody shoot. The distribution of diffusable auxin in shoots of apple and plum rootstock varieties. *Ann. Bot.* ns **23,** 409-423 (1959)

[81] Hatcher, E. S. J. The study of auxin in shoots of apple and plum. *East Malling (Kent) Research Station Ann. Rep. 1947*, 112-116 (1948)

[82] Hejnowicz, A., Tomaszewski, M. Growth regulators and wood formation in *Pinus silvestris. Physiol. Plant.* **22,** 984-992 (1969)

[83] Hess, T., Sachs, T. The influence of a mature leaf on xylem differentiation. *New Phytol.* **71,** 903-914 (1972)

[84] Hollis, C. A., Tepper, H. B. Auxin transport within intact dormant and active white ash shoots. *Plant Physiol.* **48,** 146-149 (1971)

[85] Horgan, R. Modern methods in plant hormone analysis. *Progress in Phytochemistry* **7,** 137-170 (1981)

[86] Houck, D. F., LaMotte, C. E. Primary phloem regeneration without concomitant xylem regeneration: its hormone control in *Coleus. Amer. J. Bot.* **64,** 799-809 (1977)

[87] Jacobs, W. P. Acropetal auxin transport and xylem regeneration—a quantitative study. *Amer. Nat.* **88,** 327-337 (1954)

[88] Jacobs, W. P. The role of auxin in differentiation of xylem around a wound. *Amer. J. Bot.* **39,** 301-309 (1952)

[89] Jacobs, W. P., Morrow, I. B. A quantitative study of xylem development in the vegetative shoot apex of *Coleus. Amer. J. Bot.* **44,** 823-842 (1957)

[90] Jacquiot, C. Plant tissues and excised organs culture and their significance in forest research. *J. Inst. Wood Sci.* **16,** 22-34 (1966)

[91] Jacquiot, C. Contribution à l'étude des factéurs déterminant le cycle d'activité du cambium chez quelques arbres forestiers. *Rev. Forst. Franç.* **2,** 605-610 (1950)

[92] Jeffs, R. A., Northcote, D. H. The influence of indol-3-yl acetic acid and sugar on the pattern of induced differentiation in plant tissue culture. *J. Cell Sci.* **2,** 146-152 (1967)

[93] Jeffs, R. A., Northcote, D. H. Experimental induction of vascular tissue in an undifferentiated plant callus. *Biochem. J.* **101,** 146-152 (1966)

[94] Jenkins, P. A. Influence of applied indoleacetic acid and abscisic acid on xylem cell dimensions in *Pinus radiata* D. Don. In *Mechanisms of Regulation of Plant Growth.* R. L. Bieleski, A. R. Ferguson, M. M. Creswell (eds.) pp. 737-742. (Wellington: Bull. 12, Royal Soc. N.Z., Wellington 1974)

[95] Jenkins, P. A., Shepherd, K. R. Seasonal changes in levels of indole-acetic acid and abscisic acid in stem tissues of *Pinus radiata. N.Z.J. For. Sci.* **4,** 511-519 (1974)

[96] Jenkins, P. A., Shepherd, K. R. Identification of abscisic acid in young stems of *Pinus radiata* D. Don. *New Phytol.* **71,** 501-511 (1972)

[97] Jost, L. Über Dickenwachstum und Jahresringbildung. *Bot. Z.* **49,** 482-499, 501-510, 522-531, 541-547, 557-563, 573-579, 589-595, 605-611, 625-627 (1891)

[98] Jost, L. Über Beziehungen zwischen der Blattentwicklung und der Gefässbildung in der Pflanze. *Bot. Z.* **51,** 89-138 (1893)

[99] Knudson, L. Observations on the inception, season, and duration of cambium development in the American larch *(Larix laricina* (Du Roi) Koch). *Bull. Torrey Bot. Club* **40,** 271-293 (1913)

[100] Kohlenbach, H. W., Schmidt, B. Cytodifferenzierung in Form einer direkten Umwandlung isolierter Mesophyllzellen zu Tracheiden. *Z. Pflanzenphysiol.* **75,** 369-374 (1975)

[101] Kopcewicz, J. Seasonal changes of auxin-like substances and growth inhibitors in the apical meristems of pine (*Pinus silvestris* L.) *Zesz. Nauk. Uniw. Mikolaja Kopernika Toruniu, Nauki Mat. Przyr. Biol.* **23,** 153-161 (1970)

[102] Kopcewicz, J. Gibberellin-like substances and growth inhibitors in pine trees (*Pinus silvestris* L.) of different age. *Zesz. Nauk. Uniw. Mikolaja Kopernika Toruniu, Nauki Mat. Przyr. Biol.* **23,** 130-138 (1970)

[103] Kopcewicz, J., Michniewicz, M., Kriesel, K. Auxins and growth inhibitors in pine (*Pinus silvestris* L.) and larch trees (*Larix decidua* Mill.) of different age. *Zesz. Nauk. Uniw. Mikolaja Kopernika Toruniu, Nauki Mat. Przyr. Biol.* **23,** 130-138 (1970)

[104] Kopcewicz, J., Michniewicz, M., Kriesel, K. Dynamics of gibberellin-like substances and growth inhibitors in pine (*Pinus silvestris* L.) and larch (*Larix decidua* Mill.) in relation to age and season. *Bull. Acad. Polon. Sci.* **15,** 427-433

[105] Kozlowski, T. T. *Growth and Development of Trees.* Vol. II. *Cambial Growth, Root Growth, and Reproductive Growth.* 514 p. (New York: Academic Press 1971)

[106] Kubowicz, D. B. The possible relation between cytokinins and secondary xylem formation in *Pinus silvestris. Acta Soc. Bot. Polon.* **48,** 295-303 (1979)

[107] Kutscha, N. P., Hyland, F., Schwarzmann, J. M. Certain seasonal changes in balsam fir cambium and its derivatives. *Wood Sci. Technol.* **9,** 175-188 (1975)

[108] LaMotte, C. E., Jacobs, W. P. A role of auxin in phloem regeneration in *Coleus* internodes. *Develop. Biol.* **8,** 80-98 (1963)

[109] Larson, P. R. Contribution of different-aged needles to growth and wood formation of young red pines. *Forest Sci.* **10,** 224-238 (1964)

[110] Larson, P. R. Some indirect effects of environment on wood formation. In *Formation of Wood in Forest Trees.* M. H. Zimmermann (ed.) pp. 345-365. (New York: Academic Press 1964)

[111] Larson, P. R. The indirect effect of drought on tracheid diameter in red pine. *Forest Sci.* **9,** 52-62 (1963)

[112] Larson, P. R. Auxin gradients and the regulation of cambial activity. In *Tree Growth.* T. T. Kozlowski (ed.) pp. 97-118. (New York: Ronald Press 1962)

[113] Larson, P. R. The indirect effect of photoperiod on tracheid diameter in *Pinus resinosa. Amer. J. Bot.* **49,** 132-137 (1962)

[114] Larson, P. R. A physiological consideration of the springwood-summerwood transition in red pine. *Forest Sci.* **6,** 110-122 (1960)

[115] Letham, D. S., Goodwin, P. B., Higgins, T. J. V. *Phytohormones and Related Compounds—A Comprehensive Treatise.* Vols. I and II. 641 and 648 pp. resp. (Amsterdam, Oxford, New York: Elsevier/North-Holland Biomedical Press 1978)

[116] Little, C. H. A. Inhibition of cambial activity in *Abies balsamea* by internal water stress: role of abscisic acid. *Canad. J. Bot.* **53,** 3041-3050 (1975)

[117] Little, C. H. A., Andrew, D. M., Silk, P. J., Strunz, G. M. Identification of cytokinins zeatin and zeatin riboside in *Abies balsamea. Phytochemistry* **18,** 1219-1220 (1979)

[118] Little, C. H. A., Bonga, J. M. Rest in the cambium of *Abies balsamea. Canad. J. Bot.* **52,** 1723-1730 (1974)

[119] Little, C. H. A., Eidt, D. C. Relationship between transpiration and cambial activity in *Abies balsamea. Canad. J. Bot.* **48,** 1027-1028 (1970)

[120] Little, C. H. A., Heald, J. K., Browning, G. Identification and measurement of indoleacetic and abscisic acids in the cambial region of *Picea sitchensis* (Bong.) Carr. by combined gas chromatography—mass spectrometry. *Planta* **139,** 133-138 (1978)

[121] Little, C. H. A., Strunz, G. M., La France, R., Bonga, J. M. Identification of abscisic acid in *Abies balsamea. Phytochemistry* **11,** 3535-3536 (1972)

[122] Little, C. H. A., Wareing, P. F. Control of cambial activity and dormancy in *Picea sitchensis* (Bong) Carr. by Indol-3yl-acetic and abscisic acids. *Canad. J. Bot.* (in press)

[123] Lorenzi, R., Horgan, R., Heald, J. K. Gibberellin A9 glucosyl ester in needles of *Picea sitchensis. Phytochemistry* **15,** 789-790 (1976)

[124] Lorenzi, R., Horgan, R., Heald, J. K. Gibberellins in *Picea sitchensis* Carriere: seasonal variation and partial characterization. *Planta (Ber.)* **126,** 75-82 (1975)

[125] Lorenzi, R., Horgan, R., Wareing, P. F. Cytokinins in *Picea sitchensis* Carriere: identification and relation to growth. *Biochem. Physiol. Pflanzen* **168,** 333-339 (1975)

[126] Margna, U. Control at the level of substrate supply—an alternative in the regulation of phenylpropanoid accumulation in plant cells. *Phytochemistry* **16,** 419-426 (1977)

[127] Michniewicz, M. Studies on the biological and chemical properties of the inhibitors isolated from Scotch pine. *Zesz. Nauk. Uniw. Mikolaja Kopernika Toruniu, Nauki Mat. Przyr. Biol.* **23,** 153-161 (1970)

[128] Minocha, S. C., Halperin, W. Enzymatic changes and lignification in relation to tracheid differentiation in cultured tuber tissue of Jerusalem artichoke *(Helianthus tuberosus). Canad. J. Bot.* **54,** 79-89 (1976)

[129] Minocha, S. C., Halperin, W. Hormones and metabolites which control tracheid differentiation, with or without concomitant effects on growth in cultured tuber tissue of *Helianthus tuberosus* L. *Planta* (Ber.) **116,** 319-331 (1974)

[130] Minocha, S. C., Halperin, W. Hormonal control of xylogenesis and its relation to enzymatic changes in cultured tuber slices of Jerusalem artichoke. In *Plant Growth Substances, Proc. 8th Int. Conf.* Tokyo.

[131] Mirov, N. T. Distribution of growth hormone in shoots of two species of pines. *J. Forest.* **39,** 457-446 (1941)

[132] Mizuno, K., Komamine, A. Isolation and identification of substances inducing formation of tracheary elements in cultured carrot-root slices. *Planta* **138,** 59-62 (1978)

[133] Morris, D. A. The effect of temperature on the velocity of exogenous auxin transport in intact chilling-sensitive and chilling-resistant plants. *Planta* **146,** 603-605 (1979)

[134] Münch, E. Untersuchungen über die Harmonie der Baumgestalt. *Jahrb. f. Wiss. Bot.* **86,** 581-673. (English translation of this paper available on microfilm from: Translation Center, John Crerar Library, 35 West 33rd St., Chicago, Ill. 60616) (1938)

[135] Murmanis, L., Sachs, I. B. Seasonal development of secondary xylem in *Pinus strobus* L. *Wood Sci. Technol.* **3,** 177-193 (1969)

[136] Nitsch, J. P. Naturally-occurring growth substances in relation to plant tissue culture. In *Plant Tissue and Organ Culture, A Symposium.* P. Maheshwari, N. S. Ranga Swamy (eds.) pp. 144-157. (Ranchi, India: Catholic Press 1961)

[137] Nix, L. E., Wodzicki, T. J. The radial distribution and metabolism of IAA-^{14}C in *Pinus echinata* stems in relation to wood formation. *Canad. J. Bot.* **52,** 1349-1355 (1974)

[138] Odani, K. The effects of indoleacetic acid and chilling on cambial activity of *Pinus densiflora. J. Jap. Forest. Soc.* **57,** 112-116 (1975)

[139] Odani, K. Indoleacetic acid oxidase in cambial zone of *Pinus densiflora* Sieb. et Zucc. *J. Jap. Wood Res. Soc.* **20,** 512-515 (1974)

[140] Odani, K. Seasonal changes in endogenous indoleacetic acid level and cambial activity in *Pinus densiflora* Sieb. et Zucc. *Mokuzai Gakkaishi* **20,** 507-511 (1974)

[141] Oechssler, G. Jahreszeitliche Schwankungen des Gehaltes an organischem Saüren

in den Nadeln von *Pseudotsuga menziesii* (Mirb.) Franco, *Picea Abies* (L.) H. Karsten und *Larix decidua* Mill. *Z. Pflanzenphysiol.* **59,** 213-225 (1968)

142 Ogasawara, R. Studies on auxins and growth inhibitors in Japanese red pine *(Pinus densiflora). Bull. Tottori Univ. Forests* **3,** 21-37 (1966)

143 Onaka, F. The effects of such treatments as defoliation, disbudding, girdling, and screening of light on growth and especially on radial growth of evergreen conifers. *Bull. 18, Kyoto Univ. Forests*, pp. 55-95. (English translation available from Biology Library, University College of Wales, Aberystwyth) (1950)

144 Onaka, F. The longitudinal distribution of radial increments in trees. *Bull. 18, Kyoto Univ. Forests*, pp. 1-53. (English translation available from Biology Library, University College of Wales, Aberystwyth) (1950)

145 Parker, J. Changes in sugars and nitrogenous compounds of tree barks from summer to winter. *Naturwiss.* **45,** 139 (1958)

146 Patrick, J. W. Hormone-directed transport of metabolites. In *Transport and Transfer Processes in Plants*. I. F. Wardlaw and J. B. Passioura (eds.) pp. 433-466. (New York: Academic Press 1976)

147 Phillips, I. D. J. The cambium. In *Cell Division in Higher Plants*. M. M. Yeoman (ed.) pp. 347-390. (London: Academic Press 1976)

148 Phillips, R., Torrey, J. G. DNA synthesis, cell division and specific cytodifferentiation in cultured pea root cortical explants. *Develop. Biol.* **31,** 336-347 (1973)

149 Philipson, W. R., Ward, J. M., Butterfield, B. G. *The Vascular Cambium. Its Development and Activity*. 182 pp. (London: Chapman and Hall 1971)

150 Priestley, J. H. Studies in the physiology of cambial activity. III. The seasonal activity of the cambium. *New Phytol.* **29,** 316-354 (1930)

151 Priestley, J. H., Scott, L. I. A note upon summer wood production in the tree. *Proc. Leeds Phil. Soc.* **3,** 235-248 (1936)

152 Rehm, S. Zur Entwicklungsphysiologie der Gefässe und des trachealen Systems. *Planta (Ber.)* **2,** 255-274 (1936)

153 Reinert, J., White, P. R. The cultivation *in vitro* of tumor tissues and normal tissues of *Picea glauca. Physiol. Plant* **9,** 177 (1956)

154 Reinders-Gouwentak, C. Physiology of the cambium and other secondary meristems of the shoot. In *Encycl. Plant Physiol.* W. Ruhland (ed.) Vol. 15, pp. 1076-1105 (1965)

155 Rier, J. P., Breslow, D. T. Sucrose concentration and the differentiation of xylem in callus. *Bot. Gaz.* **128,** 73-77 (1967)

156 Roberts, L. W. *Cytodifferentiation in Plants. Xylogenesis as a Model System.* 160 pp. (Cambridge: Cambridge Univ. Press 1976)

157 Roberts, L. W. The initiation of xylem differentiation. *Bot. Rev.* **35,** 201-250 (1969)

158 Rogozinska, J. H. The occurrence of cytokinins in Scots pine. *Bull. Acad. Polon. Sci.* **15,** 789-794 (1967)

159 Rubery, P. H., Northcote, D. H. Site of phenylalanine ammonia-lyase activity and synthesis of lignin during xylem differentiation. *Nature* **219,** 1230-1234 (1968)

160 Russow, E. I. Ueber Tüpfelbildung und Inhalt der Bastparenchym- und Baststrahlzellen der Dikotylen und Gymnospermen. II. Ueber den Inhalt der parenchymastischen Elemente der Rinde vor und während des Knospenaustriebes und Beginns der Cambiumthätigkeit in Stamm und Wurzel der einheimischen Lignosen. *Bot. Centralblatt* **13,** 271-275 (1883)

161 Sachs, T. The induction of vessel differentiation by auxin. In *Plant Growth Substances 1973. Proc. 8th Intern. Conf. Tokyo.* pp. 900-906. (Tokyo: Hirokawa Publishing Co. 1973)

162 Sachs, T. The induction of fibre differentiation in peas. *Ann. Bot.* **36,** 189 (1972)

163 Sarkar, S. K., Malhotra, S. S. Gas-liquid chromatographic method for separation of

organic acids and its application to pine needle extracts. *J. Chromatog.* **171,** 227-232 (1979)

[164] Savidge, R. A. Regulation of seasonal cambial activity and tracheid differentiation in *Pinus contorta* Dougl. *Ph.D. Thesis* Univ. Coll. Wales. Aberystwyth 1981.

[165] Shepherd, K. R., Rowan, K. S. Indoleacetic acid in cambial tissue in radiata pine. *Aust. J. Biol. Sci.* **20,** 637-646 (1967)

[166] Shininger, T. L. The control of vascular development. *Ann. Rev. Plant Physiol.* **30,** 313-337 (1979)

[167] Shininger, T. L. The regulation of cambial division and secondary xylem differentiation in *Xanthium* by auxins and gibberellins. *Plant Physiol.* **47,** 417-422 (1971)

[168] Siebers, A. M. Vascular bundle differentiation and cambial development in cultured tissue blocks excised from the embryo of *Ricinus communis* L. *Acta Bot. Neerl.* **21,** 327-342 (1972)

[169] Snow, R. Activation of cambial growth by pure hormone. *New Phytol.* **34,** 347-360 (1935)

[170] Snow, R. The nature of the cambial stimulus. *New Phytol.* **32,** 288-296 (1933)

[171] Söding, H. Weitere Untersuchungen über die Wuchstoffregulation der Kambiumtätigkeit. *Z. Bot.* **36,** 113-141 (1940)

[172] Söding, H. Wuchsstoff und Kambiumtätigkeit. *Jahrb. f. wiss. Bot.* **84,** 639-670 (1937)

[173] Söding, H. Über der Einfluss von Wuchsstoff auf das Dickenwachstum der Bäume. *Ber. Deut. Bot. Ges.* **54,** 291-304 (1936)

[174] Sorokin, H. P., Mathur, S. N., Thimann, K. V. The effects of auxins and kinetin on xylem differentiation in the pea epicotyl. *Amer. J. Bot.* **49,** 444-454 (1962)

[175] Sorokin, H. P., Thimann, K. V. The histological basis for inhibition of auxillary buds in *Pisum sativum* and the effects of auxins and kinetin on xylem development. *Protoplasma* **59,** 326 (1964)

[176] Studhalter, R. W., Glock, W. S., Agerter, S. P. Tree Growth. *Bot. Rev.* **29,** 245-365 (1963)

[177] Sussex, I. M., Clutter, M. E. Seasonal growth periodicity of tissue explants from woody plants *in vitro*. *Science* **129,** 386-387 (1959)

[178] Tepper, H. B., Hollis, C. A. Mitotic reactivation of the terminal bud and cambium of white ash. *Science* **156,** 1635-1636 (1967)

[179] Thimann, K. V., Kaufman, D. Cytoplasmic streaming in the cambium of white pine. In *Physiology of Forest Trees*, K. V. Thimann (ed.) pp. 479-492. (Harvard: Harvard University Press 1958)

[180] Thompson, N. P., Jacobs, W. P. Polarity of IAA effect on sieve-tube and xylem regeneration in *Coleus* and tomato stems. *Plant Physiol.* **41,** 673-682 (1966)

[181] Timell, T. E. Ultrastructure of the dormant and active cambial zones and the dormant phloem associated with formation of normal and compression woods in *Picea abies* (L.) Karst. Tech. Publ. No. **96,** State Univ. N.Y., College of Environmental Science of Forestry, 94 pp. (1973)

[182] Torrey, J. G. Tracheary element formation from single isolated cells in culture. *Physiol. Plant.* **35,** 158-165 (1975)

[183] Torrey, J. G. Hormonal control of cytodifferentiation in agar and cell suspension cultures. In *Biochemistry and Physiology of Plant Growth Substances.* F. Wightman and G. Setterfield (eds.) pp. 843-855. (Ottawa: Runge Press 1968)

[184] Torrey, J. G., Fosket, D. E. Cell division in relation to cytodifferentiation in cultured pea root segments. *Amer. J. Bot.* **57,** 1072-1080 (1970)

[185] Torrey, J. G., Fosket, D. E., Hepler, P. K. Xylem formation: a paradigm of cytodifferentiation in higher plants. *Amer. Scientist* **59,** 338-352 (1971)

[186] Torrey, J. G., Loomis, R. S. Auxin-cytokinin control of secondary vascular tissue formation in isolated roots of *Raphanus. Amer. J. Bot.* **54,** 1098-1106 (1967)
[187] Torrey, J. G., Loomis, R. S. Ontogenetic studies of vascular cambium formation in excised roots of *Raphanus sativus* L. *Phytomorphology (Panchanan Maheshwari Memorial Volume:* 401-409 (1967)
[188] Tsivion, Y. Possible role of cytokinins in nonspecific recognition of a host and in early growth of haustoria in the parasitic plant, *Cuscuta campestris. Bot. Gaz.* **139,** 37-31 (1978)
[189] Waisel, Y., Fahn, A. The effects of environment on wood formation and cambial activity in *Robinia pseudacacia* L. *New Phytol.* **64,** 436-443 (1965)
[190] Wangermann, E. Further localization of auxin transport through internode segments. *New Phytol.* **79,** 501 (1977)
[191] Wangermann, E. The effect of the leaf on differentiation of primary xylem in the internode of *Coleus blumei* Benth. *New Phytol.* **66,** 747-754 (1967)
[192] Wareing, P. F. Hormonal regulation of assimilate movement. In *Opportunities for Chemical Plant Growth Regulation*, British Crop Protection Council, Monograph No. 21, pp. 105-111 (1978)
[193] Wareing, P. Some aspects of differentiation in plants. In *Control Mechanisms of Growth and Differentiation.* D. D. Davies and M. J. Balls (eds.) pp. 323-344. (Cambridge: Cambridge University Press 1971)
[194] Wareing, P. F. The physiology of cambial activity. *J. Inst. Wood Sci.* **1,** 34-42 (1958)
[195] Wareing, P. F. Interaction between indoleacetic acid and gibberellic acid in cambial activity. *Nature* **181,** 1744-1745 (1958)
[196] Wareing, P. F. Photoperiodism in woody plants. *Ann. Rev. Pl. Phys.* **7,** 191-214 (1956)
[197] Wareing, P. F. Growth studies in woody species. III. Further photoperiodic effects in *Pinus silvestris. Physiol. Plant* **4,** 41-56 (1951)
[198] Wareing, P. F. Growth studies in woody species. IV. The initiation of cambial activity in ring-porous species. *Physiol. Plant* **4,** 546-562 (1951)
[199] Wareing, P. F. Growth studies in woody species. II. Effect of day-length on shoot-growth in *Pinus silvestris* after the first year. *Physiol. Plant* **3,** 300-314 (1950)
[200] Wareing, P. F. Photoperiodic control of leaf growth and cambial activity in *Pinus sylvestris. Nature* **163,** 770-771 (1949)
[201] Wareing, P. F., Hanney, C. E. A., Digby, J. The role of endogenous hormones in cambial activity and xylem differentiation. In *The Formation of Wood in Forest Trees.* M. H. Zimmermann (ed.) pp. 323-344 (1964)
[202] Wareing, P. F., Roberts, D. L. Photoperiodic control of cambial activity in *Robinia pseudacacia* L. *New Phytol.* **55,** 356-368 (1956)
[203] Warren Wilson, J. The position of regenerating cambia: auxin/sucrose ratio and the gradient induction hypothesis. *Proc. R. Soc. Lond. B.* **203,** 153-176 (1978)
[204] Warren Wilson, J., Warren Wilson, P. M. The position of regenerating cambia—a new hypothesis. *New Phytol.* **60,** 63-73 (1961)
[205] Webber, J. E., Laver, M. L., Zaerr, J. B., Lavender, D. P. Seasonal variation of abscisic acid in the dormant shoots of Douglas-fir. *Canad. J. Bot.* **57,** 534-538 (1979)
[206] Weinstein, L. H., Porter, C. A., Laurencot, H. J. Role of quinic acid in aromatic biosynthesis in higher plants. *Contr. Boyce Thompson Inst.* **21,** 201-214 (1962)
[207] Went, F. W. Wuchstoff und Wachstum. *Rec. Trav. Bot. Neerl.* **25,** 1-116 (1928)
[208] West, P. W. Alkali-labile IAA complexes in *Pinus radiata* D. Don. *Ann. Bot.* **40,** 265-268 (1977)
[209] Wetmore, R. H., Rier, J. P. Experimental induction of vascular tissues in callus of angiosperms. *Amer. J. Bot.* **50,** 418-430 (1963)

210 Wetmore, R. H., Sorokin, S. On the differentiation of xylem. *J. Arn. Arboretum* **36,** 305-317 (1955)
211 Wilcox, H. Cambial growth characteristics. In *Tree Growth.* T. T. Kozlowski (ed.) pp. 57-88. (New York: Ronald Press 1962)
212 Wilson, B. F., Wodzicki, T. J., Zahner, R. Differentiation of cambial derivatives: proposed terminology. *Forest Sci.* **12,** 438-440 (1966)
213 Wodzicki, T. J. Seasonal variation of auxin in stem cambial region of *Pinus silvestris* L. *Acta Soc. Bot. Polon.* **47,** 225-231 (1978)
214 Wodzicki, T. J. Mechanism of xylem differentiation in *Pinus silvestris* L. *J. Exp. Bot.* **22,** 670-686 (1971)
215 Wodzicki, T. J. Annual ring of wood formation and seasonal changes of natural growth inhibitors in larch. *Acta Soc. Bot. Polon.* **34,** 117-151 (1965)
216 Wodzicki, T. J. Photoperiodic control of natural growth substances and wood formation in larch (*Larix decidua* D.C.) *J. Exp. Bot.* **15,** 584-599 (1964)
217 Wodzicki, T. J. Investigation on the kind of *Larix polonica* Rac. wood formed under various photoperiodic conditions. III. Effect of decapitation and ringing on the wood formation and cambial activity. *Acta Soc. Bot. Polon.* **30,** 293-306 (1961)
218 Wodzicki, T. J. Investigation on the kind of *Larix polonica* Rac. wood formed under various photoperiodic conditions. II. Effects of different light conditions on wood formed by seedlings grown in greenhouse. *Acta Soc. Bot. Polon.* **30,** 111-131 (1961)
219 Wodzicki, T. J. Investigation on the kind of *Larix polonica* Rac. wood formed under various photoperiodic conditions. I. Plants growing in natural conditions. *Acta Soc. Bot. Polon.* **29,** 713-730 (1960)
220 Wodzicki, T. J., Witkowska, L. On the photoperiodic control of extension growth and wood formation in Norway spruce (*Picea abies* (L.) Karst). *Acta Soc. Bot. Polon.* **30,** 755-764 (1961)
221 Wodzicki, T. J., Wodzicki, A. B. Seasonal abscisic acid accumulation in stem cambial region of *Pinus silvestris*, and its contribution to the hypothesis of a late-wood control system in conifers. *Physiol. Plant.* **48,** 443-447 (1980)
222 Wodzicki, T. J., Wodzicki, A. B. Auxin stimulation of cambial activity in *Pinus silvestris* L. II. Dependence upon basipetal transport. *Physiol. Plant.* **29,** 288-292 (1973)
223 Wodzicki, T. J., Zajaczkowski, S. Effect of auxin on xylem tracheid differentiation in decapitated stems of *Pinus silvestris* L. and its interaction with some vitamins and growth regulators. *Acta Soc. Bot. Polon.* **43,** 129-148 (1974)
224 Wolter, K. E. A new method for marking xylem growth. *Forest Sci.* **14,** 102-104 (1968)
225 Worrall, J. Absence of "rest" in the cambium of Douglas-fir. *Canad. J. Forest Res.* **1,** 84-89 (1971)
226 Wort, D. J. Physiology of cambial activity. In *Tree Growth.* T. T. Kozlowski (ed.) pp. 89-96. (New York: Ronald Press 1962)
227 Young, B. S. The effects of leaf primordia on differentiation in the stem. *New Phytol.* **53,** 445-460 (1954)
228 Zajaczkowski, S. Xylem formation in isolated stem segments of *Pinus silvestris* L. *Acta Soc. Bot. Polon.* **38,** 671-675 (1969)
229 Zajaczkowski, S. Auxin stimulation of cambial activity in *Pinus silvestris* L. I. The differential cambial response. *Physiol. Plant.* **29,** 231-287 (1973)
230 Zajaczkowski, S., Wodzicki, T. J. Inhibition and requirement of natural stimulator for cambial xylem production in isolated stem segments of *Pinus silvestris*. *Physiol. Plant.* **33,** 71-74 (1975)
231 Zimmermann, W. A. Untersuchungen über die räumliche und zeitliche Verteilung des Wuchsstoffes bei Bäumen. *Z. Bot.* **30,** 209-252 (1936)

8. The Environmental Control of Xylem Differentiation

M P DENNE and R S DODD
Department of Forestry and Wood Science, University College of North Wales, Bangor, Gwynedd, Wales

With the depletion of the natural forest resources of the world, our future supplies of timber will have to come mainly from managed secondary regrowth, and from plantation forests. Much of this timber will be from young trees, grown faster than most of those harvested in the past. If we hope to maintain the quality as well as the quantity of wood produced, systems of silvicultural management will have to be devised for the new forest conditions, to produce wood tailored for the varied end-uses of timber. To select the most appropriate management system, it is crucial to quantify effects of environment on xylem differentiation, for at present many aspects of wood quality can be controlled only relatively crudely by silvicultural management. Whether the wood is to be used for structural timber, or pulped for paper, or used as a source of energy, one of the most important determinants of its quality is wood density. Density depends primarily on the proportion of wall material to cell lumen; this proportion varies with the diameter and wall thickness of each cell type, which may vary across the growth ring, down the tree, from season to season, and with the distribution of cell types. This chapter is concerned with effects of environment on fibre diameter and wall thickness—obviously this is only one aspect of the control of wood quality, but is particularly important since fibres (or tracheids) form the bulk of the wood in both hardwoods and softwoods.

There is already a vast literature concerned with effects of environment on fibre dimensions, but a confusing variety of experimental results has produced a diversity of interpretations. For example, fertilisation with nitrogen has been variously reported to increase wood density, decrease wood density, or have no significant effect upon it[1, 40], and reduction in daylength has been shown to decrease wall thickness of tracheids in *Pinus radiata*[33], but to increase wall thickness in *Larix decidua*[45]. As Larson pointed out[24, 27], these conflicting viewpoints may be reconciled by considering the growth processes of the whole tree in relation to its total environment, taking into account not only the direct effects of environmental factors on cambial activity, but also their indirect influence through effects on the growth of the crown. For this reason, we will start by attempting to summarise the main physiological interactions likely to influence cambial differentiation directly or indirectly. This rationale should provide a framework for the interpretation of experimental data—with the obvious limitation that important physiological processes may have been underestimated or omitted from our scheme.

A PHYSIOLOGICAL RATIONALE FOR INTERPRETING ENVIRONMENTAL EFFECTS ON XYLEM DIFFERENTIATION

As a working hypothesis, we may assume that the effects of environmental factors (such as temperature, light intensity, or water deficit) on xylem differentiation may be direct, or indirect through substrate availability or growth regulator balance (Fig. 8.1). For example, a change in temperature might be expected to have a direct effect on rate of expansion or wall thickening in differentiating xylem cells, or a direct effect on the duration of each phase of differentiation (arrow 1 in Fig. 8.1), and also indirect effects through substrate availability (arrow 2) or growth regulator balance (arrow 3). Figure 8.2 summarises some of the complex variety of interacting physiological processes that would be expected to affect xylem differentiation. In addition to the direct effects (arrow 1 in Fig. 8.2), a change of environment might affect shoot and root activity (arrows 4 and 5), which would perhaps involve changes in the partitioning of available substrate between shoot (arrow 6), root (arrow 7), and cambium (arrow 2). A change in environment would also be likely to affect rates of photosynthesis and respiration (arrow 8), and hence substrate availability (9), and a change in production of photosynthates would presumably also follow from increase in leaf area after increase in shoot activity (10, 11). Rate of transpiration might be affected indirectly through change in leaf area (arrow 12), as well as being affected directly by change in the environment (13). Rate of transpiration, along with changes in root activity, would influence nutrient

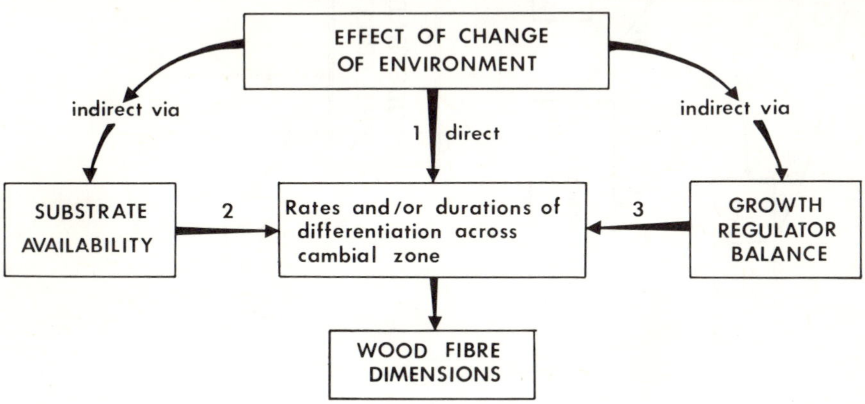

Figure 8.1: *Summary of direct and indirect effects of environment on xylem differentiation.*

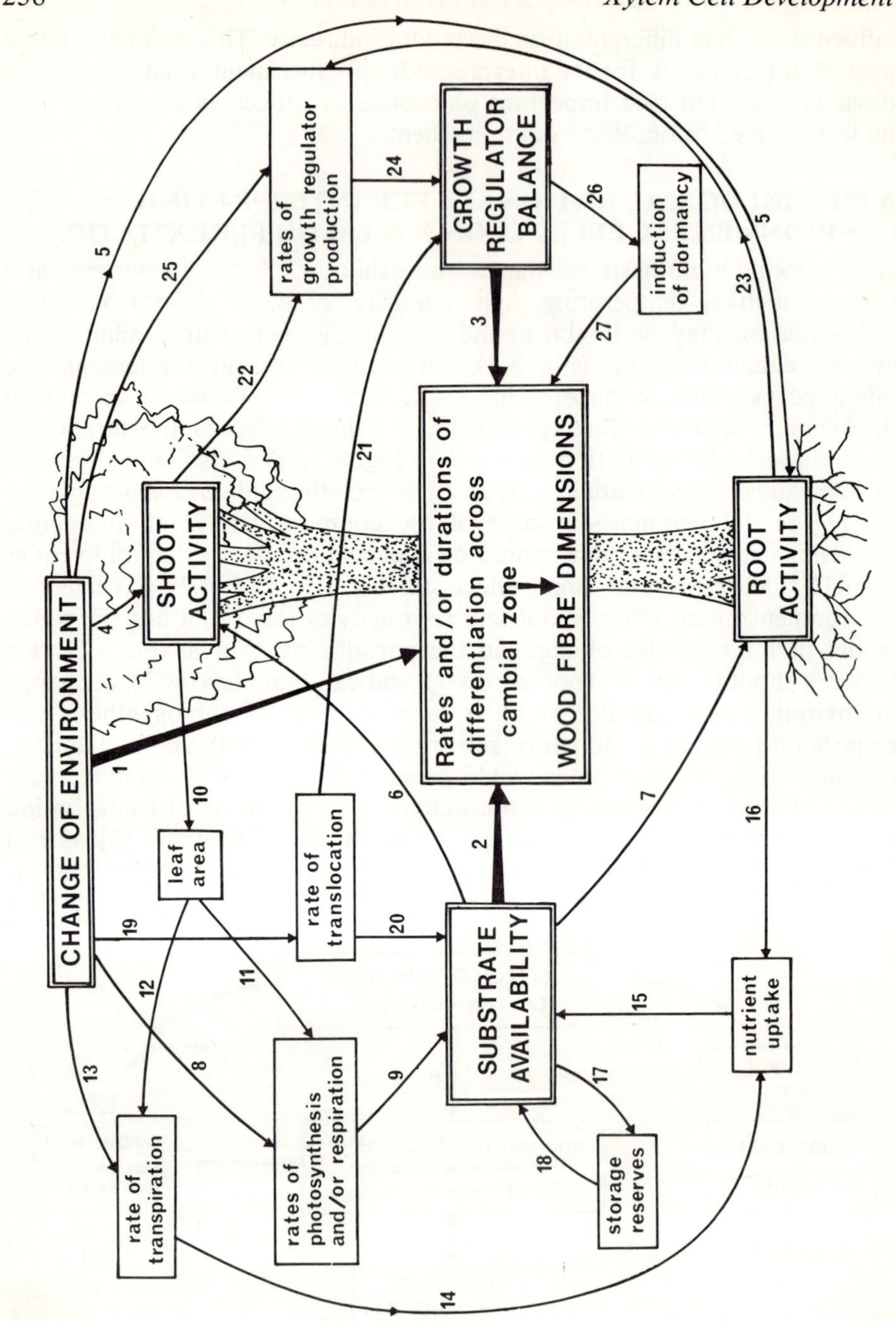

Figure 8.2: *Some of the interacting physiological processes involved in effects of environment on xylem differentiation.*

uptake (arrows 14, 16), and hence substrate availability (15). Fluctuation in storage reserves also have to be considered (arrows 17, 18). The rate of translocation would also be directly affected by environment (arrow 19), and would be expected to influence both substrate availability (20) and growth regulator balance (21). Shoot and root activity would presumably affect rates of production of growth promoters and inhibitors, and hence growth regulator balance (arrows 22, 23, 24). Direct effects of environment on growth regulator production and destruction also have to be included (25). Finally, the growth regulator balance would affect cambial differentiation directly (arrow 3), and also indirectly through effects on cambial dormancy (26, 27). This model owes much to those put forward by Fritts[17], who analysed the complex interactions between internal and external factors affecting growth ring width. As Fritts[17] pointed out, a change in one environmental factor (such as precipitation) is likely to be associated with changes in other factors (such as cloud cover, and temperature). Obviously the model shown in Fig. 8.2 is greatly oversimplified; for example the box labelled 'shoot activity' includes a wide range of developmental processes (such as apical activity, production of leaf primordia, flowers and fruit, and stem elongation), each of which might be affected differentially by environmental factors.

In temperate regions, there are likely to be regular seasonal cycles of temperature, daylength, light intensity, and water deficit, with other environmental factors, such as wind, frost, defoliation, mineral deficiency, or flooding, making occasional impacts. Seasonal cycles of external factors will interact with internal growth cycles, including periodic shoot elongation, leaf expansion, apical activity, and root growth, together with associated cycles of leaf ageing and photosynthetic efficiency, and storage reserves. Within the tree there will be variations in microclimate, including differences in light intensity, carbon dioxide concentration, humidity and temperature, interacting with position in relation to shoot and root apices and leaves. Considering effects of environment on wood production from year to year, cambial ageing also has to be taken into account, together with the changing position of the cambium in relation to crown and root, and changes in the ecology of the surroundings. The models produced by Fritts[17] also include climatic factors from previous seasons; he pointed out that water stress in previous years may reduce the number of leaf primordia, and hence the photosynthetic capacity of the crown, as well as the amount of stored substrate.

Given all these complex interactions, it is not surprising that experiments on effects of environment on wood structure sometimes appear to give conflicting results. It is obvious that such contradictory effects will only be fully understood when they can be interpreted in relation to the interacting physiological processes involved in cambial differentiation, and that these processes need to be quantified systematically throughout the tree. Furthermore, since there are adaptations to change of environment[6], as well as longer term effects on cambial dormancy[4] these interactions have to be quantified over a period of time.

THE EXPERIMENTAL EVIDENCE

Some factors affecting fibre dimensions through substrate availability

Experimental evidence suggests that environmental factors affecting substrate availability may have significant effects on fibre dimensions, though these effects are often small[4, 5, 11, 33]. For example, seedlings of *Picea sitchensis* germinated and grown at three light intensities showed considerable differences in overall growth rate, but relatively little variation in wood structure between treatments (Table 8.1). By the end of three months, plants in the lowest light intensity appeared to be only just surviving, having a single stem with very little root, while plants in the highest light intensity had a much branched stem with a vigorous root system. Yet, in spite of great differences in dry weight (Table 8.1), tracheid wall thickness did not vary significantly between treatments, and although tracheid diameter did increase with light intensity, the differences between treatments were small. Rate of wood production was slower in low light, but not significantly different between medium and high light. The calculated rates of increase in wall material per internode suggest that total increment in wall material was limited at low light intensity, but did not differ between medium and high light (Table 8.1). It seems that effects of

Table 8.1: *Effects of light intensity on seedling dry weight, tracheid production, and tracheid dimensions of Picea sitchensis.* Seedlings grown for 13 weeks in 18 hour photoperiods, in light intensities of 2,200 lx, 6,700 lx, or 20,000 lx. Sample size at least 8 plants from each treatment. Tracheid dimensions from outermost layer of fully mature tracheids, sectioned 5 mm above cotyledons. For each row, different letters indicate treatment means differ by at least P = 0.05. (These data are extracted from tables and figures published in Smith (1974) and Denne (1974), by permission of the Annals of Botany Company).

Light intensity		*2,200 lx*	*6,700 lx*	*20,000 lx*
Dry weight in mg:	total	18	200	348
	main stem	1	7	7
	branch and leaf	15	159	239
	root	2	34	102
Rate wood production in tracheids/day		0.16 a	0.27 b	0.30 b
Radial tracheid diameter in μm		10.1 a	12.4 b	13.9 c
Tracheid wall thickness in μm		2.53 a	2.64 a	2.57 a
Rate of accumulation of wall volume, per leaf internode per day, in μm^3		180 a	622 b	690 b

light intensity were partially offset by differences in both photosynthetic efficiency and distribution of assimilates. As shown by Smith[37], leaf structure and efficiency varied with light intensity, and above the medium light level any extra substrate was apparently diverted into branch and root production rather than into more wood in the main stem. At low light, differences in rate of total accumulation were nullified by differences in rate of tracheid production and tracheid diameter, resulting in similar wall thicknesses between light intensity treatments. Conversely however, Richardson[33] reported that tracheid wall thickness increased with light intensity in seedlings of *P. sitchensis*, and found no consistent effect of light intensity on lumen diameter.

Data from *Eucalyptus* suggests that light intensity also has relatively minor effects on the fibre dimensions of hardwood seedlings[11]. Doley found fibre walls to be slightly thicker in *Eucalyptus* grown in low light intensity than in those grown in high light intensity. This was associated with a considerably longer duration of the wall thickening phase of fibre differentiation in lower light intensity, in spite of a slower rate of thickening. As in *P. sitchensis*, the fibre diameter of *Eucalyptus grandis* was greater in high than in low light intensity, associated with a slightly faster rate and longer duration of cell expansion at higher light intensity. Our recent work with seedlings of *Acer pseudoplatanus* suggests that light intensity has a significant effect on total dry weight (Table 8.2), and (as with *P. sitchensis* in Table 8.1) the differences in root dry weight were particularly marked (Table 8.2). Mean fibre wall thickness was slightly greater in high light intensity in the shoot, but in the root there was no significant effect of light intensity on fibre wall thickness.

Returning to Fig. 8.2, the above data suggest that environmental effects on fibre dimensions through substrate availability (arrow 2) are minimised by differences in photosynthetic efficiency (arrows 8, 9) and distribution of assimilates (6, 7). However, experiments with change of environment suggest that this regulatory system can be temporarily upset before adaptation occurs, causing transitory changes in rates of wood production and structure[5, 8, 19, 38]. With seedlings of *P. sitchensis*, tracheid diameter and wall thickness decrease rapidly after transfer to treatments designed to limit substrate availability (decrease in light intensity or light quantity, or increase in night temperature), as shown in Fig. 8.3. But after reaching a minimum about 10 days to 3 weeks after change of environment, tracheid wall thickness tends to increase again to a maximum about 4 to 6 weeks after transfer. This recovery parallels a recovery in net assimilation rate, and appears to be associated with increase in photosynthetic efficiency, and with changes in distribution and type of assimilate produced[5, 38, 39]. Tracheid diameter also increases at the same time in some (but not all) treatments. Measurements of rates and durations of expansion and

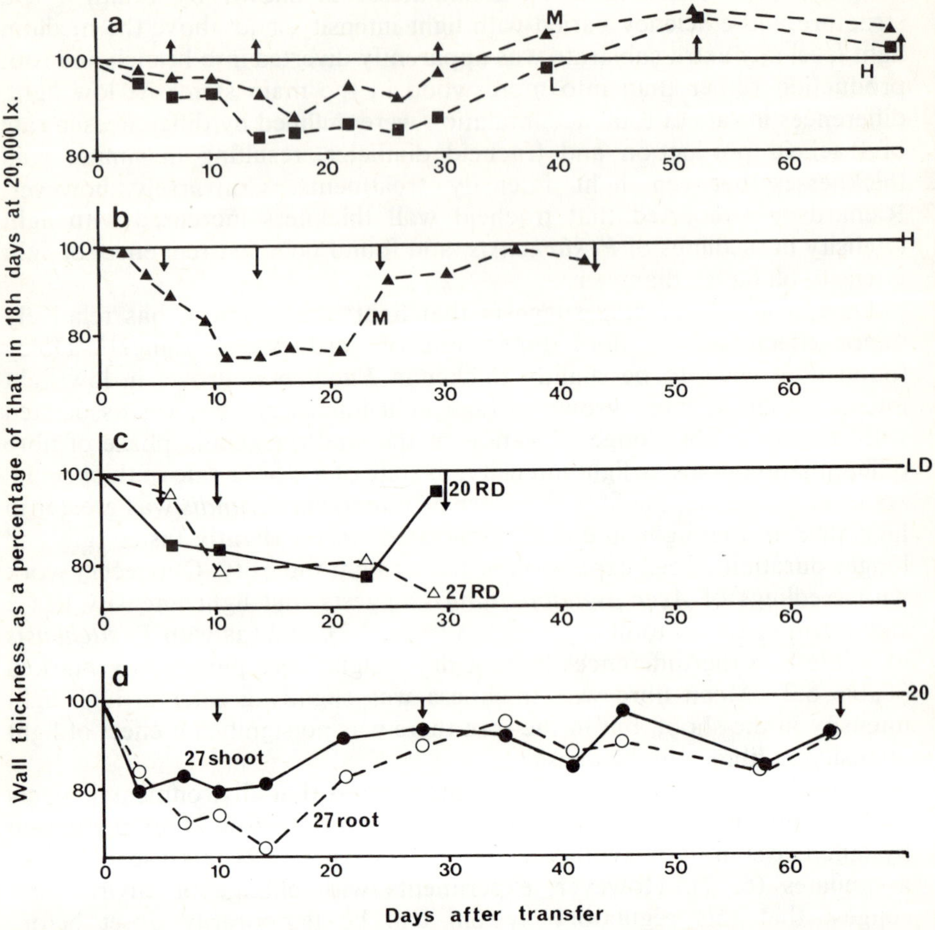

Figure 8.3: *Effects of environmental change on tracheid wall thickness in* Picea sitchensis *seedlings. Wall thickness of most recently matured tracheid expressed as a percentage of wall thickness in plants remaining in 18 hr photoperiods at 20,000 lx. Vertical arrows indicate least significant difference at P = 0.05.*

(a) Plants transferred from high light intensity (H) into: ▲M (6700 lx) or ■L (4300 lx) (experiment 1).

(b) Plants transferred from high light intensity (H) into ▲M (6700 lx) (experiment 2).

(c) Plants transferred from 18 hr photoperiods (LD) into 8 hr high light intensity supplemented with 10 hr very low light intensity (RD), at night temperatures of: ■20°C, or △27°C.

(d) Plants transferred from constant 20°C into higher night temperature (27°C): ● shoot, ○ root.

(This figure is adapted from a figure published in Denne (1976), by permission of the Annals of Botany Company).

wall thickening in the differentiating xylem suggest that the initial decline in tracheid dimensions is associated with decrease in rates of expansion and wall thickenings, while the duration of each phase of differentiation increases; the differences in rate and duration of wall thickening diminishing after the period of adaptation. It is interesting to note that the time scale for the adaptation was comparable in all experiments (Fig. 8.3), though they were done with different batches of seedlings, in a variety of environmental conditions. However, as noted by Ford *et al.*[16] this adaptive mechanism would operate too slowly to keep pace with many of the variations in environment likely to be encountered in the forest. Using older trees, Ford *et al.*[16] found a close correlation between rate of wood production and daily solar radiation in *Picea sitchensis* and considered that fluctuations in rate of cell production (resulting from environmental fluctuations) were responsible for variations in tracheid wall thickness and diameter, an increase in rate of production increasing the queue of cells waiting to pass through each phase of development. However, as shown by Creber[2], cyclical fluctuations in tracheid diameter are not necessarily synchronous in neighbouring lines of cells.

Changes in substrate availability would also be expected to follow reduction in photoperiod, though this would be complicated by other photoperiodic effects on growth regulator balance. Figure 8.4 shows tracheid diameter and wall thickness of *Picea sitchensis* seedlings in two experiments involving transfer from long (18 hr) to short (8 hr) photoperiods, the dimensions being expressed as a percentage of those in plants remaining in the long photoperiods. These data suggest an initial decrease in wall thickness to a minimum about 10 days to 3 weeks after transfer to short days, followed by an increase up to 4 weeks after transfer (maximum wall thickness being greater than in plants remaining in the long photoperiods), then a final decrease in the last few tracheids produced before cambial dormancy. The rate of tracheid production showed a similar trend[8], while tracheid diameter appeared to remain constant for about one week after transfer, then declined until dormancy[8]. Up to about 30 days after transfer to short photoperiods, the trends in wall thickness and tracheid production closely paralleled the pattern of change after transfer to other conditions expected to decrease substrate availability (Fig. 8.3). Also, in the seedlings illustrated in Fig. 8.4, these trends were accompanied by changes in rate of root growth[8, 41], rate of dry matter production[38], and photosynthetic efficiency[38]; all these seem to indicate an initial phase of substrate deficiency followed by recovery. With continued short photoperiods, however, these effects were modified, presumably through photoperiodic effects on the induction of apical dormancy—in the same seedlings bud scales were produced at the main shoot apex after about 13

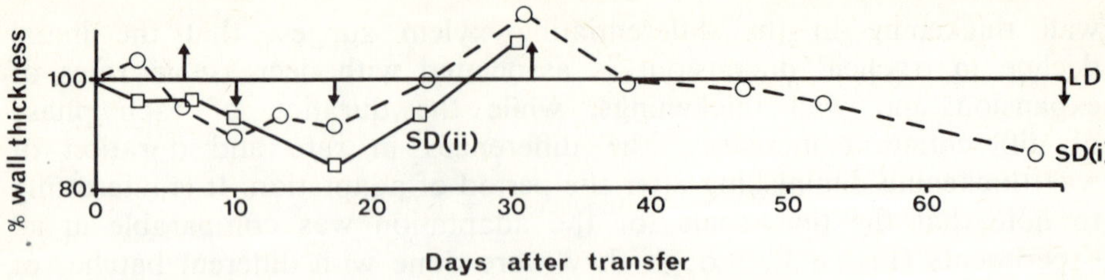

Figure 8.4: *Effects of reduction in photoperiod on tracheid wall thickness in* Picea sitchensis *seedlings. Wall thickness of most recently matured tracheid in plants transferred to 8 hr photoperiods (S) expressed as a percentage of the wall thickness of plants remaining in 18 hr photoperiods (LD): ○ experiment 1, □ experiment 2. (This figure is adapted from a figure published in Denne (1976), by permission of the Annals of Botany Company).*

to 14 short photoperiods, and stem extension ceased after 22 short photoperiods[8]. It would be logical to ascribe the increase in wall thickness after 20 short photoperiods (Fig. 8.4) to increase in substrate availability following cessation of apical activity, but similar work by Wodzicki[45] with *Larix decidua* suggested that this increase in wall thickness was independent of extension growth. It is interesting to note that the data presented by Wodzicki[45] indicate trends in wall thickness similar to those in Fig. 8.4; his data suggesting that wall thickness did not begin to increase until after 21 short photoperiods, and appears to have been preceded by a decrease in wall thickness. On the other hand, Richardson[33] reported decrease in wall thickness after several weeks of short photoperiods in *Picea sitchensis* and *Pinus radiata*, and Denne and Smith[8] found tracheid walls also remained thinner in short days in *Pinus sylvestris*—though in the latter case the situation in all treatments may have been complicated by approaching dormancy. Using cuttings and seedlings of *P. radiata*, Jenkins and co-workers[20] investigated effects of growth in long or short photoperiods applied continuously, or in combination with increasing or decreasing photoperiods. In this experiment very low light levels were used to lengthen the photoperiod so avoiding any appreciable increase in light quantity. Under these conditions the length of photoperiod appeared to have little effect on either tracheid diameter or wall thickness, though there was a slight decrease in wall/lumen and wall/diameter ratios with decreasing photoperiod and short photoperiod.

Experiments with defoliation or stem girdling should also affect substrate availability, though this is likely to be confounded by effects on transpiration and translocation. In *P. sitchensis* Richardson[33, 34] reported that though carbohydrate accumulated above a stem girdle, it did not appear to have any significant effect on wall thickness, and lumen diameter decreased above the girdle. Defoliation has been shown to decrease both tracheid diameter and wall thickness in *P. sitchensis*[33, 34], although it is obvious that defoliation is likely to have effects additional to the removal of the source of carbohydrate.

Working with *Pinus resinosa*, Gordon and Larson[18] related change in wood structure from early to latewood to the maturation of the current season's leaves. Logically, they attributed the production of the thick walled latewood tracheids to increased substrate availability, for they found that the change to latewood tracheids coincided with the time when the newly expanded leaves reached maximum photosynthetic efficiency, and began to translocate greatly increased quantities of ^{14}C to the stem. However, the increase in wall thickness from early to latewood appears to be associated with increase in duration of the wall thickening phase of tracheid differentiation, in spite of a decrease in the rate of thickening[4, 36, 46], while (as noted above) change in substrate availability has been shown to affect the rate of wall thickening in the differentiating zone rather than its duration[5]. Other work has shown that the increase in wall thickness after application of IAA is associated with increase in the duration of wall thickening[9, 31], implying that IAA delays the senescence of cells in the differentiating xylem. From this, it seems likely that the change from early to latewood may be associated with change in growth regulator balance, which might well be associated with the cessation of shoot extension, or with leaf maturation. In other words, returning to Fig. 8.2, an effect of environment on the production of latewood would seem to be determined through arrows 4, 22, 24 and 3, rather than through arrows 4, 6 and 2.

Some effects of change in temperature

Seedlings grown at lower temperatures have been reported to have thicker walls than those at higher temperatures[3, 33, 42]. The greater effect of night temperature than day temperature has been attributed to depletion of substrates by night respiration[33]. Tracheid lumen diameter is reported to increase with temperature[33, 35], and again night temperature was more effective than day. Calculations from the data supplied by Richardson and Dinwoodie[35] suggest that this effect was mainly due to change in wall thickness, radial tracheid diameter being variably affected by night temperature. Increasing the temperature of part of the stem of *Picea sitchensis* seedlings with a heating coil bridge increased both tracheid and lumen diameter, but did not affect wall thickeness[34]. In *Pinus sylvestris*, increase in temperature increased rates of both tracheid expansion and wall thickening in the differentiating xylem[3], as would be expected from a direct effect of temperature on rate processes (arrow 1 in Fig. 8.2); but since the duration of both phases of differentiation was decreased by rise in temperature, the net affect on tracheid dimensions was relatively minor, with not more than a 20% difference in dimensions with a 10°C rise in temperature. Similarly with older trees of *P. sylvestris* Wodzicki[46] found a positive correlation between mean daily temperature and rate of cell enlargement and wall thickening. In the hardwood *Terminalia superba*, cuttings grown at lower night temperature (23 °C) had a faster rate of wood production than those given warmer nights (30°C), and fibre diameter and wall thickness were both slightly greater in cooler nights[28]. Jenkins and Shepherd[21] similarly found that tracheid diameter increased with decrease in temperature in *Pinus radiata* seedlings transferred to a range of temperatures. Later work by Jenkins suggested that transfer to lower temperature initially led to decreased tracheid diameter relative to that produced after transfer to higher temperature, but tracheid diameter subsequently increased with continued low temperature, suggesting adaptation to the changed conditions[19]. Tracheid walls were thicker in lower than in higher temperature, and as in the *Pinus sylvestris* seedlings mentioned above, this was associated with a longer duration of wall thickening in the differentiating xylem[19].

Some effects of water deficit

The complex effects of water deficit on xylem differentiation have been discussed fully by Kozlowski[22] and Fritts[17]. As these authors pointed out, the influence of internal water stress may be direct, affecting the expansion and the wall thickening of the differentiating cells (arrow 1 in Fig. 8.2), or indirect through effects on rates of photosynthesis (arrow 8), transpiration (arrow 13), and translocation (arrow 19), affecting both substrate availability and growth regulator balance. Evidence for direct effects of water stress on cambial differentiation has been discussed by Zahner[47]. Whitmore and Zahner[44] found evidence that water deficits directly affected the cell wall metabolism of the differentiating xylem, and Doley and Leyton[12] showed that even slight increases in water deficit inhibited cell expansion in the cambial zone of *Fraxinus excelsior*. Working with *Liriodendron* seedlings, Doley[10] found that the effects of internal water deficit on wood structure decreased towards the apex of the stem, and he attributed this to reduced synthesis and translocation of both photosynthates and growth regulators, which consequently were more abundant near the apex. On the other hand, Larson[25, 26] emphasised the indirect effects of drought on xylem differentiation, concluding that water stress limited auxin production by restricting the growth of the terminal meristem, so affecting tracheid expansion indirectly. Larson found that although tracheid diameter normally increased again when a drought period was broken, this response could be prevented by removing the current needles, or by suppressing needle elongation by exposure to short photoperiods.

Water stress has been found to account for a large part of the ring-width variation in older trees[17]. If a water deficit persists long enough, premature dormancy will be induced[17,24], and consequently the proportion of latewood in the growth ring will be reduced[24,30]. Referring again to Fig. 8.2, this would presumably result from the effects of water deficit on shoot and root activity (arrows 4, 5), which would probably affect growth regulator production (arrows 22, 23) and hence growth regulator balance (arrow 24), affecting the onset of dormancy (arrows 26, 27).

Variation in environment around the tree

The distribution of light intensity in the crown has been shown to have considerable influence on the wood production and structure of branches of *Picea sitchensis*[4,7]. Between branches, the rate of tracheid production increased linearly over the whole range of light intensity[4,7]. This differs from the more limited effect of light intensity level on rate of wood production in the seedlings (Table 8.1), perhaps because the older trees have a higher proportion of non-photosynthesising tissue to support[37]. The time of cambial reactivation varied between branches of *P. sitchensis*, cambial

Table 8.2: *Effects of light intensity on seedling dry weight, fibre production, and fibre dimensions of Acer pseudoplatanus*. Seedlings grown for 21 days in light intensity of 10,500 lx, then grown for 46 days in either 10,500 lx or 21,000 lx. Fibre dimensions from at least 3 radii at the middle of the second internode above the cotyledons in stem, and 1.5 mm below root collar

Dry weight (g) (mean of 10 seedlings)	*Light intensity in lux*		*Treatment differences*
	10 500	*21 000*	
Total	3.3	4.7	*
Leaf	1.7	2.5	*
Stem	1.3	1.7	N.S.
Root	0.3	0.6	*

Wood production		*Light intensity in lux*		*Treatment differences*
		10 500	*21 000*	
Rate (fibre/day) (8 seedlings)	Stem	1.0	1.1	N.S.
	Root	1.2	1.5	N.S.
Fibre diameter (μm) (8 seedlings)	Stem	16.6	17.2	N.S.
	Root	19.5	19.9	N.S.
Fibre wall thickness (μm) (3 seedlings)	Stem	1.80	2.05	*
	Root	1.73	1.77	N.S.

* significant at $P = 0.05$
N.S. not significant at $P = 0.05$

Table 8.3: *Effects of light intensity on shoot growth, tracheid production, and tracheid dimensions within 8-year-old trees of Picea sitchensis*. Mean of at least six branches in each light intensity class. For each row, different letters indicate means differ by at least P=0.05. (These data are derived from tables published in Denne (1974), by permission of the Annals of Botany Company)

Light intensity as a percentage of full daylight	*up to 10%*	*11-20%*	*21-30%*	*31-50%*	*51-70%*	*71-100%*
Date shoot elongation 98% complete	June 8	June 15	June 20	June 23	June 26	July 4
Date wood production 98% complete	—	June 23	July 4	July 13	August 12	September 15
Rate of tracheid production in tracheids per day	0.10 a	0.14 a	0.18 ab	0.26 bc	0.28 c	0.30 c
Mean radial tracheid diameter in μm	11.4 a	14.4 b	14.4 b	14.5 b	16.5 c	18.1 d
Mean tracheid wall thickness across whole ring in μm	2.8 a	3.0 ab	3.2 b	3.6 c	3.6 c	3.7 c
Mean tracheid wall thickness in middle of earlywood in μm	3.0 a	3.2 a	3.1 a	3.2 a	3.4 a	3.1 a

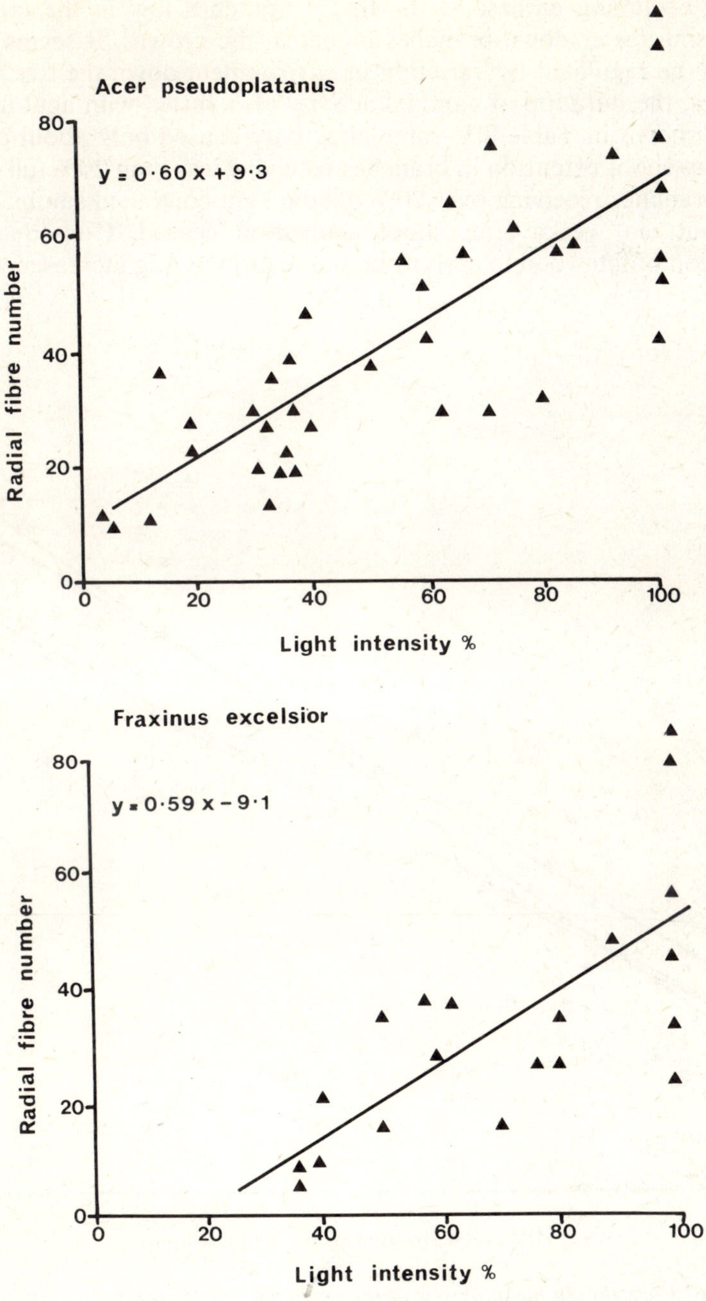

Figure 8.5: *Fitted regressions of radial fibre number across the growth ring with light intensity at the tip of the branch, in* Acer pseudoplatanus *and* Fraxinus excelsior.

activity beginning earliest at the tip of branches low in the crown, but progressing faster down branches higher in the crown[7]. It seems that this too may be regulated by variation in environment down the tree. Between branches, the duration of cambial activity also varied with light intensity[4], and, as shown in Table 8.3, cambial activity ceased only about one week later than shoot extension in branches receiving less than 20% full daylight, whilst branches receiving over 70% full daylight continued cambial activity for about ten weeks after shoot elongation ceased. Consequently the proportion of latewood to earlywood in the growth ring increased with light

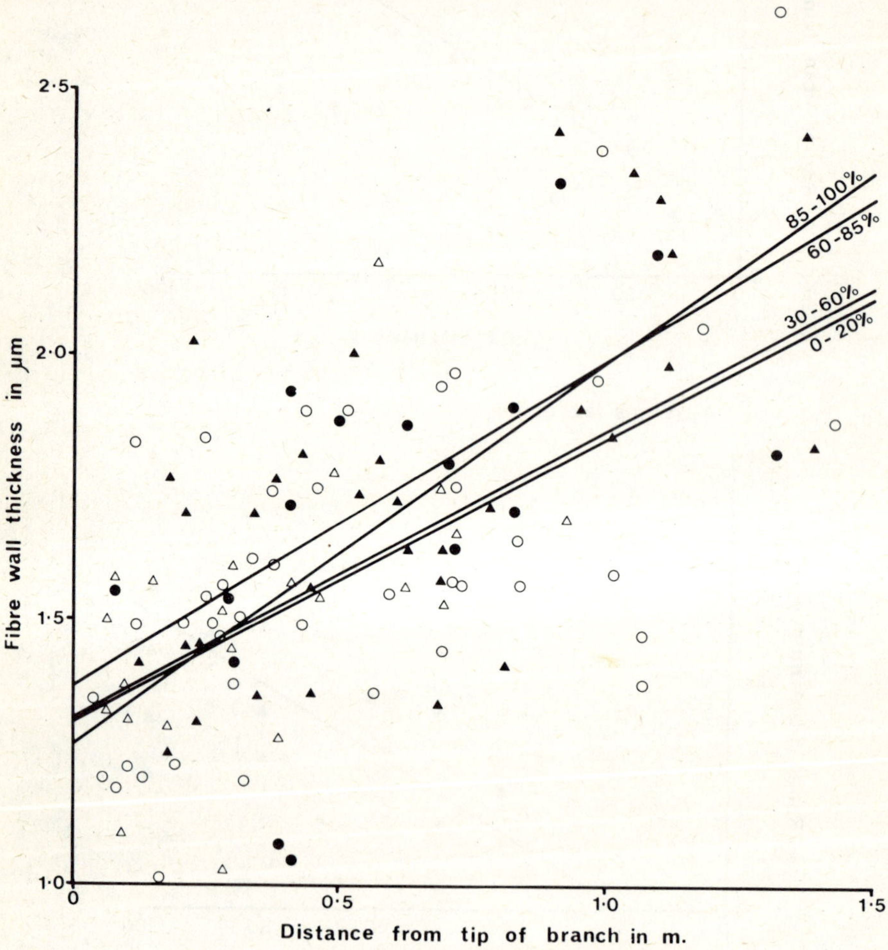

Figure 8.6: *Variation in mean fibre wall thickness with distance from the tip of branches in* Acer pseudoplatanus. *Light intensity measured at the branch tip:* △ *0-20% outside light intensity,* ○ *30-60%,* ▲ *60-85%,* ● *85-100%.*

intensity in branches of *P. sitchensis*[4]. Experimental shading has been shown to induce premature dormancy, so decreasing the amount of latewood in branches of *P. sitchensis*[6]. Mean tracheid diameter increased with light intensity between branches (Table 8.3), but the wall thickness of tracheids produced in comparable parts of the growth ring did not vary (as for example in the middle of the earlywood, Table 8.3). In these two respects the branches resemble the seedlings (Table 8.1). However, since wall thickness increases from earlywood to latewood, and since branches continued cambial activity longer in higher than in lower light intensity, mean wall thickness for the whole growth ring increased with light intensity (Table 8.3).

Similarly in the hardwoods *Acer pseudoplatanus* and *Fraxinus excelsior*, recent work suggests that the final fibre number across the growth ring of branches is closely correlated with light intensity (Fig. 8.5). Preliminary data suggest that in *A. pseudoplatanus* cambial reactivation began earliest in the lower part of the crown, while in *F. excelsior* cambial reactivation appeared to begin simultaneously throughout the tree; these trends confirm those reported for several other diffuse and ring porous hardwoods[23, 43]. Preliminary data also indicates that the onset of cambial dormancy varied with light intensity between branches of both *A. pseudoplatanus* and *F. excelsior*. This would be expected to have a greater effect on the wood structure of the ring porous *F. excelsior* than on that of the diffuse porous *A. pseudoplatanus*, the ring porous species having a far greater variation in wood structure across the growth ring. Between branches mean fibre diameter increased with light intensity in *A. pseudoplatanus* as in *P. sitchensis* (Table 8.3). Fibre wall thickness did not appear to vary with light intensity between branches of *A. pseudoplatanus*, provided they were measured at comparable distance from the branch tip (Fig. 8.6); fibre wall thickness increased linearly with distance from the tip of the branch, with no significant difference between regression coefficients at the different light intensities.

Control of wood structure within the main stem

Within the trunk of softwood trees, it has long been appreciated that both wood production and wood structure vary with position in relation to the crown[13, 14, 15, 26, 27, 29, 32]. Hence, environmental factors affecting shoot production, or leaf efficiency, should modify the distribution of wood quantity and quality within the trunk. Larson described seasonal trends in tracheid dimensions within the main trunk of conifers[26, 27], pointing out that the seasonal transition from earlywood to latewood is delayed in the crown, becoming earlier with distance below the crown. Data from *P. sitchensis* suggest that this gradient in the proportion of early to latewood is determined partly by the time of cambial reactivation, and partly by rate of change in wall thickness across the growth ring[7]. In the main trunk, cambial reactivation appears to be largely determined by the rate of progress of the reactivation stimulus down the branches, and this in turn appears to be controlled by the environment[7]. Silvicultural practices which affect the penetration of light into the crown, or the temperature profile down the crown, would presumably influence the time of reactivation in the trunk and hence the amount of earlywood. As suggested earlier (p. 9), the change in wall thickness from earlywood to latewood is likely to be associated with growth regulator balance rather than with change in substrate availability—presumably silvicultural practices affecting the form or the efficiency of the crown would affect growth regulator balance, either directly through affecting the distribution of sources of regulators, or indirectly through effects on microclimate around the crown.

In suppressed conifers, the cambium has been shown to become dormant earlier than in co-dominant trees[7, 36]. Hence in suppressed trees a silvicultural practice increasing the efficiency of the crown would be expected to have the additional effect of delaying cambial dormancy, so increasing the amount of latewood in the growth ring[7]. This may account for some of the conflicting reports of environment on wood quality, increase in crown efficiency tending to result in decreased density in co-dominant trees, but to promote increased wood density in suppressed trees.

It is clear that far more data are needed to quantify effects of environment on the gradients in wood structure within the main trunk of softwoods. Still less is known about the control of fibre dimensions within the main trunk of hardwoods, in which even the extent of the variation in wood structure is as yet largely unexplored.

SOME IMPLICATIONS FOR THE CONTROL OF WOOD QUALITY

In spite of the complexity of interacting processes involved in the environmental control of xylem differentiation, this survey suggests that it is possible to perceive some general trends—though there are still conflicting viewpoints to be reconciled. To summarise: *Firstly*, environmental factors such as light intensity, water deficit, and temperature, have been shown to influence fibre dimensions both directly (through effects on rate or duration of expansion and wall thickening in the differentiating xylem) and indirectly (through effects on substrate availability or growth regulator production). But since effects on rates often are offset by effects on duration in the differentiating zone, and effects on substrate availability tend to be diminished by adaptive changes in photosynthetic efficiency and correlative changes in distribution of substrate, the net influence on fibre dimensions usually appears to be relatively small. However, it should be noted that even minor differences in fibre dimensions may be sufficient to produce economically important variations in wood quality. *Secondly*, environmental factors have a more profound influence on fibre dimensions in species showing variation in cell dimensions across the growth ring, through affecting duration of cambial activity. Any factor tending to induce cambial dormancy (such as water deficit, or low light intensity) will reduce the amount of latewood formed, so affecting the density and the uniformity of the wood produced. *Thirdly*, the structure and amount of branchwood appears to be particularly susceptible to light intensity regimes; other variations in microclimate may be equally, or even more important, but these do not seem to have been determined as yet. Such effects on branchwood will affect the knottiness of timber in the trunk, and may affect stresses producing degrade of timber in the trunk, as well as affecting the use of the branch timber. *Fourthly*, effects of environment on xylem differentiation in the main stem have to be considered in relation to gradients in wood structure down the trunk. To understand these gradients we need far more information about rate and duration of cambial activity, cell expansion and wall thickening, in relation to the distance from shoot and root apices, and in relation to other internal growth cycles within the tree.

Ideally, to improve the quality of our future timber supplies, we need to achieve a more precise control of wood structure by silvicultural management. Although we are still a long way from that objective, there seems no reason why it should not be feasible, provided we quantify effects of management on environment, and effects of environment on xylem differentiation. To fully understand the environmental control of xylem differentiation we need to quantify each of the complex interacting physiological processes involved, systematically throughout the tree, through the season, and through the life of the tree. The task is daunting but not impossible.

ACKNOWLEDGEMENTS

The authors thank the Natural Environment Research Council for financial support, and Dr. C. J. Smith for kindly allowing them to use some of his data in Table 8.1.

REFERENCES

1 Brazier, J. D. The effect of forest practices on quality of the harvested crop. *Forestry* **50,** 49-66 (1977)

2 Creber, G. T. The effects of gravity and the earth's rotation on the growth of wood. In *Growth Rhythms and History of the Earth's Rotation*. G. D. Rosenberg, S. K. Runcorn (eds.) (London: John Wiley 1975)

3 Denne, M. P. Temperature and tracheid development in *Pinus sylvestris* seedlings. *J. Exp. Bot.* **22,** 362-370 (1971)

4 Denne, M. P. Effects of light intensity on tracheid dimensions in *Picea sitchensis*. *Ann. Bot.* **38,** 337-345 (1974)

5 Denne, M. P. Effects of environmental change on wood production and wood structure in *Picea sitchensis* seedlings. *Ann. Bot.* **40,** 1017-1028 (1976)

[6] Denne, M. P. Wood production and structure in relation to bud activity in some softwood and hardwood species. In *Wood Structure in Biological and Technological Research*. P. Baas, A. J. Bolton, D. M. Catling (eds.) Leiden Botanical Ser. 3, pp. 204-211 (1976)
[7] Denne, M. P. Wood structure and production within the trunk and branches of *Picea sitchensis* in relation to canopy formation. *Canad. J. Forest. Res.* **9,** 406-427 (1979)
[8] Denne, M. P., Smith, C. J. Daylength effects on growth, tracheid development, and photosynthesis in seedlings of *Picea sitchensis* and *Pinus sylvestris*. *J. Exp. Bot.* **22,** 347-361 (1971)
[9] Denne, M. P., Wilson, J. E. Some quantitative effects of indoleacetic acid on the wood production and tracheid dimensions of *Picea*. *Planta* **134,** 223-228 (1977)
[10] Doley, D. Effects of simulated drought on shoot development in *Liriodendron* seedlings. *New Phytol.* **69,** 655-673 (1970)
[11] Doley, D. Effects of shade on xylem development in seedlings of *Eucalyptus grandis* Hill ex. Maiden. *New Phytol.* **82,** 545-555 (1979)
[12] Doley, D., Leyton, L. Effects of growth regulating substances and water potential on the development of secondary xylem in *Fraxinus*. *New Phytol.* **67,** 579-594 (1968)
[13] Duff, G. H., Nolan, N. J. Growth and morphogenesis in the Canadian forest tree species. I. The controls of cambial and apical activity in *Pinus resinosa* Ait. *Canad. J. Bot.* **31,** 471-513 (1953)
[14] Elliott, G. E. Wood density in conifers. *Commonwealth Forest. Bur. Tech. Commun.* **8** (1970)
[15] Farrar, J. L. Longitudinal variation in the thickness of the annual ring. *Forest. Chron.* **37,** 323-330 (1961)
[16] Ford, E. D., Robards, A. W., Piney, M. D. Influence of environmental factors on cell production and differentiation in the earlywood of *Picea sitchensis*. *Ann. Bot.* **42,** 683-692 (1978)
[17] Fritts, H. C. *Tree rings and climate*. (London: Academic Press 1976)
[18] Gordon, J. C., Larson, P. R. Seasonal course of photosynthesis, respiration, and distribution of ^{14}C in young *Pinus resinosa* trees as related to wood formation. *Plant Physiol.* **43,** 1617-1624 (1968)
[19] Jenkins, P. A. Influence of temperature change on wood formation in *Pinus radiata* grown in controlled environments. *N.Z.J. Bot.* **13,** 579-592 (1975)
[20] Jenkins, P. A., Hellmers, H., Edge, E. A., Rook, D. A., Burdon, R. D. Influence of photoperiod on growth and wood production of *Pinus radiata*. *N.Z.J. Forest. Sci.* **7,** 172-191 (1977)
[21] Jenkins, P. A., Shepherd, K. R. Influence of temperature on cambial activity and cell diameter in *Pinus radiata* D. Don. *J. Inst. Wood Sci.* **6,** 36-39 (1972)
[22] Kozlowski, T. T. *Growth and development of trees. Volume II. Cambial growth, root growth, and reproductive growth*. (New York: Academic Press 1971)
[23] Ladefoged, K. The periodicity of wood formation. *Biol. Skr.* **7,** 1-98 (1952)
[24] Larson, P. R. A biological approach to wood quality. *Tappi* **45,** 443-448 (1962)
[25] Larson, P. R. The indirect effect of drought on tracheid diameter in red pine. *Forest Sci.* **9,** 52-62 (1963)
[26] Larson, P. R. Some indirect effects of environment on wood formation. In *The Formation of Wood in Forest Trees*. pp. 345-365. M. H. Zimmermann (ed.) (New York: Academic Press 1964)
[27] Larson, P. R. Wood formation and the concept of wood quality. *Yale University School of Forestry Bull.* **74,** 54 pp. (1969)
[28] Longman, K. A., Leakey, R. R. B., Denne, M. P. Genetic and environmental effects on shoot growth and xylem formation in a tropical tree. *Ann. Bot.* **44,** 377-380 (1979)

[29] Myers, C. A. Vertical distribution of annual increment in thinned ponderosa pine. *Forest Sci.* **9,** 394-404 (1963)
[30] Nicholls, J. W., Wright, J. P. The effect of environmental factors on wood characteristics. 3. The influence of climate and site on young *Pinus radiata* material. *Canad. J. Forest. Res.* **6,** 113-121 (1976)
[31] Nix, L. E., Wodzicki, T. J. The radial distribution and metabolism of IAA-^{14}C in *Pinus echinata* stems in relation to wood formation. *Canad. J. Bot.* **52,** 1349-1355 (1974)
[32] Richardson, S. D. A biological basis for sampling in studies of wood properties. *Tappi* **44,** 170-173 (1961)
[33] Richardson, S. D. The external environment and tracheid size in conifers. In *The Formation of Wood in Forest Trees*. pp. 367-388. M. H. Zimmermann (ed.) (New York: Academic Press 1964)
[34] Richardson, S. D. Studies on the physiology of xylem development. III. Effects of temperature, defoliation and stem girdling on tracheid size in conifer seedlings. *J. Inst. Wood Sci.* **12,** 3-12 (1966)
[35] Richardson, S. D., Dinwoodie, J. M. Studies on the physiology of xylem development. I. The effect of night temperature on tracheid size and wood density in conifers. *J. Inst. Wood Sci.* **6,** 3-13 (1960)
[36] Skene, D. S. The kinetics of tracheid development in *Tsuga canadensis* Carr and its relation to tree vigour. *Ann. Bot.* **36,** 179-187 (1972)
[37] Smith, C. J. Light intensity and substrate availability in relation to tracheid development in *Picea sitchensis*. *Ann. Bot.* **38,** 347-358 (1974)
[38] Smith, C. J. Substrate availability in relation to daylength effects on tracheid development in *Picea sitchensis* (Bong.) Carr. *Ann. Bot.* **39,** 101-111 (1975)
[39] Smith, C. J. Adaptations to environmental change in *Picea sitchensis* seedlings—changes in photosynthesis, growth and substrate availability. *Ann. Bot.* **40,** 1003-1015 (1976)
[40] Smith, C. J., Wellwood, R. W., Elliott, G. K. Effects of nitrogen fertilizer and current climate on wood properties of corsican pine (*Pinus nigra* var *maritima* (Ait) Melv.). *Forestry* **50,** 117-138 (1977)
[41] Stahel, J. B. The effect of daylength on root growth of sitka spruce. *Forest Sci.* **18,** 27-31 (1972)
[42] Van Buijtenen, J. P. Experimental control of environmental factors and their effect upon some aspects of wood anatomy in loblolly pine. *Tappi* **41,** 175-178 (1958)
[43] Wareing, P. F. Growth studies in woody species. 4. The initiation of cambial activity in ring-porous species. *Physiol. Plant.* **4,** 546-562 (1951)
[44] Whitmore, F. W., Zahner, R. Evidence for a direct effect of water stress on tracheid cell wall metabolism in pine. *Forest Sci.* **13,** 397-400 (1967)
[45] Wodzicki, T. J. Photoperiodic control of natural growth substances and wood formation in larch (*Larix decidua* D.C.). *J. Exp. Bot.* **15,** 584-599 (1964)
[46] Wodzicki, T. J. Mechanism of xylem differentiation in *Pinus sylvestris* L. *J. Exp. Bot.* **22,** 670-687 (1971)
[47] Zahner, R. Water deficits and growth of trees. In *Water deficits and Plant Growth II*. pp. 191-254. T. T. Kozlowski (ed.) (New York: Academic Press 1968)

9. Spiral Grain Formation

J MADDERN HARRIS
Forest Research Institute, Rotorua, New Zealand

The term 'spiral grain' refers to the alignment of the main conducting and structural elements of wood at an angle to the axis of a tree stem. The direction of the spiral may be to the left or right of the upper extremity of the longitudinal axis of the tree as viewed by an observer on the ground, and is thus described as left or right spiral grain (or left or right grain angle).

Even casual inspection of a line of wooden telegraph or transmission poles will show that spiral grain is a very common phenomenon, and it may well be the normal growth pattern in trees as suggested by Northcott[35]. Fortunately, grain angles seldom becomes so extreme as to make logs unusable, although quite small grain angles can introduce problems in timber utilisation[5, 32]. In sawn timber cross-grain causes twist on drying, and comparatively small angles give rise to marked loss of strength in comparison with straight-grained timber. Spirally grained poles tend to twist and untwist with changing moisture content and the resultant movement of attached cross-arms can severely stress and even break wires strung between them. When used as plywood, veneers from spirally grained logs create dimensional instability because successive sheets cannot be laid up with their grain directions strictly at right angles to one another.

All of these considerations have made spiral grain a matter of considerable interest to foresters, botanists, and wood technologists. Many papers have been published over the last hundred years or so on the possible causes and on the incidence and effects of spiral grain[36]. However, there has been no clear agreement as to which of the various forms of grain deviation can properly be described as spiral grain. Indeed several workers have suggested that the ways in which spirality develops are so unlike in different species that each distinct growth pattern should be treated separately.

For example, when dye is injected along a radius at the bottom of a living conifer stem there are various pathways along which dye can be conducted according to the different patterns of spiral grain development. At least five patterns of dye distribution have been described from examination of cross-sections at the top of injected stems[46, 47]; each pattern is said to be specific to certain groupings of species. For this reason it has been suggested that spiral grain should not be regarded as a defect common to all trees, but rather as a natural phenomenon which emerges from a variety of normal growth patterns.

Similarly, it has been contended[41] that differences between hardwoods and softwoods with respect to the way that grain inclinations develop over the first year are so fundamental that the two phenomena should be treated as being quite different and should be placed in separate categories.

In contrast to these views, a considerable body of recent work dealing mainly with the morphogenetic phenomena giving rise to grain deviations in hardwoods[23] suggests that similar structural mechanisms can be used to explain 'simple' spiral grain, interlocked grain, and wavy grain, as well as the complex combinations of these. For the sake of completeness, therefore, all aspects of grain deviation from the strictly axial direction will be considered in this chapter whether they occur in hardwoods or in softwoods.

TYPICAL SPIRAL PATTERNS

Noskowiak[36] lists more than 20 authors who have reported the same general pattern of spiral grain development in conifers. In this, the tracheids of the first formed xylem adjacent to the pith are laid down roughly parallel with the stem axis, but subsequent growth layers develop increasing left spiral grain which commonly reaches its maximum development within the first 10 years of growth. With further growth the left grain angle decreases until the xylem becomes straight grained once again, but in some cases this change of grain angle may continue through to a right spiral which increases in magnitude with age.

If this generalised description is analysed in terms of the grain direction that might be observed down a single growth sheath from stem apex to the base of a middle-aged tree, it will be realised that straight grain in the apical shoot passes into increasing left spirality in the upper stem after which the grain angle diminishes in mid-stem and passes once more through straight grain until a right-hand spiral develops in the lower stem. A nearly complete wave pattern of spirality will have been developed within a single growth sheath.

However, significant deviations from this pattern have been described by several authors. Northcott[35] observed that 19 per cent of the Douglas fir stems which he examined passed directly from a small left spiral in the early years to an increasing left spiral in later life without changing to a right spiral at all. In certain species spirality continues to develop so strongly in occasional trees that angles as high as 40° have been recorded (e.g. in *Pinus balfouriana*[36]) and in *Pinus roxburghii* tracheids are sometimes orientated almost horizontally around the stem[45].

Marked variations of grain angle can occur within an annual growth layer, and across different radii in the same stem section[31, 38]. In some cases a pattern of interlocked grain may be superimposed upon the more generalised spiral pattern[31, 14].

Turning now to the developmental pattern of spiral grain in hardwoods, Noskowiak[36] comes to the conclusion that it is frequently the reverse of that found in softwoods, so that grain angles that develop to the right in early years change to the left in maturity. However, exceptions are much more numerous amongst hardwoods and the whole question of the relationship between 'simple' spirality and other grain deviations such as interlocked grain, wavy grain, and so on, will be considered later in terms of the morphogenetic phenomena that produce them.

There remains one spiral pattern which appears to differ from all others, and which may represent a distinctive development peculiar to certain hardwoods as suggested by Priestley[41]. In *Sambucus nigra*, *Syringa vulgaris*, and *Castanea sativa* Priestley described a form of spirality in which the pitch of the spiral is determined in the first year's wood and this then increases purely as a result of the geometry of increase in girth (Fig. 9.1). This apparently 'passive' development of spirality would seem to arise through total inability to restore axial polarity once some small initial twist has developed in the apical shoot. In all other forms of spiral grain there seems to be at least some response by way of cambial reorientation, otherwise the geometric effect of increasing girth would be a much more widely observed component of spiral development than it is.

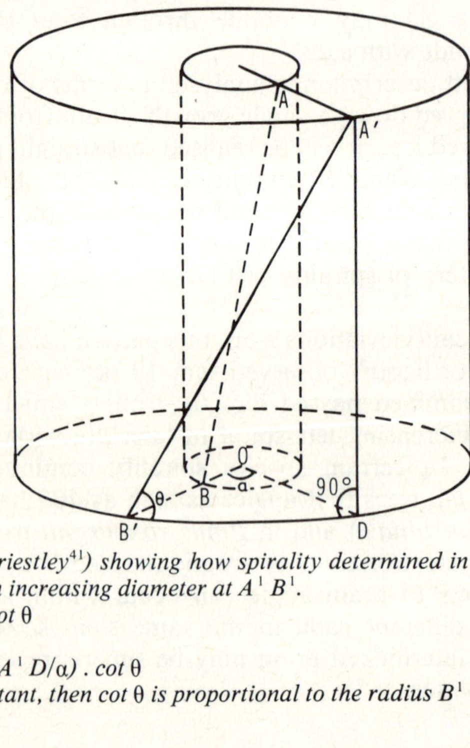

Figure 9.1: *(From Priestley[41]) showing how spirality determined in the first year wood at AB will increase with increasing diameter at $A^1 B^1$.*

$B^1D/A^1D = \cot\theta$

$B^1D/B^1O = \alpha$

Hence $B^1O = (A^1D/\alpha) \cdot \cot\theta$

Since A^1D/α is constant, then $\cot\theta$ is proportional to the radius B^1O.

EXTERNAL CAUSES OF SPIRAL GRAIN

Almost every imaginable feature of the external environment has been invoked at one time or another as being a possible cause of spiral grain. Altitude, aspect, soil, stem injuries, solar movement, the earth's rotation, and wind action have all been suggested, and detailed reviews have been presented by Noskowiak[36] and Nicholls[34]. Yet as Nicholls[34] pointed out, none of these factors can adequately account for those changes of direction in the developing patterns of spiral grain which are so often observed in nature.

On the other hand, since the external environment is capable of influencing or modifying any or all of the growth processes within a tree, which in turn modify the morphological features arising from those growth processes, it is almost axiomatic that external environment can influence development of spiral grain. Whereas external factors seem by their nature inadequate to account solely for the variable patterns of spirality, there are therefore good *a priori* reasons to believe that they might modify patterns established by other means and determine the level of their development. Unfortunately very little experimental work has been undertaken to find out how or to what extent factors of the environment may modify the expression or development of basic spiral patterns.

An exception to this is the work of Pyszynski[42] who studied the effects of torsion on the stems of *Aesculus hippocastanum*. In this species he reported that the basic direction of spiral grain is always to the right though wavy grain may sometimes be superimposed on this. After subjecting stems to alternate right and left-handed torsion, residual torsions to the right were found to be larger than those to the left. As the predominant orientation of the fibrillar helix in cell walls is also to the right, Pyszynski deduced that asymmetry of the stem's mechanical properties may be responsible for the formation of right spiral grain in the stem of a tree subjected to various torsions under the action of wind.

This is the most convincing evidence to date of the direct influence of external factors on the development of spirality in one constant direction.

EFFECT OF GROWTH RATE

Any effect of growth rate on the development of spirality is obviously of interest to foresters who will usually be seeking to optimise tree volume increment. On the purely geometrical grounds discussed by Priestley[41] it would appear that, for any initial grain deviation, rapid diameter growth will tend to produce wider grain angles than slow growth per unit of time. If the 'geometrical effect' is an important feature of developing spirality it might be anticipated therefore that rapid diameter growth would be closely associated with severe spirality. That this is not so is evident from the very extensive literature review by Sachsse[44]. Almost as many authors have in the past associated intensification of spiral grain with decreasing growth rate as have reported the reverse situation, and a number of authors have plumped for total independence of spiral growth and growth rate.

Elliott[12] points out that many of the reports referred to in literature surveys fail to differentiate between the independent influence of age (rings from the pith) and growth rate, that few are based on a biologically unequivocal sampling system, and that most were carried out before today's more sophisticated approaches to numerical analysis were readily available. All this is quite true of course, and in view of the success achieved in examining other wood properties by these methods it is highly desirable that developing spiral grain should be subjected to equally rigorous scrutiny. Nevertheless, it is probably fair to say that if there were any close association between growth rate and spirality in any of the major timber species which have been repeatedly examined over the years it would have become apparent by now. Whilst not dismissing the possibility that growth rate may sometimes influence the development of spiral grain it therefore seems unlikely that such influences are consistent or of major consequence in the majority of instances.

GENETIC CONTROL

Early observations of spiral grain among open-pollinated progenies (families) and in vegetatively propagated individuals gave rise to the impression that it is a highly heritable feature of certain species. This impression was reinforced by the results from a small but important series of tree breeding experiments in India using *Pinus roxburghii*—a species that often produces severely spiralled grain. These studies were summarised in the classical papers by Champion[7, 8, 9]. There is then a gap of some years in the literature and it is only in relatively recent times that attempts have been made to initiate critical well-designed genetic experiments. These have mainly been concerned with fast-growing conifers subject to intensive plantation silviculture such as *Pinus radiata* and *Pinus taeda*.

Three studies demonstrate the essentially additive nature of the inheritance of spiral grain in *P. radiata*. Dadswell *et al.*[10] reported broad-sense heritability for spirality in *P. radiata* clones of 0.66 for rings 2 to 8 from the pith, but this was reduced to only 0.28 when rings 7 to 8 were considered by themselves. Nicholls[33] found that broad-sense heritability decreased slowly from 0.6 for the first ring from the pith to 0.1 for the twenty-first ring in the same species. In practical terms the low heritability outside the central core of wood may not be a serious disadvantage in a species such as *P. radiata*. Spirality in this species usually reaches its maximum value within the first three growth layers and is quite small outside the tenth growth layer from the pith. Moreover, plantation grown trees will normally be harvested between 25 and 40 years of age so that increasing right-hand spirality past 'middle age' will have no opportunity to develop. Pederick[39] measured spiral grain in 4-year-old *P. radiata* seedlings of 22 full-sib families. Heritability of spiral grain in the second annual growth layer was calculated to be 0.40, and the non-additive component of the genetic variance was found to be very small, if not zero.

Zobel *et al.*[50] analysed the data from 1,043 four-year-old trees belonging to 52 control-pollinated *P. taeda* families. Among these there were definite family differences in grain spirality at 20 cm height: the smallest average grain angle for a family of 21 trees was 1.7° while the largest average angle for a family was 4.8°. Further analysis of these results suggested that, in sharp contrast to the case of *P. radiata*, the genetic variation for spiral grain was non-additive. Consequently, there would be little to be gained from uncontrolled mating of selected individuals (mass selection), and in order to reduce the average level of spirality a tree breeder would have to take advantage of specific combining abilities between selected parents.

Mikami *et al.*[30] examined 400 trees belonging to 80 clones of 8- and 9-year-old *Larix leptolepis* and obtained gross heritability values ranging from 0.35 to 0.42 for grain angles in individual growth layers and 0.49 for maximum grain angle.

There is also a considerable body of evidence, from limited progeny trials and cases of vegetative propagation, that variously figured woods (including interlocked grain, wavy grain, and curly grain) are under a measure of genetic determination[49].

Even though the weight of reported evidence from past observations and experiments may indicate a considerable measure of genetic control over the development of spiral grain (and related phenomena) in a wide range of tree species, there remains an urgent need for better understanding of inheritance before the most appropriate breeding methods can be applied and suitable intensities of parent tree selection determined. In the meantime potential breeding trees are often excluded from use in seed orchards and the like because of excessive spirality. This is usually done out of deference to the penalties that might arise if such precautions were not taken, rather than being based on any precise understanding of what is involved.

ANATOMY OF SPIRAL GRAIN DEVELOPMENT

Many writers discussing other aspects of spiral grain have stressed the need for a better understanding of the anatomy of spiral development. It is generally acknowledged that spirality arises in most instances from reorientation of cambial elements, and although it is often convenient to follow cambial reorientation by reference to changes in successive xylem layers it must always be remembered that it was in the cambium that most of these changes first took place.

The relatively simple anatomy of gymnosperm wood usually ensures that a continuous record of cambial reorientation is left behind in the secondary xylem. In dicotyledonous wood the complexities of vessel formation which are often combined with intermittent production of axial parenchyma make for a highly discontinuous record. Nevertheless, it is possible to use the terminal zone of xylem in the annual growth layer to interpret changes that have taken place in the cambium from year to year[21].

It will be convenient first of all to consider cambial reorientation purely in terms of the events that can be detected in the xylem, i.e. without reference, to causes. These are the so-called 'morphogenetic events' of Hejnowicz[18] and they comprise (a) orientation of the new cell wall formed during pseudo-transverse anticlinal division, (b) the direction of overlap when cell tips move past each other in sliding growth, (c) the direction of ray splitting by intrusive growth of fusiform cell ends, (d) change of direction arising from union of wood rays when fusiform cells between them are 'withdrawn' (Fig. 9.2).

Hejnowicz[19] cites work by Braun as early as 1854 in which it was suggested that spiral grain arises from preferred orientation of pseudo-transverse divisions in the cambium, and from elongating cell ends always overlapping the cells immediately above them to the same side. Similar beliefs were expressed by later workers but these phenomena were not fully quantified until the pioneering studies by Bannan[1, 2, 3, 4] and Hejnowicz[17, 18]. The results of their studies on conifers can be summarised as follows:

1. Neighbouring cells within a particular cambial zone, or 'domain', tend

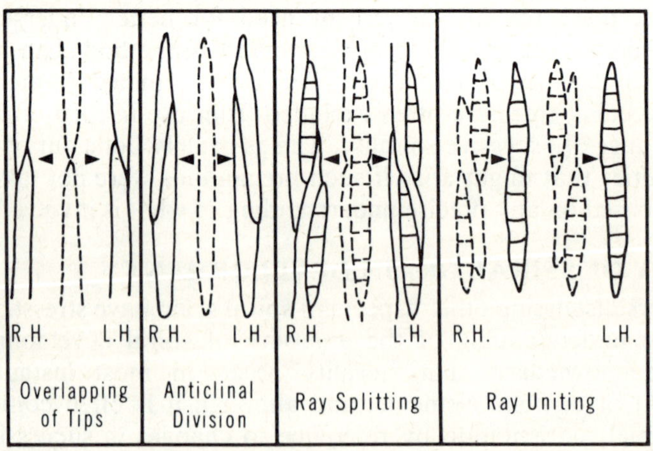

Figure 9.2: *(From* Z. Hejnowicz *Domain pattern in cambium and formation of spiral grain (undated)). Four different classes of morphogenetic events giving rise to right or left spiral grain.*

to produce anticlinal divisions in the same direction, rather than at random.
2. This is a general phenomenon found in stems producing either straight or spiral grain.
3. The same direction of division is not maintained indefinitely, so that the domain pattern within the cambium changes with time.
4. If there are frequent periodic reversals in the preferred orientation of anticlinal division, changes in grain angle will be kept to low values. A low frequency of morphogenetic events and extensive losses of cambial initials also tend to keep changes of grain angle low[4].
5. When grain angles do change consistently this arises because intervals between reversals are longer in one preferred direction—that of the changing grain angle.
6. Orientation of the overlap achieved by elongating cell tips and of related morphogenetic events in wood rays is usually the same as that of the anticlinal divisions.

It should be noted that the direction of morphogenetic events determines only the *change* of direction in the cambium. Left-handed events in an area where grain angles are to the left will produce increasing spirality in that direction. Left handed events in an area where grain angles are to the right will result in decreasing spirality tending towards the initiation of a left grain angle.

Wood rays appear to play a completely passive role in changing grain angles no matter how violent these may be. Their reorientation may lag behind that of the fusiform initials[15,16]. As the cambium grows outwards each grouping of ray initials appears to rotate about its centre in adjustment to the changing orientation of fusiform initials rather than behaving as a separate dynamic component of changing grain angles[16,19,25]. The relative positions of wood rays remain remarkably constant across successive growth layers despite the fact that a ray may be split by intrusive growth of fusiform initials, or that two rays may become united either by loss, retraction, or parenchymatisation of a part or the whole of a fusiform initial.

ANATOMY OF COMPLEX GRAIN FIGURES

Following the initial studies on spiral grain in conifers, further investigations by Hejnowicz and his co-workers at Wroclaw University have covered many aspects of other more complex developing grain patterns in both conifers and hardwoods. All of these have been shown to arise from the systematically changing orientation of cambial fusiform initials.

In order to understand the unifying features of grain patterns as apparently dissimilar as interlocked grain (ribbon figure) and wavy grain[37] it is necessary to consider the effects on wood structure of moving cambial domains in which morphogenetic events lead either to left-handed or to right-handed grain deviations.

The starting point is the established fact that the cambium at any given site will from time to time be involved in events leading to either left or right grain angles[4, 18, 19]. It can be shown that these changes are brought about by the migration of cambial domains along the stem in the course of wood formation[20, 22, 27].

The boundaries of the domain types lie predominantly transversely to the stem axis[28], and the direction of migration is usually upwards. However, Pyszynski[43] noted downward movement of cambial domains in *Aesculus*. In either case a given site in the cambium will successively be occupied by domains of alternate type. The length of time during which a domain type determines the direction of change in grain angle at a particular site in the cambium depends on the height of the domain and the speed with which it moves along the stem (Fig. 9.3). There appears to be a correlation between domain height and speed of movement (Table 9.1).

Movement of domains along the cambium prevents excessive inclination of cells by producing cyclical changes of domain type at each site. Each domain opposes the effects of the previous one so that successive sequences of cambial cells along a radial file show oscillations to the left and to the right with respect to the stem axis. The size of the oscillations will also depend on the frequency of the events leading to reorientation and to the survival rate of cells in which they occur[4, 20, 22].

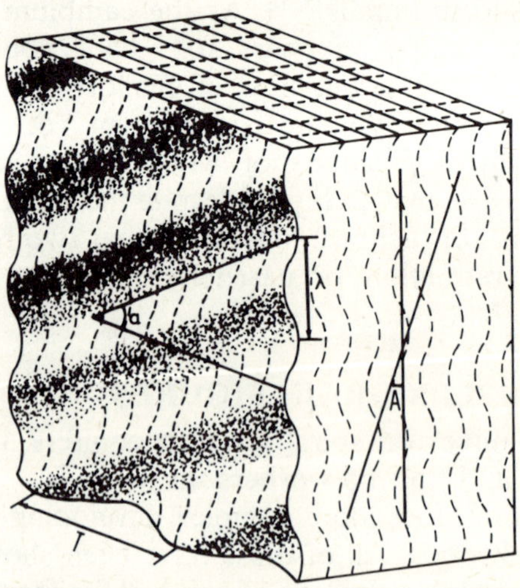

Figure 9.3: *(From Pyszynski[43]) Spatial model of wood sample with wavy grain. A—amplitude. λ—wavelength. α—angle between bands of different light reflections and wood rays. T—period.*

When domain height is short, and domain movement along the stem is slow, morphogenetic events of sufficient frequency will produce grain deviations that appear on a radial wood surface as nearly horizontal bands of differing light reflection. This is the basis for the development of curly grain in *Acer* and *Fraxinus* (Table 9.1). When domain height is high and movement along the stem is rapid, grain deviations are seen on the radial face as the typical ribbon figure of interlocked grain as in *Platanus* (Table 9.1). In the first case the radial bands appear to be horizontal but in fact always subtend a small angle to the wood rays. In the second the ribbon figure appears to be vertical but in fact forms a small angle with the growth ring boundaries.

In extreme cases the wavelength of grain deviations as they appear on the outside of a log (tangential surface) can be more than 3 m as in *Entandophragma* spp.[24] If the wavelength exceeds the length of a log in which the figure is being observed, surface grain deviations may give the impression of being a simple spiral. Shorter wave lengths appear on the log surface as serpentine grain deviations whose peaks would be seen on a radial surface to be developed upwards when traced growth layer by growth layer from the centre of the stem outwards. Hejnowicz and Romberger[23] provide illustrations of a continuum of grain patterns between the two extremes represented by curly grain and the ribbon figure of interlocked grain.

The slope of the stripes seen on a radial face can be calculated from distance moved by a domain pattern along the stem/unit of radial growth in the xylem, which is the tangent of the angle between the stripes and the rays on a radial section. From Fig. 9.4 it can be seen that the difference between an almost horizontal (curly) figure and an almost vertical (ribbon) figure is brought about by increasing domain velocity from 0.1 to 5.0 mm/mm of wood laid down[28]. Typically waves move forward slowly at about one wavelength in 10 to 15 years. Thus in wavy grained *Fraxinus*, for example, the migration rate is usually less than 1 mm/year[23].

Although complex grain patterns develop more commonly in hardwoods

Table 9.1: *Relationship between domain height and velocity of domain movement (from Krawczyszyn[28])*

Grain type in wood	*Tree*	*Domain height (mm)*	*Velocity of domain movement (mm/mm radial growth)*
Curly	*Acer, Fraxinus*	2–8	0.1 Hejnowicz and Romberger[22]
Wavy	*Betula, Picea*	10	1.0 Hejnowicz[20]
Interlocked	*Platanus*	80–280	6.5–12 Krawczyszyn[28]

than in softwoods, fluctuating grain angles may sometimes be superposed on spiral grain in pine[14,35]. Occasionally quite pronounced wavy grain may develop in patches on softwood logs. Wavy grain in 34 trees of *Picea excelsa* was investigated by Hejnowicz[20] who found that the cambium producing wavy grained wood differed from normal cambium only in the higher frequency of anticlinal division and higher rate of intrusive growth of fusiform initials. Cambial domains in these trees averaged about 4 mm in height and their rate of movement along the stem was about 0.7 mm/mm of xylem laid down. Thus the existence of cambial domains giving rise to left-handed or right-handed morphogenetic events was confirmed in this softwood, and the influence of event frequency and domain movement were also clearly implicated in the development of complex grain patterns as compared with the development of normal wood. In purely anatomical terms, therefore, close similarities can be demonstrated between softwoods and hardwoods with respect to changing cambial orientation.

Certain of the morphogenetic events giving rise to changing grain angles in trees with non-storeyed cambium are obviously placed under a degree of constraint by the presence of a storeyed structure in the cambium. The cambium of *Entandophragma cyclindricum* is storeyed, and the wood rays too, which are multiseriate, are usually limited to one storey in height Axial strand parenchyma, which is mostly terminal and four or more cells wide, accurately reflects the condition of the cambium at the time it was laid down. Hejnowicz and Zagorska-Marek[24] used tangential sections through

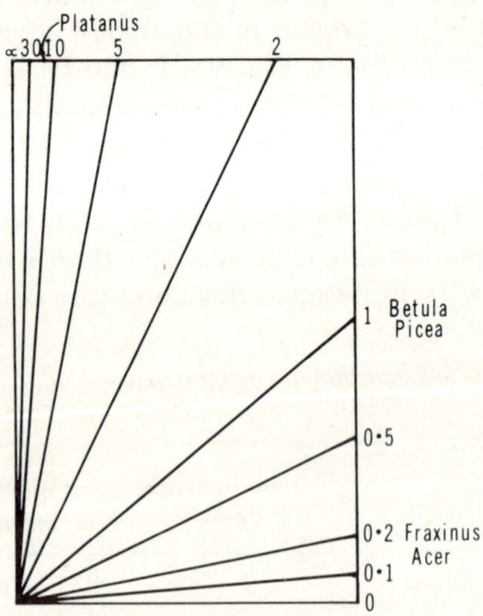

Figure 9.4: *(From Krawczyszyn[28]) Relation between velocity of domain movement (mm/mm of radial growth) and the inclination (relative to wood rays) of stripes on a radial face.*

successive terminal parenchyma bands to trace cambial events during the formation of the interlocked grain that is characteristic of this species. They found that changes in grain angle result from changes in the intrusive growth of fusiform cells which repeatedly produce a new pointed tip to one side of the existing tip which then disappears. This causes a sort of creeping movement of the cells of one storey past those of the adjoining storey. Moreover the opposite ends of fusiform cells creep in opposite directions so that successive cells along a radial file appear to rotate about their centres. During the period of change of grain orientation in one direction cell ends of one storey were observed to creep past nine ends of the next storey. The authors concluded that stratification of rays in *Entandophragma* represents an adaptation to the changes in cell orientation involved in the formation of interlocked grain.

The slowly moving morphogenetic waves in the cambia of trees have the same attributes as waves in other physical systems. They have measurable lengths, amplitudes, and periods. Under suitable conditions these waves can also exhibit superposition and interference phenomena. These have been illustrated by Hejnowicz and Romberger[23] using the wood of *Guibourtia ehie*. In their illustration 'pure' waves of similar length but different direction can be seen to become superposed to produce moiré, quilted, or chequered figures, lending considerable support to their contention that the common basis of wood grain figures is the systematically changing orientation of cambial fusiform cells.

Pyszynski[43] also discovered superposed long and short grain patterns arising from migrating domains in *Aesculus*. Whereas the long waves moved up the axis at an angle of 5° to the stem axis, the short waves moved very slowly down the stem at an average angle of 86° to the axis. All of these events were in turn superposed on a basic spiral pattern.

In summarising the considerable body of work now assembled on migrating domain patterns, Hejnowicz and Romberger[23] speculate that the morphogenetic waves which form the basis of wood grain figures are themselves interference patterns of more fundamental and much more rapidly moving waves. These primary waves may be several orders removed from the events that we can directly observe. The nature of the fundamental phenomena that may give rise to morphogenetic waves will be discussed in the final section of this chapter.

PHYSIOLOGICAL BASIS

It will be helpful first of all to review what has been discovered about the structure of trees containing spiral grain and other more complex grain patterns. It appears that there are at least two sources of spirality which are not primarily dependent on active reorientation of cambial elements. Indeed the case of *Aesculus*[42] was investigated because morphogenetic events in the cambium indicated a tendency towards left spiral grain whereas exclusively right spiralled grain was encountered[21]. The explanation of residual stem torsion to the right in response to wind stress was advanced to meet this apparent anomaly. However, Pyszynski[43] has shown that even in this species migrating domain patterns are responsible for complex wavy grain superposed on spiral grain.

The second type of spirality is that described by Priestley[41] in *Sambucus*, *Syringa*, and *Castanea*. This he explained as being a purely geometrical consequence of circumferential growth magnifying some chance spiral developed early in wood formation. To what extent geometrical projections may be valid when applied to repeated cell divisions in the cambium—involving as they do competition between, and loss of, individual elements—is perhaps a matter for further debate. There is, however, one important question to be answered in the present context: if the geometrical explanation appears valid for one set of observations why does it not appear more regularly and prominently as a component of developing spirality among the vast majority of species in which grain deviations are known to develop? Why should the cambium in a few instances show no evidence of active reorientation, whereas in all others the development of grain deviations, and even the absence of deviations, can be explained in terms of the frequency and orientation of morphogenetic events in the cambium, and the migration of domains in which the great majority of such events are in the same direction? The answer to this question might well provide a key to understanding the fundamental phenomena behind the events 'several orders removed'[23] that can be observed in tree cambia.

For the present the leads to such understanding are few. Several earlier hypotheses have failed to stand the test of time, or have at least received little over the years by way of confirmation. Nicholls[34] and Elliott[12] have reviewed the many theories that have been advanced to explain the physiological basis of spiral grain. For the most part they have little experimental justification, but certain concepts deserve to be examined to see how far they may be used to explain the known anatomical bases of spiral grain development.

Preston[40] postulated that an elongating fusiform initial will tend to roll about its longitudinal axis as a consequence of interaction between the forces producing elongation and the cell's spiral wall structure. Elliott[12] disputed this because elongating fusiform cells have only primary cell walls and these do not have the spiral structure so characteristic of the secondary walls that develop later. Furthermore, the spiral structure of cell walls remains in the same direction throughout the life of a tree so that this feature could not account for the changes in spirality that are so often in evidence. However, Preston was one of the first to point out a very important feature of developing spirality: only minute changes are required in individual fusiform cells to produce quite major changes in cambial reorientation. He calculated that the displacement needed to produce a 13° change of angle over 16 mm radial growth was 0.2° or barely 0.5% of the girth of a xylem cell.

Elongation of daughter cells produced by uni-directional pseudo-transverse division of fusiform initials, together with intrusive growth and 'retraction' of cell ends provide the basis of the series of morphogenetic events which have been used so successfully to explain changing grain angles in a wide range of species. However, Harris[15, 16] suggested that these phenomena are only consequential to the more fundamental changes that occur in cambial reorientation. By spiral girdling of young *Pinus radiata* Harris was able to change the alignment of fusiform cambial initials by 40° over 9 weeks' growth. The observed pseudo-transverse divisions and intrusive growth of cell ends by themselves were inadequate to account for changes of this magnitude, being sufficient to produce 0.2°/mm of radial growth compared with the more than 10°/mm observed. Detailed measurements of fusiform initials indicated that major changes took place between successive periclinal divisions.

In this respect it is as well to remember that the cambium is not a static tissue that simply cuts off phloem to one side and xylem to the other. A natural consequence of cambial division is that the cambium must also renew *itself* in successively widening concentric sheaths. After each periclinal division the cell that retains the meristematic function must grow, mainly by extension of its radial walls, before redividing. Moreover cambial cells are highly plastic, totally lacking any of the rigidity that derives from secondary wall formation and lignification.

Measurements made on reorientating fusiform initials showed that the first changes manifested themselves as curvature of cell ends. This gave the fusiform cell a slightly sigmoid conformation during the period of most active reorientation[16]. Adjustment in the alignment of the main cell axis was envisaged as taking place in the wake of subsequent periclinal divisions since the total cambial alignment changed with progressive growth but the average angle between cell tips and the body of the cell never exceeded 5°.

From these experiments it was concluded that grain deviations arise as a result of differential growth of highly plastic fusiform initials, that differential growth induces cytoplasmic stress, and that cytoplasmic stress determines the direction in which the mitotic spindle is skewed from the cell axis during anticlinal divisions and hence the orientation of the new cell wall.

The orientation of pseudotransverse divisions is therefore seen as being symptomatic of differential growth and is not in itself a primary cause of spirality. The type of creeping movement of storeyed initials relative to cell ends in the adjacent storey as described for *Entandophragma*[24] would seem to confirm that this sort of plastic realignment of cambial cells provides the essential basis for morphogenetic events that can be observed in the xylem record. Here again it is necessary to repeat that the required readjustments between periclinal divisions are very small. Even the change of 40° over 3 mm radial growth induced experimentally in *Pinus radiata* requires the cell axis to move through less than half a degree for each periclinal division.

Influence of growth substances

That auxin is implicated in differential growth is apparent from several studies[15, 16, 26]. Girdling experiments in *Robinia pseudacacia* designed to produce non-axial bark bridges with or without direct contact with root and shoot[26], indicated that the stimuli which control normal vascular differentiation can also be used to change the orientation of the vascular system. When a bark bridge was cut off from the root only, regenerative vessels were formed in the new orientation and these vessels connected the shoot with areas of callus growth. When a bridge was detached from the shoot but had direct contact with the root it showed even less xylem differentiation, but the regenerative vessels that did form also connected the root with callus growth. The influence of the detached shoot could be partially replaced by a source of auxin. There were also some indications that separate centres of callus growth can partially replace both the shoot and the root in their effect on vascular differentiation.

In girdling experiments with *Pinus radiata*[15], a source of auxin could also be used to maintain stem polarity. In one experiment, originally designed to reduce the production of endogenous auxin by removal of all stem apices, the opposite effect of increasing the relative amounts of auxin to total photosynthate was postulated to be the result. This arose from repeated and massive production of epicormic shoots and from the partial defoliation which occurred in attempting to remove them. All young pine stems so treated reverted to straight grain within one growing season.

It could be speculated that the inconclusive results of experiments to examine the effect of radial growth rate on spirality may arise from the multiple functions of auxin. High auxin levels may be required for rapid growth but in normal circumstances may also be the means of maintaining stem polarity.

The classical experiments with seedling hypocotyls suggest that polar transport of auxin is a result, and not the cause, of longitudinal polarity in plants[13]. Nevertheless, other experiments with young plants, e.g. with *Pinus sylvestris*[11], have shown that auxin exerts a controlling influence in xylem differentiation.

Several types of structures have been associated with the initiation of grain deviations including spiral parastichies of young pine stems[15], as well as the more obvious effects of branching and stem wounds. The initial development of a patchwork pattern of cambial domains in *Platanus* has been associated with phyllotaxy[29], and it is only later in development that the domains resolve themselves into circumferential bands.

If such structures induce non-polar movement of metabolites including auxin, it is still difficult to interpret the means by which auxin cross-flow might induce differential growth in fusiform initials because so little is known about auxin transport. In analysing the sigmoid curvature that develops in fusiform initials during reorientation, it has been pointed out[16] that curvature in the upper end of the cell must proceed in the opposite sense to curvature in the lower end—'into' the auxin flow *versus* 'with' it—and curvature of both must respond to the direction of flow. These considerations support the idea that curvature of fusiform initials arises from movement of auxin through their cytoplasm. The attractive feature of symplastic transport in this situation is that it would contain vector components related both to cell polarity and to the direction of auxin flow, i.e. in by the up-stream side at the top of a cell, and out by the down-stream side at the bottom.

Wave phenomena in morphogenesis

The final feature of reorientating fusiform initials that must be considered is the very widespread phenomenon of wave patterns arising from moving cambial domains in which morphogenetic events are predominantly in one direction. It is well established that auxin flow is a periodic phenomenon rising to a maximum in relation to extension growth in stem and branches. Stimuli originating in the root may also be implicated in cambial reorientation[26]. It is possible to envisage interactions between such stimuli (possibly compounded by developing bark pressures[6] or by the geometrical effect of circumferential expansion on developing spirality[41], but these hardly add up to the 'much more rapidly moving waves' needed to create the interference patterns postulated by Hejnowicz and Romberger[23].

This role seems much more likely to be supplied by shorter waves such as those measured in the amounts of diffusable auxin along stem sections of *Pinus sylvestris*[48]. Wavelengths were about 20–40 mm and are interpreted as evidence of wave-front mobility in the stem. The observed pattern can be seen as a two-dimensional wave in which tangentially neighbouring cells are at the same phase of oscillation even though their ends do not lie in horizontal rows in non-storeyed cambium. Zajaczkowski and Wodzicki[48] suggest that cells may be integrated into a supracellular oscillating system. This would have great significance in all matters relating to wave-like morphogenetic patterns.

Studies of spiral grain have made considerable advances over recent years, especially by way of observations leading to a better understanding of morphogenetic events at an anatomical level. The dim outlines of a physiological system can now be perceived which may well provide crucial pieces of the jig-saw pattern in future. Indeed, from a physiological viewpoint, it is now becoming apparent that developing grain patterns in trees are capable of providing excellent material towards understanding some very fundamental and important growth processes.

REFERENCES

[1] Bannan, M. W. The frequency of anticlinal divisions in fusiform cambial cells of Chamaecyparis. *Amer. J. Bot.* **37,** 511-519 (1950)

[2] Bannan, M. W. Tracheid size and anticlinal divisions in the cambium of Pseudotsuga. *Canad. J. Bot.* **42,** 603-631 (1964)

[3] Bannan, M. W. Tracheid size and anticlinal divisions in the cambium of lodgepole pine. *Canad. J. Bot.* **42,** 1105-1118 (1964)

[4] Bannan, M. W. Spiral grain and anticlinal divisions in the cambium of conifers. *Canad. J. Bot.* **44,** 1515-1538 (1966)

[5] Brazier, J. D. An assessment of the incidence and significance of spiral grain in young conifer trees. *Forest Product. J.* **15,** 308-312 (1965)

[6] Brown, C. L., Sax, K. The influence of pressure on the differentiation of secondary tissues. *Amer. J. Bot.* **49,** 683-691 (1962)

[7] Champion, H. G. Contributions towards a knowledge of twisted fibre in trees. *Ind. Forest Records* **11,** part II, 11-80 (1925)

[8] Champion, H. G. An interim report on the progress of investigations into the origin of twisted fibre in *Pinus longifolia* Roxb. *Ind. Forester* **53,** 18-22 (1927)

[9] Champion, H. G. Second interim report on the progress of investigations into the origins of twisted fibre in *Pinus longifolia* Roxb. *Ind. Forester* **56,** 511-520 (1930)

[10] Dadswell, H. E. and others. Tree-to-tree variations and the gross heritability of wood characteristics of *Pinus radiata. TAPPI* **44,** 174-179 (1961)

[11] Egierszdorff, S., Tomaszewski, M. Auxin dependent accumulation of photosynthates in cambium and wood formation in Scots pine. *Trans. 3rd Symp. Accumulation and Translocation of Nutrients and Regulators in Plant Organisms* (1973)

[12] Elliott, G. K. Some problems of spiral grain—with special reference to conifers. *Proc. XIV IUFRO Congress, Munich,* **IX,** 413-435 (1967)

[13] Goldsmith, M. H. M. Transport of plant growth regulators. M. B. Wilkins (ed.) *Physiology of Plant Growth and Development.* (London: McGraw-Hill 1969)

[14] Harris, J. M. Preliminary studies of spiral grain in radiata pine. Proceedings of a meeting of Section 41, IUFRO, Melbourne, Vol. 1 (1965)

[15] Harris, J. M. On the causes of spiral grain in corewood of radiata pine. *N.Z.J. Bot.* **7,** 189-213 (1969)

[16] Harris, J. M. Spiral grain and xylem polarity in radiata pine: microscopy of cambial reorientation. *N.Z.J. Forest. Sci.* **3,** 363-378 (1973)

[17] Hejnowicz, Z. Anticlinal division, intrusive growth, and loss of fusiform initials in non-storeyed cambium. *Acta Soc. Bot. Polon.* **30,** 729-748 (1961)

[18] Hejnowicz, Z. Orientation of partition in pseudotransverse division in cambia of some conifers. *Canad. J. Bot.* **42,** 1685-1691 (1964)

[19] Hejnowicz, Z. The structural mechanism involved in the changes of grain in timber. *Acta Soc. Bot. Polon.* **37,** 347-365 (1968)

[20] Hejnowicz, Z. Upward movement of the domain pattern in the cambium producing wavy grain in *Picea excelsa. Acta Soc. Bot. Polon.* **40,** 499-512 (1971)

[21] Hejnowicz, Z., Krawczyszyn, J. Oriented morphogenetic phenomena in cambium of broadleaved trees. *Acta Soc. Bot. Polon.* **38,** 547-560 (1969)

[22] Hejnowicz, Z., Romberger, J. A. Migrating cambial domains and the origin of wavy grain in xylem of broadleaved trees. *Amer. J. Bot.* **60,** 209-222 (1973)

[23] Hejnowicz, Z., Romberger, J. A. The common basis of wood grain figures is the systematically changing orientation of cambial fusiform cells. *Wood Sci. Technol.* **13,** 89-96 (1979)

[24] Hejnowicz, Z., Zagorska-Marek, B. Mechanism of changes in grain inclination in wood produced by storeyed cambium. *Acta Soc. Bot. Polon.* **43,** 381-398 (1974)

[25] Jones, B. E. Cell adjustments accompanying the development of spiral grain in a specimen of *Pseudotsuga taxifolia. Brit. Com. Forest. Rev.* **42,** 151-158 (1963)

[26] Kirschner, Hanna et al. Secondary xylem reorientation as a special case of vascular tissue differentiation. *Israel Journal Bot.* **20,** 184-198 (1971)

[27] Krawczyszyn, J. Unidirectional splitting and uniting of rays in the cambium of Platanus accompanying the formation of interlocked grain in wood. *Acta Soc. Bot. Polon.* **40,** 57-79 (1971)

[28] Krawczyszyn, J. Movement of the cambial domain pattern and mechanism of formation of interlocked grain in Platanus. *Acta Soc. Bot. Polon.* **41,** 443-461 (1972)

[29] Krawczyszyn, J. Domain pattern in cambium of young Platanus stems. *Acta Soc. Bot. Polon.* **42,** 637-648 (1973)

[30] Mikami, S. et al. Clonal variation in spiral grain of *Larix leptolepis* Gord. *J. Japan. Forest. Soc.* **54,** 213-217 (1972)

[31] Misra, P. Observations on spiral grain in the wood of *Pinus longifolia* Roxb. *Forest.* **13,** 118-133 (1939)

[32] Nicholls, J. W. P. The relation of spiral grain to wood quality. Paper presented to a meeting of Section 41, IUFRO, Madison (1963)

[33] Nicholls, J. W. P. Preliminary observations on the change with age of the heritability of certain wood characteristics in *Pinus radiata* clones. Proceedings of a meeting of Section 41, IUFRO, Melbourne, **1** (1965)

[34] Nicholls, J. W. P. The possible causes of spiral grain. Proceedings of a meeting of Section 41, IUFRO, Melbourne, **1** (1965)

[35] Northcott, P. L. Is spiral grain the normal growth pattern? *Forest. Chron.* **33,** 335-352 (1957)

[36] Noskowiak, A. F. Spiral grain in trees—a review. *Forest. Product. J.* **13,** 266-275 (1963)

[37] Panshin, A. J., De Zeeuw, C. *Textbook of wood technology*, Third edition, Vol. 1. (New York: McGraw-Hill 1970)

[38] Paul, B. H. Importance of wood quality in tree breeding. *J. Forest.* **53,** 659-661 (1955)

[39] Pederick, L. A. Inheritance of spiral grain in young radiata pine. Presented to XVth Congress IUFRO, Gainesville, Florida (1971)

[40] Preston, R. D. Spiral structure and spiral growth. The development of spiral grain in conifers. *Forestry.* **23,** 48-55 (1949)

[41] Priestley, J. H. Observations on spiral grain in timber. *Amer. J. Bot.* **32,** 277-284 (1945)

[42] Pyszynski, W. Mechanism of formation of spiral grain in *Aesculus* stems: dissymetry of deformation of stems caused by cyclic torsion. *Acta Soc. Bot. Polon.* **46,** 501-522 (1977)

[43] Pyszynski, W. Complex wavy grain in the stem of *Aesculus. Acta Soc. Bot. Polon.* **46,** 231-249 (1977)

[44] Sachsse, H. The effect of the rate of growth on the occurrence of spiral grain. Proceedings of a meeting of Section 41, IUFRO, Melbourne, **1** (1965)

45 Venkataramanan, S. V. Spiral grain in chir *(Pinus roxburghii Sargent)*. Proceedings of XIV IUFRO Congress, Munich, **IX,** 484-497 (1967)
46 Vité, J. P. Water conduction and spiral grain: causes and effects. Proceedings of XIV IUFRO Congress, Munich, **IX,** 338-351 (1967)
47 Vité, J. P., Rudinsky, J. A. The water conducting systems in conifers and their importance to the distribution of trunk-injected chemicals. *Contributions of the Boyce Thompson Institute*, **20,** 27-38 (1959)
48 Zajaczkowski, S., Wodzicki, T. J. On the question of stem polarity with respect to auxin transport. *Physiol. Plant.* **44,** 122-126 (1978)
49 Zobel, B. J. Inheritance of spiral grain. Proceedings of a meeting of Section 41, IUFRO Melbourne, **1** (1965)
50 Zobel, B. J. et al. Inheritance of spiral grain in young loblolly pine. *For. Sci.* **14,** 376-379 (1968)

10. The Development of Growth Strains and Stresses in Reaction Wood

BRAYTON F WILSON
Department of Forestry and Wildlife Management, University of Massachusetts, Amherst, Massachusetts, USA

INTRODUCTION

The term 'reaction wood' includes two very different types of wood commonly found in leaning tree stems, compression wood in gymnosperms and tension wood in angiosperms. The most important characteristic of reaction wood *to people*, is that it generally degrades the value of logs and lumber. Perhaps its worst characteristic is that it shrinks excessively when dried, causing boards to warp and split. The most important characteristic of reaction wood *to trees*, is its mechanical action whereby it generates growth strains that cause bending moments within branches and stems of living trees. These bending moments are the mechanism by which trees can re-orient their branches or tilted stems, usually upwards toward vertical[43]. Reaction wood has evolved in trees because of its mechanical action, but the structural characteristics that produce the mechanical action in the tree have undesirable side effects when the wood is dried outside the tree. This chapter will be concerned with the process of strain development in reaction wood in living trees. The basic question is, how does cell development in reaction wood produce the high levels of strain that are observed. There are many other aspects of reaction wood that will not be considered in this chapter. Interested readers are referred to review articles[12, 18, 22, 27, 30, 35, 36, 40, 41, 42].

It seems axiomatic that the mechanical action of reaction wood results from anatomical, structural characteristics. Some of these characteristics are obvious macroscopically but others only microscopically or submicroscopically. What, then, are the differences between normal and reaction wood? A knowledge of the differences and how they originate should provide clues about the mechanism of strain development. Characteristics that seem to be related to strain development are presented in the first section of this chapter.

For years most studies were concerned only with the most obvious examples of reaction wood. There were occasional references to 'weak' or 'mild' reaction wood. Evidence has, however, been accumulating which suggests that there is a continuum of wood types from normal through to the extremes of reaction wood. Not only do anatomical characteristics follow a continuum, but also the strain generated by the cells. It is quite clear that normal wood also can generate growth strains, although these are not as extreme as in reaction wood. From a developmental point of view, the significance of a continuum of wood types would be that reaction wood may be only quantitatively different from normal wood. No new characteristics have been evolved, just different levels of the same characteristics present in normal wood. The second section of this chapter presents some of the evidence for the existence of a continuum of wood characters.

Several hypotheses have been advanced to explain the development of growth strains in normal and reaction wood. These hypotheses are presented and discussed in the final section of this chapter. Assessment of these hypotheses is limited to a large extent by the difficulty of measuring strain levels in cells. There are increasingly more data on anatomical features, which are relatively easy to measure, but relatively few strain measurements. At present, accurate strain measurements can only be made on fairly large woody stems and branches.

Throughout this chapter there will be references to strain and stress. Reaction wood develops growth strains and stresses. Strains are changes in length per unit length (cm/cm). Stresses are forces per unit area (kg/cm^2, or pascals, 1 pascal = 1 $newton/m^2$). Strains and stresses are related to each other linearly at low values. The slope of this 'stress–strain curve' is the elastic modulus of the material (E), so that stress = strain $\times$ E. Stress causes strain and strain causes stress, the two are inextricably related.

In most respects strains are more biologically significant than stresses because: (1) It is not possible to measure stresses directly, they must be determined from strain measurements. This applies to experimental methods and, presumably, also applies to organisms that respond to a stress–strain stimulus. The organisms detect strain, not stress; (2) It is strains that are important in elongation and, through differential elongation, curvatures. A growth strain develops when a cell tends to shrink (tensile strain) or expand (compressive strain). If this tendency is resisted by surrounding tissue then growth stresses develop. The standard way to measure growth strains is to remove the resistance of the surrounding tissue (strain relief) and measure the resulting change in length. Fairly precise methods have been developed for measuring growth strains in trees by Nicholson[24] and by Saurat and Geuneau[29]. When the strains have been measured, the stresses are calculated by using the elastic modulus which has been determined experimentally using a testing machine. Data in literature on wood are often given as stress and the phenomenon is called growth stress[19], even though it might be more apt to call it growth strain. Therefore this chapter will refer to both growth stress and growth strain, but the reader should be aware of the relationship between the two parameters.

CHARACTERISTICS OF REACTION WOOD

Each type of reaction wood has specific characteristics not found in normal wood which are helpful for its identification. Compression wood is made up of tracheids that are rounded in cross section so that the tissue has intercellular spaces, and whose tips tend to be distorted and bifurcated. The secondary wall usually lacks an S3 layer. The thicker S2 layer has an outer highly lignified zone and an inner less lignified zone with helical cavities which run parallel to the microfibril angle of the S2 layer and form ribs, or striations visible at low magnifications in longitudinal section.

Tension wood is a heterogeneous tissue with relatively few vessels, some normal fibres and also 'gelatinous' fibres. The presence of gelatinous fibres is usually used to define tension wood. They are similar to normal fibres except that they have an additional layer of secondary wall, the gelatinous (Sg) layer, formed usually as the innermost layer. The Sg layer is convoluted during development in some species (Wardrop[39]), but by maturity it is adpressed to the rest of the wall. During sectioning it often pulls away from the rest of the wall, suggesting that it is not tightly bonded to the other layers. The Sg is neutral to many stains and so is often difficult to see in sections. As a result, tension wood is usually more difficult to identify than compression wood. Within tension wood Trenard and Gueneau[32] found that the level of tensile stress was related to the proportion of gelatinous fibres in the tissue.

Some characteristics of reaction wood are different from those of normal wood in amount or intensity, but are not unique to it. Perhaps the most interesting are those characteristics which are opposite to each other in compression and tension wood. Compared with normal wood compression wood tracheid walls have relatively high lignin (40% *vs* 30%) and low cellulose (40% *vs* 50%), while tension wood gelatinous fibre walls have relatively low lignin (15% *vs* 20%) and high cellulose (65% *vs* 55%). The S2 layer of compression wood tracheids has an unusually high (45°) microfibril angle (MFA) whereas the Sg layer of tension wood fibres has an unusually low (5°) MFA. Compression wood cells are shorter than normal, but tension wood fibres are about normal length.

The most important characteristic of tension wood and compression wood is that both bend leaning stems upwards toward vertical. Yet, this identical end result is accomplished by totally opposite action and position of the reaction wood. Tension wood forms on the upper side of leaning stems and contracts to pull the stem up. Compression wood forms on the lower side and expands to push the stem up. The strains developed by reaction wood are much greater than in normal wood; tension wood develops tensile strains 2–3 times those of normal wood while compression wood develops compressive stresses of <100 kg/cm^2 compared to the tensile stresses of 40 kg/cm^2 in normal wood[28].

There are various minor differences between reaction wood and normal wood. One that has received attention is the presence in compression wood of a β 1–3 glucan called laricinan[17]. Laricinan occurs in normal wood, but in much lower concentrations. There are other variations, for instance in type and amount of hemicellulose and the type of lignin[2]. Of these minor variations, so far only laricinan has been linked to a strain generating mechanism[10]. In general, small variations in composition are unlikely to produce the large observed differences in strain generation.

Although the chemical composition of wood is usually expressed on a per cent weight basis, from a developmental point of view it should be presented on a per cell basis. For example, by weight, compression wood is roughly 40% lignin and 40% cellulose compared to a 30:50 ratio for normal gymnosperm wood. It is, however, about 25% denser than normal wood because the secondary walls are thicker. On a per weight basis one concludes that the lignin went up 10% and the cellulose went down 10%. On a per cell basis one would conclude that the cellulose stays about the same, but that lignin increases 20%. Similarly, in tension wood, lignin is low on a per cent weight basis compared with normal wood, but the amount per cell appears to be about the same[2]. There is no need to hypothesize about a mechanism for decreased production of these constituents if, in fact, the decrease is not real.

THE CONTINUUM FROM NORMAL TO REACTION WOOD

The most important initial point is that normal wood, in both gymnosperms and angiosperms, develops biologically significant growth strains. These growth strains have been measured by many investigators in all species tested for more than 40 years[42], yet they still seem to be a well kept secret outside of the group of researchers interested in them. In general, new wood added to the outside of a tree develops tensile strains and tends to contract. The strain is about 3×10^{-4} in gymnosperms and 8×10^{-4} in angiosperms (0.03 and 0.04% decrease in length). The stresses are of the order of 10–100 kg/cm^2.[19]. Thus, year after year more tensile strain is added to the outside of the tree and the entire cylinder of older wood is gradually compressed as the outside shrinks. The distribution of residual strains inside trees has been measured and also calculated using incremental models[42].

Jacobs[19] reported that young (1–3 year-old) portions of vertical pine stems developed compressive strains. To my knowledge these measurements have not been repeated with modern methods, primarily because of the difficulty of measuring strains in small stems. Compressive growth strains have also been reported for some angiosperm wood[28]. If normal wood can generate either tensile or compressive strains, then it seems that reaction wood could be merely an extreme expression of this normal tendency.

The most direct evidence for the existence of a continuum of growth strains comes from measurements taken around the periphery of leaning stems. Measurements in Australia have shown a continuum of strain in *Pinus radiata*[9] and *Eucalyptus regnans*[25]. The strain continuum is not a smooth gradient. In both cases there is a fairly rapid change from normal levels to those in reaction wood, but in both cases there are intermediate levels of strain between reaction wood and normal wood. It would be desirable to have more measurements in these transitional areas.

It is possible to have peripheral strain gradients in leaning trees with no anatomically distinguishable reaction wood. In these cases the sides are normal, but the underside has less tensile strain than the upper side. The difference in strain levels from under to upper side generate a bending stress even without reaction wood[4]. Nicholson *et al.*[25] interpret this peripheral continuum as a whole-tree response, 'In this context, all wood is "reaction wood" in that it is formed under a closely controlled system which is designed to achieve a particular intensity and arrangement of longitudinal stresses within newly formed wood so as to enable optimum positioning of the tree crown in relation to its surroundings and environment'. Although the reader may not agree with this interpretation, it seems clear that differential strains can act to bend stems. Reaction wood may be the most dramatic, but it is not the only growth strain in developing wood.

Characters other than growth strains vary around the periphery of a leaning stem. In the same *Pinus radiata* in which strains were measured, Boyd[7] also measured MFA, lignin content, density, elastic modulus, crushing strength, wall thickness and cell diameter. He found that the measured strains matched an inverse curve of MFA at the maximum and minimum, but poorly in between. The MFA was greatest on the under side and least on the upper side. The strength properties paralleled the MFA, but the other measured characteristics were not related to either strain or MFA. Nicholson *et al.*[25], in the same *Eucalyptus regnans* used for strain determinations, measured the proportion of different kinds of fibres, the elastic modulus, density and volumetric shrinkage of wood around the periphery of the stem. All these parameters were positively associated, with the maximum values on the underside of leaning stems, and they all showed transitions from normal through to reaction wood.

Timell[33] has made anatomical studies of 'opposite wood' in conifers, the wood on the upper side of leaning stems. He did not measure growth strains, but he consistently found that characters of opposite wood differed from normal wood in opposite directions from the way that compression wood differed from normal wood; for instance, the walls are thinner and less lignified in opposite wood, and in compression wood they are thicker and more lignified. These studies of opposite wood support the presence of a continuum of wood characteristics, but the continuum probably goes all the way from opposite wood, through normal wood to reaction wood.

Other anatomical studies have found gradients of characters due to partial induction of reaction wood. Kennedy and Farrar[20] tilted *Pinus banksiana* seedlings for short periods and found that compression wood characters were formed only while the stem was tilted. Cells stimulated during enlargement might have rounded shapes with normal secondary walls. Cells stimulated after enlargement could have normal angular shape in cross section, but compression wood characters in the secondary wall. It was concluded that cells needed constant stimulation by tilting to develop all the compression wood characteristics. Tension wood of *Acer saccharinum* from a stem with a 10° lean had 18% gelatinous fibres with S1+S2+Sg, while in stems with a 20° lean there were 45% gelatinous fibres with S1+S2+S3+Sg and the Sg layer was thicker with a honeycomb structure[23]. Thus it seemed that the tension wood became more pronounced at greater leans, both through more gelatinous fibres and through more highly developed gelatinous fibres. Scurfield and Wardrop[31] rotated horizontal, potted *Tristania conferta* and *Eucalyptus melliodora* through 180° for a series of intervals. They found that the treatment induced a sudden transition back to normal wood 'with gradual elimination of Sg' when the stem was rotated so that the previously stimulated side was now on the bottom. Therefore, both gymnosperms and angiosperms seem to form transitional cells between normal and reaction wood when the level of stimulus is changed.

The apparent existence of a continuum of wood from opposite wood to reaction wood may finally solve an old problem. Both Westing[40] and Timell[34] quote Robert Hartig from 1901: 'Above all, it remains unexplained why the gravitational stimulus, which after all acts on the cambial cells in the entire branch, elicits its remarkable reaction only on the cambial cells of the lower side'. Both authors point out that Hartig's statement is still true. Of course, all cambial cells are acted on by a gravitational stimulus. In fact, however, all the cells *do* show a response to the gravitational stimulus, but they respond differently according to their position around the branch or stem. Only a few of the cells develop that set of characters which we classify as reaction wood, but they all respond, even those cells on the sides which remain 'normal'.

MECHANISM OF GROWTH STRAIN DEVELOPMENT

It seems reasonable to assume that there is a mechanism for strain development that is common both to reaction wood and normal wood. The main support for this assumption is the apparent existence of a gradient of strains, and other characteristics, from normal wood to reaction wood. There is no sudden change from one to the other, implying that there is no developmental 'switch' that suddenly changes pathways of differentiation so that a totally new type of wood is produced. Instead there are probably modifications in structure and amounts of cellular components which affect the strain producing mechanism.

There are a few possible exceptions to the assumption above; structural aspects of reaction wood that seem to be unique and might require quite separate developmental pathways. Foremost of these is the Sg layer of gelatinous fibres in tension wood. The Sg layer can vary in thickness, but there does not seem to be any transitional state. It is either present or absent. On the other hand, gelatinous fibres are present scattered in normal wood of vertical stems in some species[42], although in a much lower percentage than in tension wood. The Sg layer may be unique to gelatinous fibres, but gelatinous fibres are not unique to reaction wood. The high proportion of gelatinous fibres in tension wood seems to be an extreme expression of a normal tendency rather than a unique feature of this tissue.

A second, unique feature of reaction wood might be the pattern of helical cavities and ribs present in the inner portion of the S2 layer of compression wood tracheids. The question here is whether the cavities are somehow related to the cause of strain development, or whether they result from it. If they are formed by the high strains in compression wood then they would be found only in compression wood. Most investigators favour the 'schizogenous' origin of the cavities as a result of strains in the cell wall[6, 13, 39]. Two main observations support this theory: (1) the lamellae of the S2 layer are present in the ribs of the wall and they run mostly straight out to the edge of the cavity, as though they had been sheared by the formation of the cavity. Côté *et al.*[13] found that some of the innermost lamellae appeared to be added to the wall after the cavities had formed. (2) The cavities are not visible until the S2 layer is almost fully formed. The S2 can be quite thick without any cracks, but later in development the cracks extend deep into the S2. Both observations are consistent with a sudden failure of the wall rather than a gradual development of ribs and cavities as the wall is formed.

Even these two features, which appear to be unique to reaction wood, cannot explain the strains in compression wood through mechanisms different from those present in normal wood. There are, however, differences between gymnosperm and angiosperm reaction wood. The direction of the growth strains is an obvious one. Another major difference seems to be the Sg layer, which is not present in gymnosperm wood. It still might be possible that there are different strain developing mechanisms in gymnosperm and angiosperm wood, even if within each wood type the basic mechanisms are the same in normal and reaction wood.

There is general agreement that strains in both types of wood originate from the tendency of the cells to change shape during development.

Tension wood and normal wood tracheids and fibres tend to contract longitudinally while compression wood tracheids tend to expand. As the cells contract longitudinally, they also expand transversely, to produce radial tensile growth strains and tangential compressive growth strains[39]. These non-longitudinal strains are important for many wood properties, but the associated stresses are of relatively low values and do not contribute significantly to longitudinal bending moments, so they will not be considered here.

It is generally assumed that the cells of most significance in strain generation are the tracheids in gymnosperms and the fibres in angiosperms. Although rays, vessels and parenchyma may vary between normal and reaction wood, they are not considered to be mechanically significant.

There is now general agreement that growth strains are developed during the development of the secondary wall of tracheids and fibres. There have been theories in which it was proposed that strains originated during cell enlargement when the cells were under high turgor pressure, but only had primary walls. For instance, the high turgor of compression wood cells might lead to their rounded outline and the presence of distorted cell tips might suggest that the tips developed strains from trying to elongate between cells above and below. Sasaki *et al.*[28] proposed that 'transverse swelling by cell division is restrained by the thick walled cells with extreme lignification and high enzyme activity. Hence it can be supposed that the transverse obstructed stress is converted into longitudinal compressive stress'. The main support for the idea that strain originates in secondary walls is that the appropriate strain levels can be measured in mature wood cells by stress relief methods. If cell enlargement and turgor cause growth strains, then the deposition of the secondary wall would fix the cells at their new length, but when the cell died the growth strains would be lost. In fact, mature wood has high residual growth strains and they have been measured many times[42].

Therefore, the major questions about strain development are (1) what causes dimensional changes in the secondary wall as it develops? (2) Does the same mechanism apply to compression, normal and tension woods? There are several different hypotheses that try to answer at least parts of these questions and these will be presented in the next section. The main barrier to the experimental testing of these hypotheses is the difficulty of making strain measurements. As a result there are relatively few strain measurements and a great many anatomical and compositional observations. There are technological problems in measuring growth strains in wet wood in standing trees. Techniques exist for measuring small strains in metal beams, but they are difficult to adapt to variable, living tissue. The techniques are particularly difficult to apply to small stems and there are consequently almost no direct strain measurements on such stems.

Undoubtedly these technological difficulties will be solved. Another problem is that growth strains have to be measured on young tissue at the outside of the stem or branch. The inner layers have been subjected to stress from the outer layers and cells may well have changed their dimensions, particularly if the stem or branch has actually bent. As a result of these difficulties the various hypotheses are supported largely by circumstantial evidence and, although they may be correct, none of them has been tested rigorously.

Strain origin from lignification

J. D. Boyd has worked out a carefully reasoned brief for his hypothesis in a series of papers on growth strains in normal and reaction wood, particularly compression wood[3, 4, 5, 6, 7, 8]. I will summarise the hypothesis, but clearly such a summary leaves out detail and much supporting evidence.

Boyd believes that lignification of the secondary wall, after the formation of the framework of cellulose microfibrils, causes the wall to swell. He emphasizes the S2 layer in tracheids and fibres because it is the thickest and most variable layer of the secondary wall. Other investigators concur with this emphasis on the S2 layer. He cites numerous references to support the assumption that lignification causes swelling, putting particular reliance on his calculations based on data from Grozdits and Ifju[15], showing that tracheids in *Tsuga canadensis* increase in radial diameter during secondary wall formation. A model developed by Barber and Meylan[1] is used to explain the anisotropic shrinkage of wood, and it is assumed that lignin and water occupy the same types of sites in the cell wall, so that shrinkage from loss of water should be essentially the reverse of swelling that occurs during lignification. Boyd's calculations show that swelling from lignification would cause tracheids to contract at MFA less than about 40° and to expand at MFA greater than 40°. The strains in either direction would increase as the MFA changed away from 40°. His calculated stresses based on the model are essentially the same as the mean stresses reported in the literature. Thus in this hypothesis the two main factors in strain generation are lignification and the MFA. Boyd feels a third factor influencing the total amount of strain is the thickness of the wall, particularly the S2.

This idea is neat, simple, and it accounts for both tensile and compressive strains through changes in MFA. In his papers he systematically refutes alternative hypotheses. Boyd's hypothesis, however, does not seem to cover the high tensions in the Sg layer in tension wood which, although it has very low MFA, is scarcely lignified. Boyd[6] does extend his hypothesis to the origin of the helical cracks in the outer S2 of compression wood cells. He calculates that the S2 layer of cells with greater than 45° MFA would develop transverse strains great enough to cause failures between the helices of microfibrils. These failures would produce the helical cavities.

The conclusion that cells increase in radial diameter during lignification, calculated from Grozdits and Ifju[15], may be correct, but it would be desirable to have corroborative evidence from a study specifically designed to investigate the change in cell dimensions during secondary wall formation. Grozdits and Ifju[15] compare cells formed at different times during a growing season. If final cell size was constant this would be acceptable, but if final diameter decreases over this period then it would result in an apparent radial enlargement during lignification from measurements taken at intervals from the cambial zone. It is also possible that there was shrinkage during dehydration for paraffin embedding. This shrinkage would be greatest in the least lignified cells so that the apparent increase in diameter would be really a gradient of shrinkage from mature through to enlarging cells.

Boyd's theory about a critical MFA for the change in direction of strains seems sound qualitatively. Similar assumptions have been used by others[36]. In support of the hypothesis, it has been found that in some conifer species the MFA is greater than 45° in normal wood near the pith of the tree and MFA decreases in rings further from the pith[21]. Jacobs[19] found that this normal wood with large MFA near the pith might have compressive strains, just as Boyd's hypothesis would predict. Compression wood, however, may form in cells with MFA of only 25–30°. Boyd has dismissed reports of compression wood with small angles, saying that x-ray techniques for measuring MFA include angles from normal wood with low MFA. My own measurements in *Pinus strobus* (unpublished) of MFA in compression wood on the underside of a branch that was 60 years old were made using the angle of the helical ribs and cavities in the S2 and averaging 50 cells. By 30 rings from the pith the MFA had dropped to 31° in the early wood and 25° in the late wood. Cockrell[11] measured MFA of 21–25° in the latewood of compression wood in *Sequoia gigantea*. It is not known whether these compression wood tracheids with low angles actually generate compressive strain, but they have the other major compression wood characteristics. I would assume that they do generate compressive strains at low MFA. Despite these exceptions, most compression wood does have MFA greater than 40°. There is an appealing extension of Boyd's hypothesis. If we assume that an increase in MFA causes a shift from tensile to compressive strains, then the induction of compression wood action would depend simply on the reorientation of the MFA. Some investigators find that microtubules are oriented parallel to developing microfibrils[14, 26]. If the orientation of MFA was determined by the orientation of microtubules then the key developmental change would be in the orientation of the microtubules. To continue this speculation, if compression wood is stimulated by high auxin, then the auxin could be acting to change microtubule orientation.

There is a major objection to any hypothesis suggesting that there can be rapid changes in MFA with the induction of compression wood. This is the high correlation between tracheid length and MFA, $r = -0.956$ in *Pinus elliottii*[21]. This correlation is also found in compression wood where cells are shorter and MFA higher than in normal wood. Wardrop and Dadswell[38] concluded that 'micellar orientation in compression wood tracheids does not differ significantly from that existing in normal wood tracheids of similar length'. Therefore, the increase in MFA in compression wood may merely reflect the decrease in cell length usually, but not always, associated with compression wood. Compression wood is usually associated with increased radial growth rate[40], which usually increases the rate of pseudotransverse divisions and thus reduces the average tracheid length[38]. Wardrop and Dadswell[38] followed cell length and MFA changes in a few radial files after sudden induction of compression wood. Unfortunately there were so few files that 'the values plotted must be considered only as trends'. Their data do not seem to show a sudden increase in MFA before cell length decreased.

If MFA is truly tied to cell length then there are cases where compression wood develops without changes in cell length and there should be no accompanying increase in MFA. Kennedy and Farrar[20] showed that in cells with normal characteristic length, compression wood secondary walls could be induced, even though primary walls, and thus cell length, were unaffected. They did not measure MFA. In general, when a gymnosperm stem is tilted to induce compression wood there must be a lag period before all the radial files undergo pseudotransverse divisions to reduce cell length. More measurements are needed to follow the change in MFA during this lag period.

It seems likely that the low MFA in compression wood reflects the shorter cell lengths associated with fast growth. In cases where cell length did not decrease with compression wood formation, in old or slow grown wood, then the MFA would be less than Boyd's postulated 40°. Whether these cells generate compressive strains cannot be answered because the strains have not been measured. They may have helical cavities, which Boyd[6] argues come from the transverse strains associated with compression wood. Perhaps compressive strains can come from heavy lignification at angles even lower than 40°.

Scurfield[30] proposed an hypothesis, based on swelling from lignification, to account for the extreme shrinkage of the Sg layer in tension wood fibres. He speculated that as the wall layers external to the Sg swell from lignification, the high resistance to longitudinal expansion offered by the Sg would cause cell expansion to occur predominantly in the transverse direction.

This transverse expansion would then result in a marked longitudinal contraction and cause the observed compression of vessels in tension wood.

A problem with this hypothesis is that most investigators find that the outer wall layers of gelatinous fibres are the same as comparable layers in normal fibres and that they differentiate the same way[36]. Therefore the amount of lignification in the outer layers of gelatinous fibres is about the same as in normal fibres. In angiosperms both normal wood and tension wood contract to develop tensile strains. It is hard to see how the gelatinous fibres could generate 2–3 times as much tensile strain without some extra strain being contributed by the Sg layer. In any case, it would seem that the orientation of the microfibrils in Sg would offer resistance to longitudinal contraction as well as to longitudinal expansion, so if Scurfield's hypothesis is correct the Sg layer should reduce tensile strains.

Strains from the crystallisation of cellulose

Wardrop[36] proposed that cellulose is laid down in microfibrils in an uncrystallised form. As portions of the cellulose crystallise the microfibrils tend to shrink. He hypothesised that lignification inhibits crystallisation and therefore reduces the tendency of the cell to shrink. The least lignified walls, the Sg layer, would contract the most and generate the greatest tensile strains. The most lignified, in compression wood tracheids, would contract least. In addition, Wardrop cites Preston's work on drying of wood which showed that at MFA less than 40° the contracting microfibrils would cause the whole cell to contract, but at MFA greater than 40° the same microfibril contraction would cause the whole cell to expand. The shrinkage of the microfibrils could therefore cause either tensile strains in normal and tension wood, or compressive strains in compression wood (MFA = 45°).

Both Wardrop's[36] and Boyd's[4] hypotheses suggest a switch from tension to compression at MFA of 40–45°. The difference is that Boyd postulates a swelling of the whole cell wall and Wardrop postulates a shrinking of the cellulose helical framework, both basing their conclusions on models derived from the behaviour of wood when dried. The problems about the relationship of MFA to cell length discussed in relation to Boyd's hypothesis apply equally well to Wardrop's hypothesis and may mean that a rigid concept of some cross-over angle would have to be modified.

Some combination of Wardrop's and Boyd's hypotheses could explain most aspects of growth strains. Crystallisation of the unlignified Sg layer could cause its high tensile strains. Swelling of the walls during lignification could cause the high compressive strains in compression wood cells with large MFA. In normal wood the two mechanisms might both act to induce normal tensile growth strains. If both have a tendency to switch from tensile to compressive strains as the MFA increases then the problem of normal wood with high MFA having compressive strains is also solved. Neither of the mechanisms alone seem adequate to explain all the observed growth strain phenomena. There is, however, no basis for combining the two hypotheses other than the ability of the combined hypothesis to explain so many different growth strains.

Strains due to changes in water content

Wood which contains reaction wood and normal wood will bend when it dries because reaction wood shrinks longitudinally more than does normal wood. Hejnowicz[16] proposed that changes in water content of fresh wood are responsible for the development of growth strains. He tilted *Populus tremula* and *Pinus sylvestris* trees to induce reaction wood. He then cut out longitudinal strips that contained reaction wood on one side and normal wood on the other. Fresh strips with compression wood bent away from the compression wood because of stress release of the compressive strains, although strips from the lower stem did not bend even though there was compression wood in them. Hejnowicz concluded that this compression wood was ‘neutral’. It did not expand even though its MFA was 56.60°. Strips containing tension wood and normal wood did not bend if cut in August, but in October bent towards the tension wood because of its tensile growth strains. He then dried the strips and found that the compression wood in the bent strip contracted by the same amount as it had expanded after stress relief so that it became straight. The neutral compression wood also contracted on drying to bend the strips even though it apparently had no growth strains. This change in compression wood length on drying was reversible upon rewetting. The *Populus* strips bent on drying whether or not the strips had shown active tension wood. The curvature after drying in the strips without *active* tension wood (although it was present in the strips) was about equal to the curvature in the green state produced by active tension wood in October. The curvature in *Populus* strips was not reversible on rewetting.

Hejnowicz[16] concluded that: (1) Compression wood develops its strains as it is formed; (2) Tension wood does not develop its strains until the leaves fall off; (3) Compression wood develops its strains by imbibing water as the wall develops; (4) Tension wood develops its strains by losing water from the wall causing shrinkage. The parallels between dimensional changes of wood on drying and the dimensional changes with growth strains have been the basis of other models for growth strain action[4, 36]. There is, however, no indication that there is a significant change in water content of reaction wood cells as they develop. Boyd[4] suggests that the lignin occupies sites similar to those occupied by water, so lignification may effectively reduce water content by replacing it in the cell wall.

Other investigators have not found neutral compression wood. In *Pinus thunbergii*, for instance, the level of growth strain in compression wood is high at the base of a tree[28]. In the case of Hejnowicz's results, strips would not bend if the growth strains were equal in the compression and normal wood. He concluded that the compression wood was neutral, with no compressive growth strains, but the normal wood in the strip had a high MFA (52.1°) and might have generated its own compressive strains, so that the strip did not bend because both types of wood expanded. Neutral compression wood does, however, seem to be a case where drying strains and growth strains are independent.

Strains due to laricinan deposition

Brodzki[10] investigated the distribution of a substance he called 'callose' in the cell walls of compression wood tracheids. The substance is probably the β 1–3 glucan identified by Hoffman and Timell[17] which they call 'laricinan'. Brodzki reported that the substance occurred within the helical cavities of the compression wood in 17 of the 20 gymnosperm species he studied. He postulated that the swelling of the substance as it imbibes water within the cracks caused the compressive strains of compression wood. Boyd[8] devoted a whole paper to refuting this theory and concluded that the effect of the laricinan was of little, if any, importance in strain generation. It is present in many tissues and is not present in all compression wood. Boyd also calculated that the laricinan could not generate large strains. There has also been no suggestion that laricinan causes strain generation in normal wood or tension wood. It seems surprising, however, given the absence of lignin in the Sg layer, that no one has postulated that some substance like laricinan, or some hemicellulose, swells within the cellulose matrix just as lignin is supposed to swell in secondary walls. If some substance swelled markedly in the Sg layer it should cause longitudinal contraction if the theories about MFA and the direction of strain generation are correct.

SUMMARY

The important functional aspect of reaction wood is the strains it develops and the bending moments that these strains generate. In fact, all types of wood seem to generate strains, but not such large ones as in reaction wood. It seems that compression and tension woods are not qualitatively different from normal wood. They do not have new features not present in normal wood. Therefore, one might expect general strain generating mechanisms that act in all developing wood cells, but are extreme in reaction wood cells.

Of the proposed strain generating mechanisms none seem adequate to explain all the strain phenomena. Microfibril angle does seem to be a key in explaining the direction, compression or tension, of the strains generated. It seems unlikely that microfibril angle can be varied independently of cell length, so the developmental regulation of the direction of growth strains may be complex. A combination of several mechanisms seems most likely to explain strain phenomena, for instance some combination of the swelling of the secondary walls during lignification and the contraction of cellulose microfibrils during their crystallisation. We are coming closer to satisfactory explanations of strain development, but critical testing of hypotheses will probably have to wait for the development of better techniques for measuring growth strains in developing wood cells.

REFERENCES

[1] Barber, N. F., Meylan, B. A. The anisotropic shrinkage of wood, a theoretical model. *Holzforschung* **18,** 146-156 (1964)

[2] Bland, D. E. The chemistry of reaction wood Part 1. The lignins of *Eucalyptus goniocalyx* and *Pinus radiata. Holzforschung* **12,** 36-43 (1958)

[3] Boyd, J. D. Tree growth stresses. III. The origin of growth stresses. *Austral. J. Sci. Res.* **B3,** 294-309 (1950)

[4] Boyd, J. D. Compression wood force generation and functional mechanics. *N.Z.J. Forest. Sci.* **3,** 240-258 (1973)

[5] Boyd, J. D. Tree growth stresses—Part V: evidence of an origin in differentiation and lignification. *Wood Sci. Technol.* **6,** 251-262 (1972)

[6] Boyd, J. D. Helical fissures in compression wood cells: causative factors and mechanics of development. *Wood Sci. Technol.* **7,** 92-111 (1973)

[7] Boyd, J. D. Relating lignification to microfibril angle differences between tangential and and radial faces of cell wall layers in wood cells. *Drévarsky Výskum* **19,** 41-54 (1974)

[8] Boyd, J. D. Significance of laricinan in compression wood tracheids. *Wood Sci. Technol.* **12,** 25-35 (1978)

[9] Boyd, J. D., Foster, R. C. Tracheid anatomy changes as responses to changing structural requirements of the tree. *Wood Sci. Technol.* **8,** 91-105 (1974)

[10] Brodski, P. Callose in compression wood tracheids. *Acta. Soc. Bot. Polon.* **41,** 321-327 (1972)

[11] Cockrell, R. A. A comparison of latewood pits, fibril orientation and shrinkage in normal and compression wood of giant sequoia. *Wood Sci. Technol.* **8,** 197-206 (1974)

[12] Côté, W. A., Day, A. C. Anatomy and ultrastructure of reaction wood. In *Cellular Ultrastructure of Woody Plants*. pp. 391-418. W. A. Côté (ed.) (Syracuse, New York: Syracuse University 1965)

[13] Côté, W. A., Kutscha, N. P., Timell, T. E. Studies on compression wood VIII. Formation of cavities in compression wood tracheids of *Abies balsamea* (L.) Mill. *Holzforschung* **22,** 138-144 (1968)

[14] Cronshaw, J. Cytoplasmic fine structure and cell wall development in differentiating xylem elements. In *Cellular Ultrastructure of Woody Plants*. pp. 99-124. W. A. Côté (ed.) (Syracuse, New York: Syracuse University 1965)

[15] Grozdits, G. A., Ifju, G. Development of tensile strength and related properties in differentiating conifer xylem. *Wood Sci.* **1,** 137-142 (1969)

[16] Hejnowicz, Z. Some observations on the mechanism of orientation movement of woody stems. *Amer. J. Bot.* **54,** 684-689 (1967)

[17] Hoffman, G. C., Timell, T. E. Isolation of a β 1–3 glucan (laricinan) from compression wood of *Larix laricinia. Wood Sci. Technol.* **4,** 159-162 (1970)

[18] Hughes, J. F. Tension wood: a review of the literature. *Forestry Abst.* **26,** 1-16 (1965)

[19] Jacobs, M. R. The growth stresses of woody stems. *Comm. Forest. Bur. Austral. Bull.* **28,** 67 pp. (1945)

[20] Kennedy, R. W., Farrar, J. L. Tracheid development in tilted seedlings. In *Cellular Ultrastructure of Woody Plants*. pp. 419-453. W. A. Côté (ed.) (Syracuse, New York: Syracuse University 1965)

[21] Koch, P. Utilization of Southern Pines. *U.S. Dept. Agric. Handbook 420,* **1** (1972)

[22] Low, A. J. Compression wood in conifers: a review of the literature. *Forest. Abst.* **25,** 1-14 (1964)

[23] Manwiller, F. G. Tension wood anatomy of silver maple. *Forest. Prod. J.* **17,** 43-48 (1967)

[24] Nicholson, J. E. A rapid method for estimating longitudinal growth stresses in logs. *Wood Sci. Technol.* **4,** 40-48 (1971)

[25] Nicholson, J. E., Hillis, W. E., Ditchburne, N. Some tree growth-woody property relationships of eucalypts. *Canad. J. Forest. Res.* **5,** 424-432 (1975)

[26] Nobuchi, T., Fujita, M. Cytological structure of differentiating tension wood fibres of *Populus euroamericana (P. × canadensis). Mokusai Gakkaishi* **18,** 137-144 (1972)

[27] Robards, A. W. The effect of gravity on the formation of wood. *Sci. Progr. Oxford* **57,** 513-532 (1969)

[28] Sasaki, Y., Okuyama, T., Kikita, Y. The evolution process of growth stresses on the tree. *Mokusai Gakkaishi* **24,** 149-157 (1978)

[29] Saurat, J., Gueneau, P. Growth stresses in Beech. *Wood Sci. Technol.* **10,** 111-123 (1976)

[30] Scurfield, G. Reaction wood: its structure and function. *Science* **179,** 647-655 (1973)

[31] Scurfield, G., Wardrop, A. B. The nature of reaction wood. VII. Lignification in reaction wood. *Austral. J. Bot.* **11,** 107-116 (1963)

[32] Trenard, Y., Gueneau, P. Relations entre contraintes de croissance longitudinales et bois de tension, dans le hêtre (*Fagus sylvatica* L.) *Holzforschung* **29,** 217-223 (1975)

[33] Timell, T. E. Studies on the opposite wood in conifers. Part II: histology and ultrastructure. *Wood Sci. Technol.* **7,** 79-91 (1973a)

[34] Timell, T. E. Ultrastructure of the dormant and active cambial zone and the dormant phloem associated with the formation of normal and compression wood in Picea abies (L.) *Karst. SUNY College Env. Sci. Forest. Tech. Publ.* **60,** 94 pp. (1973b)

[35] Wardrop, A. B. The reaction anatomy of arborescent angiosperms. In *The Formation of Wood in Forest Trees*. pp. 405-456. M. H. Zimmermann (ed.) (New York: Academic Press 1964)

[36] Wardrop, A. B. The formation and function of reaction wood. In *Cellular Ultrastructure of Woody Plants*. pp. 371-390. W. A. Côté (ed.) (Syracuse, New York: Syracuse University Press 1965)

[37] Wardrop, A. B., Dadswell, H. E. The nature of reaction wood. II. The cell wall organization of compression wood tracheids. *Austral. J. Biol. Sci.* **B3,** 1-13 (1950)

[38] Wardrop, A. B., Dadswell, H. E. The nature of reaction wood. III. Cell Division and cell wall formation in conifer stems. *Austral. J. Sci. Res.* **B5,** 385-392 (1952)

[39] Wardrop, A. B., Davies, G. W. The nature of reaction wood. VIII. The structure and differentiation of compression wood. *Austral. J. Bot.* **12,** 24-38 (1964)

[40] Westing, A. H. Formation and function of compression wood in gymnosperms. *Bot. Rev.* **31,** 381-480 (1965)

[41] Westing, A. H. Formation and function of compression wood in gymnosperms. II. *Bot. Rev.* **34,** 51-78 (1968)

[42] Wilson, B. F., Archer, R. R. Reaction wood: induction and mechanical action. *Ann. Rev. Plant Physiol.* **28,** 23-43 (1977)

[43] Wilson, B. F., Archer, R. R. Tree design: some biological solutions to mechanical problems. *BioScience* **29,** 293-298 (1979)

Index

Figures in italics indicate pages on which figures or tables occur.

B

C

D

E

F

G

H

I

J

K

L

P

Q

R

S